T0262226

An Introduction to Photosynthesis

An Introduction to Photosynthesis

Edited by **Agatha Wilson**

New York

Published by Callisto Reference,
106 Park Avenue, Suite 200,
New York, NY 10016, USA
www.callistoreference.com

An Introduction to Photosynthesis
Edited by Agatha Wilson

International Standard Book Number: 978-1-63239-067-7 (Hardback)

Printed in the United States of America.

Contents

Preface

In my initial years as a student, I used to run to the library at every possible instance to grab a book and learn something new. Books were my primary source of knowledge and I would not have come such a long way without all that I learnt from them. Thus, when I was approached to edit this book; I became understandably nostalgic. It was an absolute honor to be considered worthy of guiding the current generation as well as those to come. I put all my knowledge and hard work into making this book most beneficial for its readers.

The most basic and significant aspect of life process on earth is linked to the process of photosynthesis. Photosynthesis is the most researched field amongst the scientific community. The present book examines the fundamentals of photosynthesis, and its impact on different life forms. The book contains important sections analyzing light and photosynthesis, the importance of carbon in photosynthesis, and discusses other significant topics related to the process of photosynthesis. The chapters are well-structured and are contributed by experts in the field. The readers will gain ample knowledge from the new findings documented in the book.

I wish to thank my publisher for supporting me at every step. I would also like to thank all the authors who have contributed their researches in this book. I hope this book will be a valuable contribution to the progress of the field.

                                                                                    Editor

Part 1

Introduction

Photosynthesis: How and Why?

Mohammad Mahdi Najafpour* and Babak Pashaei

Chemistry Department, Institute for Advanced Studies in Basic Sciences (IASBS), Zanjan, Iran

1. Introduction

The total solar energy absorbed by Earth is approximately 3,850,000 exajoules per year. This was more energy in one hour than the world used in one year! *Nature* uses very wonderful and interesting strategies to capture the energy in an interesting process: *Photosynthesis*.

To know more about photosynthesis, the first we should know about phototrophy. Phototrophy is the process by which organisms trap photons and store energy as chemical energy in the form of adenosine triphosphate (ATP). ATP transports chemical energy within cells for metabolism. There are three major types of phototrophy: Oxygenic and Anoxygenic photosynthesis, and Rhodopsin-based phototrophy. Photosynthesis is a chemical process that converts carbon dioxide into different organic compounds using solar energy. Oxygenic and anoxygenic photosynthesis undergo different reactions in the presence and absence of light (called light and dark reactions, respectively). In anoxygenic photosynthesis, light energy is captured and stored as ATP, without the production of oxygen. This means water is not used as primary electron donor. Phototrophic green bacteria, phototrophic purple bacteria, and heliobacteria are three groups of bacteria that use anoxygenic photosynthesis. Anoxygenic phototrophs have photosynthetic pigments called bacteriochlorophylls. Bacteriochlorophyll a and b have maxima wavelength absorption at 775 nm and 790 nm, respectively in ether. Unlike oxygenic phototrophs, anoxygenic photosynthesis only functions using a single photosystem. This restricts them to cyclic electron flow only, and they are therefore unable to produce O_2 from the oxidization of H_2O. In plants, algae and cyanobacteria, the photosynthetic processes results not only in the fixation of carbon dioxide (CO_2) from the atmosphere but also release of molecular oxygen to the atmosphere. This process is known as oxygenic photosynthesis.

Photosynthesis captures approximately 3,000 EJ per year in biomass and produces more than 100 billion tons of dry biomass annually (Barber, 2009). Photosynthesis is also necessary for maintaining the normal level of oxygen in the atmosphere.

It is believed that the first photosynthetic organisms evolved about 3,500 million years ago. In that condition, the atmosphere had much more carbon dioxide and organisms used hydrogen or hydrogen sulfide as sources of electron (Olson, 2006). Around 3,000 million years ago, Cyanobacteria appeared later and changed the Earth when they began to oxygenate the atmosphere, beginning about 2,400 million years ago. This new atmosphere was a revolution for complex life. The chloroplasts in modern plants are the descendants of

---

* Corresponding Author

these ancient symbiotic cyanobacteri (Gould et al., 2008). In plants and algae, photosynthesis takes place in chloroplasts. Each plant cell contains about 10 to 100 chloroplasts (Fig. 1).

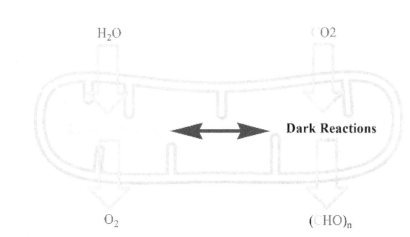

Fig. 1. In photosynthesis, organic synthesis and oxygen evolution reactions performs in two distinct enzymatic systems.

The chloroplast is composed of two membranes (phospholipid inner and outer membrane) and an intermembrane space between them. Within the membrane is an aqueous fluid called the stroma contains stacks (grana) of thylakoids, which are the site of photosynthesis. The thylakoids are flattened disks, bounded by a membrane with a lumen or thylakoid space within it. The site of photosynthesis is the thylakoid membrane, which contains integral and peripheral membrane protein complexes, including the pigments that absorb light energy, which form the photosystems. The first step in photosynthesis is the absorption of light by a pigment molecule of photosynthetic antenna resulting in conversion of the photon energy to an excited electronic state of pigment molecule. Plants absorb light primarily using the pigment chlorophyll. Besides chlorophyll, organisms also use pigments such as, phycocyanin, carotenes, xanthophylls, phycoerythrin and fucoxanthin (Fig. 2).

The most useful decay pathway is "energy transfer" to a photochemical reaction centers, and it is important to photosynthetic reactions. Excitons trapped by a reaction center provide the energy for the primary photochemical reactions. Subsequent electron transfer reactions occur in the dark which results in accumulation of chemical bound energy. In the other words, photosynthesis occurs in two stages. In the first stage, *light-dependent reactions* or *light reactions* capture the energy of light and use it to make the energy-storage molecules (ATP). During the second stage, the *light-independent reactions* use these products to capture and reduce carbon dioxide (Govindjee et al., 2010). The dark reaction doesn't directly need light, but it does need the products of the light reaction.

In the light reactions, a chlorophyll molecule of reaction center absorbs one photon and loses one electron. This electron is passed to a modified form of chlorophyll called pheophytin, which passes the electron to a quinone molecule, allowing the start of a flow of electrons down an electron transport chain that leads to the ultimate reduction of NADP to NADPH.

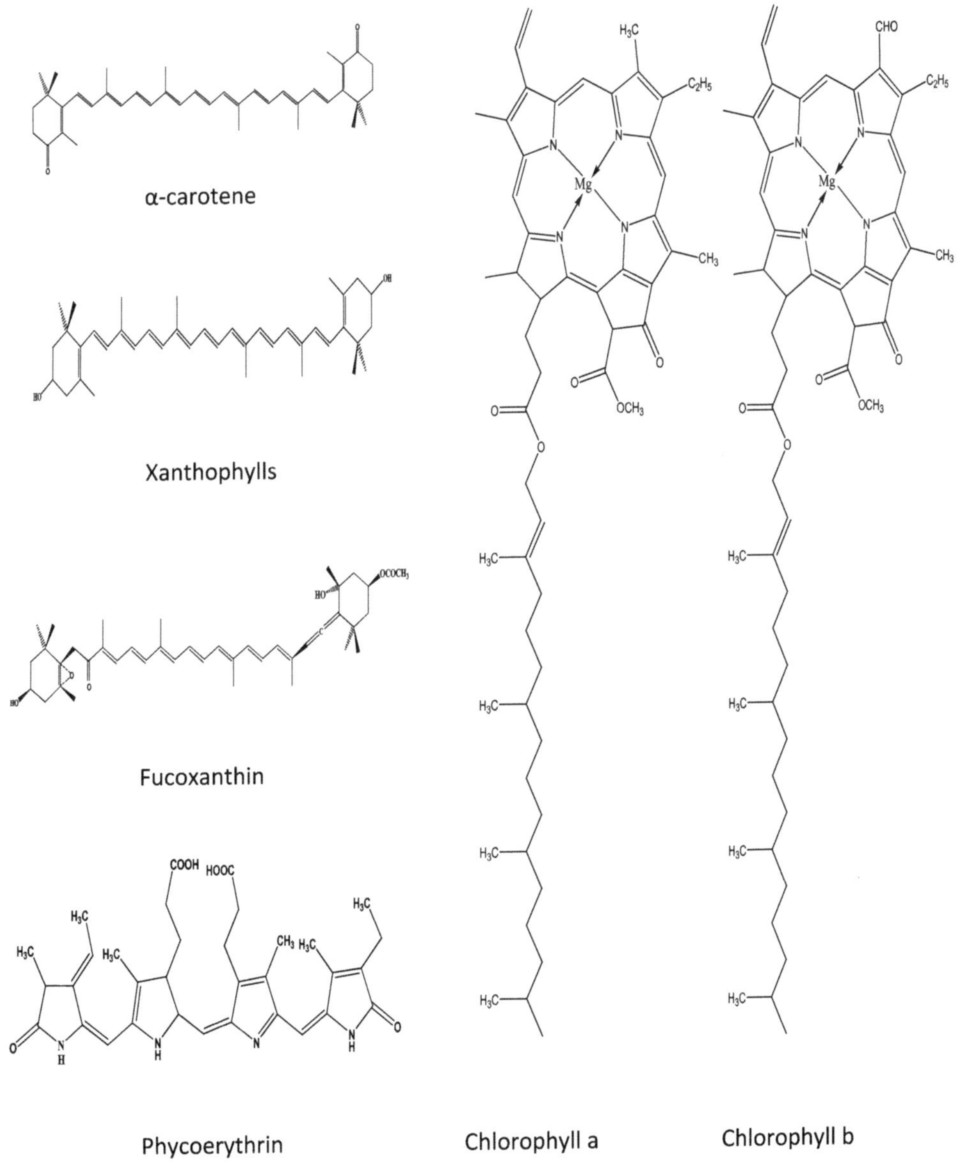

Fig. 2. Plants absorb light primarily using some pigments.

The proton gradient across the chloroplast membrane is used by ATP synthase for the concomitant synthesis of ATP. The chlorophyll molecule regains the lost electron from a water molecule and oxidizes it to dioxygen (O_2):

$$2H_2O + 2NADP^+ + 3ADP + 3P_i + light \rightarrow 2NADPH + 2H^+ + 3ATP + O_2$$

A good method to study of oxygen evolution in this process is to activate a photosynthetic system with short and intense light flashes and study of oxygen evolution reaction. Joliot's experiments in 1969 showed that flashes produced an oscillating pattern in the oxygen evolution and a maximum of water oxidation occurred on every fourth flash (Satoh et al., 2005). These patterns were very interesting because splitting of two water molecules to produce one oxygen molecule requires the removal of also four electrons. In 1970, Kok proposed an explanation for the observed oscillation of the oxygen evolution pattern (Kok et al., 1970). Kok's hypothesis (Kok et al., 1970) is that in a cycle of water oxidation succession of oxidizing equivalents is stored on each separate and independent water oxidizing complex, and when four oxidizing equivalents have been accumulated one by one an oxygen is spontaneously evolved (Kok et al., 1970). Each oxidation state of the water oxidizing complex is known as an "S-state" and S_0 being the most reduced state and S_4 the most oxidized state in the catalytic cycle (Fig. 3) (Kok et al., 1970). The S_1 state is dark-stable. The $S_4 \rightarrow S_0$ transition is light independent and in this state oxygen is evolved. Other S-state transitions are induced by the photochemical oxidation of oxidized chlorophyll (P_{680}^+) (Satoh et al., 2005).

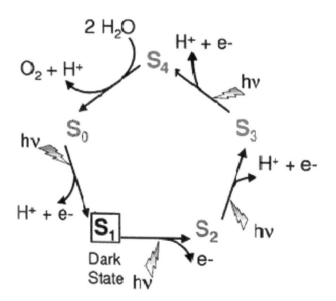

Fig. 3. Catalytic cycle proposed by Joliot and Kok for water oxidation, protons and electrons at photosystem II. The figure was reproduced from Sproviero et al., 2008.

Recently, Umena et al. (Umena et al., 2011) reported crystal structure of this calcium-manganese cluster of photosystem II at an atomic resolution. In this structure one calcium and four manganese ions are bridged by five oxygen atoms. Four water molecules were found also in this structure that two of them are suggested as the substrates for water oxidation (Fig. 4).

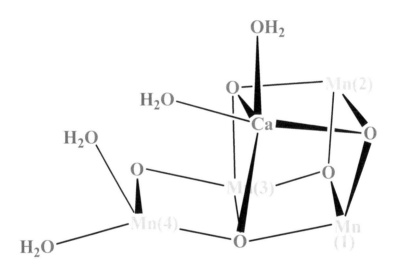

Fig. 4. The structure of water oxidizing complex (WOC) (Umena et al., 2011).

Light-dependent reactions occur in the thylakoid membranes of the chloroplasts in plants and use light energy to synthesize ATP and NADPH. Cyclic and non-cyclic are two forms of the light-dependent reaction. In the non-cyclic reaction, the photons are captured in the light-harvesting antenna complexes of photosystem II by different pigments (Fig. 5 and Fig. 6).

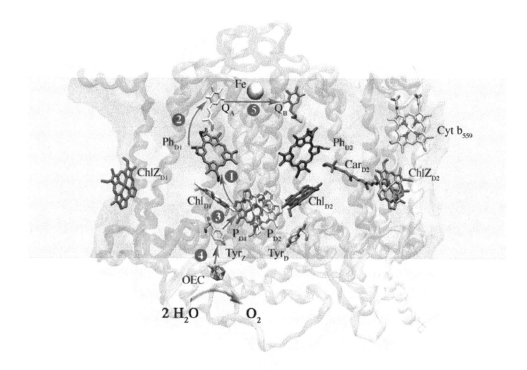

Fig. 5. Map of the main cofactors of PSII. The arrows show the electron transfer steps and the numbers indicate the order in which they occur. The figure was reproduced from Herrero et al., 2010.

When a chlorophyll molecule in reaction center of the photosystem II obtains sufficient excitation energy from the adjacent antenna pigments, an electron is transferred to the primary electron-acceptor molecule, pheophytin, through a process called photoinduced charge separation. These electrons are shuttled through an electron transport chain, the so-called Z-*scheme* shown in Fig. 7, that initially functions to generate a chemiosmotic potential across the membrane.

Z-scheme diagram of oxygenic photosynthesis demonstrates the relative redox potentials of the co-factors in the linear electron transfer from water to NADP$^+$.

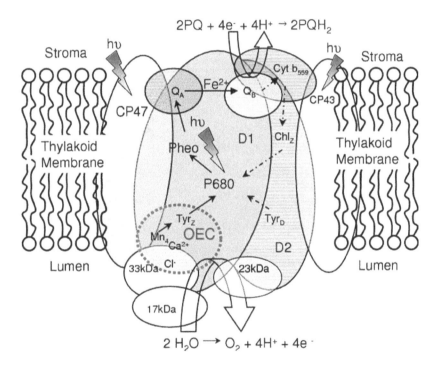

Fig. 6. Schematic representation of photosystem II and its components embedded in the thylakoid membrane. The figure was reproduced from Sproviero et al., 2008.

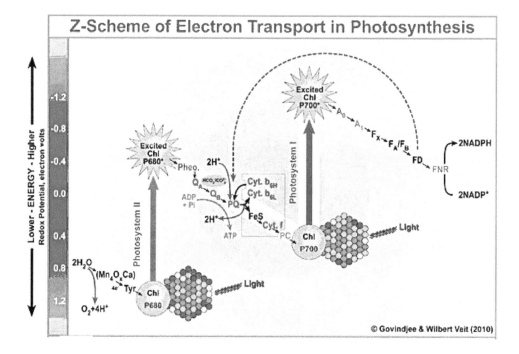

Fig. 7. Z-Scheme of Electron Transport in Photosynthesis (the picture provided by Govindjee and Wilbert Veit in http://www.life.illinois.edu/govindjee/photoweb/subjects.html#antennas).

An ATP synthase enzyme uses the chemiosmotic potential to make ATP during photophosphorylation, whereas NADPH is a product of the terminal redox reaction in the *Z-scheme*. Photosystem I operates at the final stage of light-induced electron transfer. It reduces $NADP^+$ via a series of intermediary acceptors that are reduced upon excitation of the primary donor P_{700} and oxidize plastocyanin. The cyclic reaction is similar to that of the non-cyclic, but differs in the form that it generates only ATP, and no reduced $NADP^+$ (NADPH) is created. The stored energy in the NADPH and ATP is subsequently used by the photosynthetic organisms to drive the synthesis in the Calvin - Benson cycle in the light-independent or dark reactions (Fig. 8).

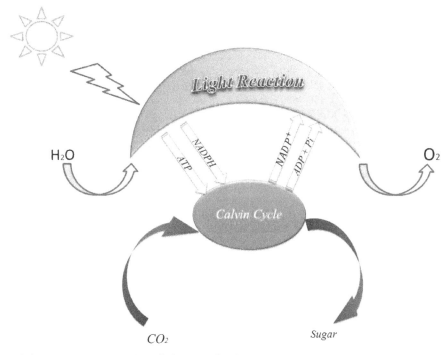

Fig. 8. Schematic representation of photosynthesis.

In these reactions, the enzyme RuBisCO captures CO_2 from the atmosphere and in a process that requires the newly formed NADPH, releases three-carbon sugars, which are later combined to form sucrose and starch. The overall equation for the light-independent reactions in green plants is:

$$3\ CO_2 + 9\ ATP + 6\ NADPH + 6\ H^+ \rightarrow C_3H_6O_3\text{-phosphate} + 9\ ADP + 8\ P_i + 6\ NADP^+ + 3\ H_2O$$

2. Why is photosynthesis important?

It is believed that photosynthesis is the most important biological process on earth. Our food, energy, environment and culture, directly or indirectly, depend on the important process. Really, the relationship between living organisms and the balance of atmosphere and life on earth needs knowledge of the molecular mechanisms of photosynthesis. The process also provides paradigms for sustainable global energy production and efficient energy transformation. Research into the nature of photosynthesis is necessary because by understanding photosynthesis, we can control it, and use its strategies for the improvement of human's life.

3. References

Barber, J. (2009). Photosynthetic energy conversion: natural and artificial. Chem. Soc. Rev, Vol. 38, pp. 185-196.

Gould, S.; Waller, R. & Mcfadden, G. (2008). Plastid evolution. Annu. Rev. Plant. Biol, Vol. 59, pp. 491-517.

Govindjee; Kern, J.F.; Messinger, J.& Whitmarsh, J. (2010) Photosystem II. In: Encyclopedia of Life Sciences (ELS). John Wiley & Sons, Ltd: Chichester.

Herrero, Ch.; Lassalle-Kaiser, B.; Leibl, W.; Rutherford, A.W.; Ally Coord, A. (2008). Artificial systems related to light driven electron transfer processes in PSII. Chem. Rev, Vol. 252, pp. 456-468.

Kok, B.; Forbush, B. & McGloin, M. (1970). Cooperation of charges in photosynthetic O_2 evolution: I. A linear four-step mechanism. Photochem. Photobiol, v. 11, p. 457 – 475.

Olson, J. (2006). Photosynthesis in the archean. Photosyn. Res, Vol. 88(2), pp. 109-117.

Satoh, K.; Wydrzynski, T.J. & Govindjee, (2005). Introduction to photosystem II. In: Wydrzynski, T.J. & Satoh, K. (eds) Photosystem II: The light-driven water: plastoquinone oxidoreductase. Advances in Photosynthesis and Respiration, v. 22. Springer, Dordrecht, p. 111-227.

Sproviero, E.M.; Gascón, J. A.; McEvoy, J.P.; Brudvig, G.W. & Batista V.S. (2008) Coord. Chem. Rev., v. 252, p. 395-415.

Umena, Y.; Kawakami, K.; Shen, J.R. & Kamiya, N. (2011). Crystal structure of oxygen-evolving photosystem II at a resolution of 1.9Å. Nature, v. 473, p. 55-60.

http://www.life.illinois.edu/govindjee/photoweb/subjects.html#antennas).

Part 2

Light and Photosynthesis

2

Carotenoids and Photosynthesis - Regulation of Carotenoid Biosyntesis by Photoreceptors

Claudia Stange and Carlos Flores
Universidad de Chile
Chile

1. Introduction

Carotenoids are isoprenoid molecules of 40 carbons which are synthesized in a wide variety of photosynthetic (plants, algae) and non photosynthetic (some fungi and bacteria) organisms. So far, over 750 carotenoid structures are known, and these are divided into nonoxygenated molecules designated as carotenes and into oxygenated carotenoids referred to as xanthophylls.

In photosynthetic organisms, carotenoids are synthesized in the plastids, such as chloroplasts. They are localized and accumulated in the thylakoid membranes of chloroplasts (Cunningham & Gantt, 1998), near the reaction center of photosystem II in the light harvesting complexes (LHC), along with other pigments such as chlorophyll *a* and *b*. Carotenoids act as accessory pigments in the LHC, where they absorb light in a broader range of the blue spectrum (400-500 nm) than chlorophyll. Carotenoids transfer the absorbed energy to chlorophyll *a* during photosynthesis (Britton, 1995). Carotenoids also protect plant cells from photo-oxidative damage as a result of their antioxidant characteristic giving by the conjugate bonds of the polyene chain (Britton, 1995; Britton et al., 1998). In this context carotenoids absorb the excess of energy from reactive oxygen species (ROS) and quench singlet oxygen produced from the chlorophyll triplet in the reaction center of photosystem II (Telfer, 2005). Carotenoids also protect the plant from photo-oxidative damage through thermal dissipation by means of the xanthophyll cycle (Baroli & Nigoyi, 2000). This process occurs when excessive light increases the thylakoid ΔpH, which activatates the enzyme violaxanthin de-epoxidase (VDE), converting violaxanthin to zeaxanthin. Zeaxanthin molecules and protons may change the conformation in the LHC, favoring the thermal dissipation.

Carotenoids are also synthesized and accumulated in chromoplasts, plastids that accumulate pigments in flowers, fruits and storage roots. Carotenoids are stored in lipid bodies or in crystalline structures inside the chromoplasts where they are more stable because they are protected from light (Vishnevetsky et al., 1999). In addition, carotenoids are precursors for apocarotenoids such as the phytohormones abscisic acid (ABA) and stringolactones. ABA is involved in dormancy, development and differentiation of plant embryos, stomata open-closure and in tolerance to abiotic stress (Crozier et al., 2000). The stringolactones act as shoot branching inhibitor hormones. Also they are involved in plant signaling to both harmful (parasitic weeds) and beneficial (arbuscular mycorrhizal fungi) rhizosphere residents (Walter et al, 2010).

In flowers and fruits, the presence of carotenoids serve also to attract pollinators and seed dispersal agents by the intense yellow, orange and red colors that they provide to these organs (Grotewold, 2006).

Animals are not able to synthesize carotenoids, so they have to be included in their diet. In animals, carotenoids are precursors of vitamin A (retinal) and retinoic acid, which play essential roles in nutrition, vision and cellular differentiation, respectively (Krinsky et al., 1994). These molecules have also antioxidant properties (Bartley & Scolnik, 1995) and therefore, oxidative damage, associated with several pathologies, including aging (Esterbauer et al., 1992), carcinogenesis (Breimer, 1990) and degenerative processes in humans, among others, can be resisted by the ingestion of carotenoids (Rao and Rao, 2007; von Lintig 2010).

2. Biosynthesis of carotenoid in plants

Carotenogenic genes are encoded in the nuclear genome and the synthesized proteins are targeted as preproteins to the plastids, where they are post-translationally processed.

Chlorophyll, carotenoids, and prenylquinones are key molecules that share early steps in the biosynthesis and directly derive from the plastidic isoprenoid biosynthetic pathway. This pathway starts within the 2-C-methyl-D-erythritol-4-phosphate (MEP) which provides isopentenylpyrophosphate (IPP) for the synthesis of the primal intermediate geranylgeranyl diphosphate (GGDP). The MEP pathway is involved in the IPP biosynthesis for plastidial isoprenodid, and the mevalonate (MEV) pathway is required for the synthesis of IPP for cytoplasmic sterols (brassinoesteroids, cytoquinins, ubiquinones, Figure 1). Despite these biosynthetic routes appear as independent and compartmentalized, a regulated metabolic cross-talk has been reported between them (Flügge & Gao, 2005).

The first step of the MEP pathway condenses glyceraldehyde- 3-phosphate and pyruvate — a reaction catalyzed by 1-deoxy-D-xylulose 5-phosphate synthase (DXS)- to produce deoxy-D-xylulose 5-phosphate (DOXP). Then, a reductive isomerization by a DOXP reductoisomerase (DXR) yields MEP; the introduction of a cytidyl moiety by 2-C-methyl-D-erythritol 4-phosphate cytidylyl transferase (CMS) produces 4-(cytidine 5′-diphospho)-2-C-methyl-D-erythritol that is further phosphorylated by 4-(cytidine 5′-diphospho)- 2-C-methyl-D-erythritol kinase (CMK) and then cyclised by 2-C-methyl-Derythritol 2,4-cyclodiphosphate synthase (MCS) to form 2C-methyl-D-erythritol 2,4-cyclo-diP. The final two reactions leading to IPP and DMAPP are carried out by (E)-4-hydroxy-3-methylbut-2-enyl diphosphate synthase (HDS) and reductase (HDR), respectively. All the enzymes of the MEP pathway reside in the stroma. Functional data suggest that the enzymes responsible for the biosynthesis of IPP and DMAPP via the MEP pathway in plants are soluble and localized to plastids (Lange & Ghassemian, 2003). IPP molecules synthesized in the plastids are isomerized to the allylic isomer, dimethylallyl pyrophosphate (DMAPP) through IPP isomerase (IPI). Three molecules of IPP condense with DMAPP to generate geranylgeranyl pyrophosphate (GGPP), in a process involving GGPP synthase (GGPPS, Figure 1). GGPPS is a central intermediate in the synthesis of plastidic isoprenoids: chlorophylls (phytyl side-chain), carotenoids and prenylquinones (isoprenoid side-chains, Figure 1).

For chlorophyll biosynthesis, the enzyme geranylgeranyl reductase (GGDR) catalyzes the formation of phytyl pyrophosphate (Phytul-PP) from GGPP and chlorophyll synthase (CHLG) catalyses the synthesis of chlorophyll *a* from Phytyl-PP and chlorophyllide (Figure 1). Chlorophyll a and b are precursors for tocopherols (Joyard et al. 2009).

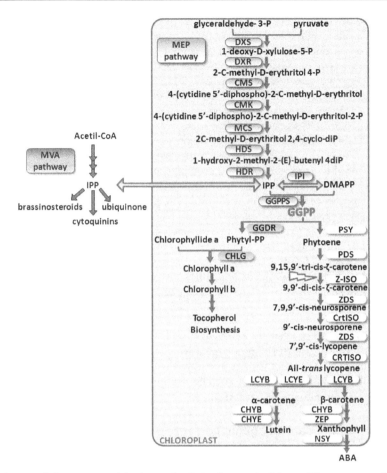

Fig. 1. **Scheme of the Isoprenoid Biosynthetic Pathways in Plants**. The non-mevalonate pathway (MEP) takes place in plastids and the mevalonate route (MEV) ocurrs in the cytoplasm of the cell. Isopentenylpyrophosphate (IPP) and geranylgeranyl pyrophosphate (GGPP) are key metabolites in the biosynthesis of chlorophylls and carotenoids. Abbreviations: 2-C-methyl-D-erythritol-4-phosphate (MEP), 1-deoxy-D-xylulose-5-phosphate synthase (DXS), 1-deoxy-D-xylulose 5-phosphate reductoisomerase (DXR), 2C-methyl-D-erythritol 4-phosphate cytidyltransferase (CMS), 4-(cytidine 5#-diphospho)-2-C-methyl-D-erythritol kinase (CMK), 2C-methyl-D-erythritol 2,4-cyclodiphosphate synthase (MCS), 1-hydroxy-2-methyl-2-(E)-butenyl 4-diphosphate synthase (HDS), 1-hydroxy-2-methyl-2-(E)-butenyl 4-diphosphate reductase (HDR), isopentenyl pyrophosphate isomerase (IPI), dimethylallylpyrophosphate (DMAPP), geranylgeranyl pyrophosphate syntase (GGPPS), phytoene synthase (PSY), phytoene desaturase (PDS), ζ-carotene isomerase (Z-ISO), ζ-carotene desaturase (ZDS), carotenoid isomerase (CrtISO), lycopene β cyclase (LCYB), lycopene ε cyclase (LCYE), β-carotene hydroxylase (CHYB), ε-carotene hydroxylase (CHYE), zeaxanthin epoxidase (ZEP), neoxanthin synthase (NSY), abscisic acid (ABA), geranylgeranyl reductase (GGDR), Chlorophyll synthetase (CHLG), light ⚡ .

With regard to the carotenoid pathway, two molecules of GGPP give rise to the colorless phytoene by means of phytoene synthase (PSY, Figure 1). The biosynthesis continues with the desaturation of phytoene to produce the pink-colored trans-lycopene. These reactions are catalyzed by two desaturases and two isomerases. The first desaturase, phytoene desaturase (PDS), catalyzes the biosynthesis of 9,15,9'-tri-cis-ζ-carotene, substrate of the 15-cis- ζ-carotene isomerase (Z-ISO) to produce 9,9'-di-cis- ζ-carotene. After the 15-cis- ζ-carotene isomerization, the second desaturase termed ζ –carotene desaturasa (ZDS) leads to the formation of 7,9,9'-cis-neurosporene and 7',9'-cis-lycopene. Finally, the carotene isomerasa (CRTISO) catalyzes the isomerization of this compound resulting in all-*trans* lycopene (Isaacson et al., 2004; Chen et al., 2010). Although isomerization can be mediated by light, carotenoid biosynthesis in "dark grown" tissues such as roots and etiolates leaves required Z-ISO and CRTISO enzymes.

Subsequently, lycopene is transformed into different bicyclic molecules by means of lycopene cyclases. Lycopene-β-cyclase (LCYB) converts lycopene into γ-carotene and afterward to β-carotene. Lycopene is also cyclized by lycopene-ε-cyclase (LCYE) and by LCYB toproduce α-carotene. The β-carotene is hydroxylated by the enzyme β-carotene hydroxylase (CβHx, CRTZ) to give rise zeaxanthin, while the hydroxylation of α-carotene by the ε-carotene hydroxylase (CεHx) and CβHx results in the formation of lutein. Abscisic acid is synthesized in the cytoplasm at the end of the pathway by the cleavage of violaxanthin and neoxanthin by carotenoid cleavage dioxygenases (CDE and NCED, Cunningham, 2002).

Some carotenoid enzymes act in multienzyme complexes in the stroma (isopentenyl pyrophosphate isomerase (IPI), geranylgeranyl pyrophosphate synthase (GGPPS) and phytoene synthase (PSY) and others are associated with the thylakoid membrane (phytoene desaturase (PDS), z-carotene desaturase (ZDS), lycopene β-cyclase (LCYB) and lycopene ε-cyclase (LCYE) (Cunningham & Gantt, 1998).

3. Regulation of the carotenogenic pathway

Due to the importance of carotenoids for plant and animal health, carotenoid biosynthesis regulation has been studied for the last 40 years both at the pure and applied levels. Nearly all carotenogenic genes in diverse plant species, algae, fungi and bacteria have been identified and characterized (Cunningham & Gantt, 1998; Cunningham, 2002; Howitt & Pogson, 2006; Cazzonelli & Pogson, 2010). The knowledge generated has been used to improve the nutritional value of several organisms, preferentially to metabolically engineer β-carotene and ketocarotenoid formation in plants (Ye et al., 2000; Davuluri et al., 2005; Aluru et al., 2008; Apel & Bock, 2009).

The regulation of carotenoid biosynthesis has been studied in photosynthetic organs (leaves) and in non-photosynthetic organs (fruits, flowers) of traditional plant models such as *Arabidopsis thaliana*, *Nicotiana tabacum* (tobacco) and *Solanum lycopersicon* (tomato) (Römer and Fraser, 2005; Howitt & Pogson, 2006).

Almost all of these studies show that carotenogenic genes are expressed in photosynthetic organs exposed to different light qualities, during the transition of etioplasts to chloroplasts (de-etiolation) which correlates with a high and concomitant increase in the carotenoid and chlorophyll levels (Römer & Fraser, 2005; Toledo-Ortiz et al., 2010).

During these processes, carotenogenic gene expression is mostly regulated at the transcriptional level mediated by photoreceptors such as the family of phytochromes

(PHYA-PHYE), cryptochromes (CRY) and phototropins. The reaction catalysed by psy has been shown to be the rate limiting step of carotenoid biosynthesis in plants and most studies on psy have been focused on the induction of its transcription by PHY and CRY during plant de-etiolation in *A. thaliana, maize,* tomato and tobacco. The expression of other carotenogenic genes such as lcyb, bhx, zep y vde is also induced in the presence of white light or during plant de-etiolation (Simkin et al., 2003; Woitsch & Römer, 2003; Briggs & Olney, 2001; Franklin et al., 2005; Briggs et al., 2007, Toledo-Ortiz et al., 2010).

3.1 Carotenoid gene activation mediated by photoreceptors in plants

Plant photoreceptors, include the family of phytochromes (PHYA-PHYE) that absorb in the red and far red range and cryptochromes (CRY) and phototropins that absorb in the blue and UV-A range (Briggs and Olney, 2001; Franklin et al., 2005; Briggs et al., 2007). Phytochrome (PHY) is the most characterized type of photoreceptor and their photosensitivity is due to their reversible conversion between two isoforms: the Pr isoform that absorbs light at 660 nm (red light) resulting in its transformation to the Pfr isoform that absorbs light radiation at 730 nm (far red). Once Pr is activated, it is translocated to the nucleus as a Pfr homodimer or heterodimer (Franklin et al., 2005; Sharrock & Clack, 2004; Huq et al., 2003;) where it accumulates in subnuclear bodies, called speckles (Nagatani, 2004). PHY acts as irradiance sensor through its active Pfr form, contributing to the regulation of growth and development in plants (Franklin et al., 2007). A balance between these two isoforms regulates the light-mediated activation of signal transduction in plants (Bae and Choi, 2008), Figure 2.

The signal transduction machinery activated by PHYA and PHYB promotes the binding of transcription factors such as HY5, HFR1 and LAF1 and the release of PIFs factors from light responsive elements (LREs) located in the promoter of genes that are up regulated during the de-etiolation process, such as the psy gene. The most common type of LREs that are present in genes activated by light are the ATCTA element, the G box1 (CACGAG) and G box (CTCGAG). PHYA, PHYB and CRY1, can also activate the Z-box (ATCTATTCGTATACGTGTCAC), another LRE present in light inducible promoters (Yadav et al., 2002). In *A.thaliana*, it has been shown that PHYA, but not PHYB, plays a role in the transcriptional induction of psy by promoting the binding of HY5 to white, blue, red and far red light responsive elements (LREs) located in its promoter (von Lintig et al., 1997). The involvement of the b-zip transcription factor HY5 in tomato carotenogenesis was proven with LeHY5 transgenic tomatoes that carry an antisense sequence or RNAi of the HY5 transcription factor gene. The transgenic Lehy5 antisense plants contained 24–31% less leaf chlorophyll compared with non-transgenic plants (Liu et al., 2004), while, immature fruit from Lehy5 RNAi plants exhibited an even greater reduction in chlorophyll and carotenoid accumulation.

Photosynthetic development and the production of chlorophylls and carotenoids are coordinately regulated by phytochrome –interacting factor (PIF) family of basic helix-loop-helix transcription factors (bHLH, Shin et al., 2009; Leivar et al. 2009) PIFs are negative regulators of photomorphogenesis in the dark. In darkness, PIF1 directly binds to the promoter of the psy gene, resulting in repression of its expression. Once etiolated seedlings are exposed to R light, the activated conformation of PHY, the Pfr, interacts and phosphorylates PIF, leading to its proteasome-mediated degradation (Figure 2). Light-triggered degradation of PIFs results in a rapid de repression of psy gene expression and a burst in the production of carotenoids in coordination with chlorophyll biosynthesis and chloroplast development, leading to an optimal transition to the photosynthetic metabolism (Toledo-Ortíz et al., 2010).

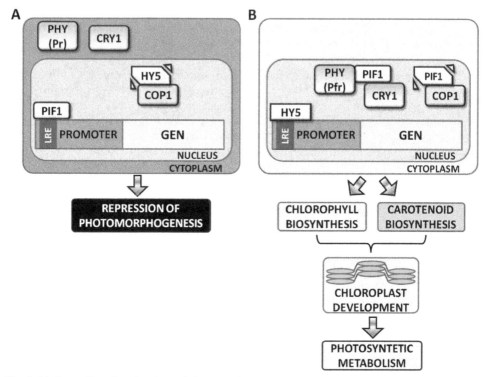

Fig. 2. **Ligh-mediated activation of the signal transduction involved in photomorphogenesis in plants.** The transition from dark conditions (A) to light conditions (B) allows the photosynthetic metabolism. Abbreviations: activated phytochromoe (PHY-Pr), cryptochrome 1 (CRY1), transcription factor LONG HYPOCOTYL 5 (HY5), constitutive photomorphogenic 1 (COP1), phytochrome interacting factor (PIF1), light response element (LRE).

Microarray transcriptome analysis during seedling deetiolation indicated that the majority of the gene expression changes elicited by the absence of the PIFs in dark grown pifq seedlings (pif1 pif3 pif4 pif5 quadruple mutants) are normally induced by prolonged light in wild-type seedlings, such as the induction of numerous photosynthetic genes related to the biogenesis of active chloroplasts, auxin, gibberellins (GA), cytokinin and ethylene hormone pathway-related genes, potentially mediating growth responses and metabolic genes involved in the transition from heterotrophic to autotrophic growth.

Besides, other functions associated with PIFs have been described as: i) regulating seed germination; dormant Arabidopsis seeds require both light activation of the phytochrome system and cold treatment (stratification) to induce efficient germination. PIF1 repress germination in the dark and exerts this function, at least in part, by repressing the expression of the key GA-biosynthetic genes GA3ox1 and GA3ox2 and promoting the expression of the GA catabolic genes. PIF1 also promotes the expression of the abscisic acid (ABA)-biosynthetic genes, and represses the expression of the ABA catabolic gene, resulting in high ABA levels. PIF4 and PI5 also promote ii) Shade Avoidance Syndrome (SAS); the abundance of these proteins increases rapidly upon transfer of white-light grown seedlings to simulated shade. Pif4, pif5 and pif4 pif5 mutants have reduced hypocotyl-elongation and marker-gene responsiveness to this signal compared with wild type (Leivar & Quail, 2011).

The cryptochrome CRY, another type of photoreceptor, is also involved in carotenoid light mediated gene activation. Phytochrome and cryptochrome signal transduction events are coordinated (Casal, 2000); PHYA phosphorylates cryptochrome *in vitro* (Ahmad et al., 1998) and blue and UV-A light trigger the phosphorylation of CRY1 and CRY2 (Shalitin et al., 2002; Shalitin et al., 2003). CRY1 localizes in the cytoplasm during darkness and when plants are exposed to light, CRY1 is exported to the nucleus (Guo et al., 1999; Yang et al., 2000; Schepens et al., 2004). CRY2 which belongs to the same family as CRY1, is localized in the nucleus of plant cells during both light and dark periods (Guo et al., 1999). Overexpression of cry2 in tomato causes repression of lycopene cyclase genes, resulting in an overproduction of flavonoids and lycopene in fruits (Giliberto et al. 2005). It has been reported that zeaxanthin acts as a chromophore of CRY1 and CRY2, leading to stomatal opening when guard cells are exposed to light (Briggs, 1999). The blue/green light absorbed by these photoreceptors induces a conformational change in the zeaxanthin molecule, resulting in the formation of a physiologically active isomer leading to the opening and closing of stomata (Talbott et al., 2002).

CRY and PHY bind and inactivate COP1 through direct protein-protein contact (Wang et al., 2001; Seo et al., 2004). COP1 is a ring finger ubiquitin ligase protein associated with the signalosome complex involved in protein degradation processes via the 26S proteasome (Osterland et al., 2000; Seo et al., 2003). During darkness, COP1 triggers degradation of transcription factors committed in light regulation, such as HY5 and HFR1 (Yang et al., 2001; Holm et al., 2002; Yanawaga et al., 2004) whose colocalize with COP1 in nuclear bodies and are marked for post-translational degradation during repression of photomorphogenesis (Ang et al., 1998; Jung et al., 2005). Light promotes conformational changes of COP1, inducing the release of photomorphogenic transcription factors. Once these factors are released, they accumulate and bind to LREs located in the promoters of genes activated by light (Wang et al., 2001;Lin & Shalitin, 2003, Figure 2). Transgenic tomatoes over expressing a Lecop1 RNAi have a reduced level of cop1 transcripts and significantly higher leaf and fruit chlorophyll and carotenoid content than the corresponding non-transformed controls (Liu et al. 2004),.

The UV-damaged DNA binding protein 1 (DDB1) and the de-etiolated-1 (DET1) factors are also negative regulators of light-mediated gene expression, they interact with COP1 and other proteins from the signalosome complex, and lead to ubiquitination of transcription factors (Osterlund et al., 2000; Yanawaga et al., 2004). Post transcriptional gene silencing of det1 leads to an accumulation of carotenoids in tomato fruits (Davuluri et al., 2005). Highly pigmented tomato mutants, *hp1* and *hp2* display shortened hypocotyls and internodes, anthocyanin accumulation, strongly carotenoid colored fruits and an excessive response to light (Mustilli et al., 1999). HP1 and HP2 encode the tomato orthologs of DDB1 and DET1 in *A. thaliana*, respectively (Liu et al., 2004). Carotenoid biosynthesis in *hp2* mutants increased during light treatments, due to the inactivation of the signalosome, decreasing the ubiquitination of transcription factors involved in phytochrome/cryptochrome transduction mechanisms.

The involvement of other photoreceptors such as phototropins, phytochrome C and E or CRY2 in the activation of carotenogenic genes has been evaluated through mutants. PhyC mutants, revealed that PHYC is involved in photomorphogenesis throughout the life cycle of *A. thaliana* playing a role in the perception of day length and acting with PHYB in the regulation of seedling de-etiolation in response to constant red light (Monte et al., 2003). As outlined above, regulation of light-mediated gene expression at the transcriptional level is the key mechanism controlling carotenogenesis in the plastids. Nonetheless, Schofield & Paliyath (2005) demonstrated post-translational control of PSY mediated by phytochrome. In red light exposed seedlings, PHY is activated which lead to an increase in PSY activity (Schofield & Paliyath, 2005). Therefore, light by means of photoreceptors, regulates carotenoid biosynthesis through transcriptional and post-transcriptional mechanisms.

3.2 Carotenoid and chlorophyll biosynthesis are simultaneously regulated

As mentioned previously, carotenoids carry out an essential function during photosynthesis in the antennae complexes of chloroplasts from green organs. Therefore, the regulation of the biosynthesis of chlorophyll and carotenoid biosynthesis are associated in photosynthetic organs (Woitsch & Römer, 2003; Joyard et al., 2009).

The photosynthetic machinery is composed of large multisubunit protein complexes composed of both plastidial and nuclear gene products, therefore a proper coordination and regulation of photosynthesis-associated nuclear genes (PhANG) and photosynthesis-associated plastidic genes is thought to be critical for proper chloroplast biogenesis. Light and plastidial signals trigger PhANG expression using common or adjacent promoter elements. A plastidial signal may convert multiple light signaling pathways, that perceive distinct qualities of light, from positive to negative regulators of some but not all PhANGs. Part of this remodeling of light signaling networks involves converting HY5, a positive regulator of PhANGs, into a negative regulator of PhANGs. In addition, mutants with defects in both plastid-to-nucleus and CRY1 signaling exhibited severe chlorophyll deficiencies.

Thus, the remodeling of light signaling networks induced by plastid signals is a mechanism that permits chloroplast biogenesis through the regulation of PhANG expression (Rucke et al., 2007)

White light induces a moderate stimulation of the expression of ppox, that encodes for protophorphirine oxidase (PPOX), an enzyme involved in chlorophyll biosynthesis, and simultaneously induces the expression of several carotenogenic genes (lcyβ, cβhx, violaxanthin de-epoxidase (vde) and zeaxanthin epoxidase (zep) genes). In addition, the psy gene, the fundamental gene that controls the biosynthesis of carotenoids, is co-expressed with photosynthetic genes that codify for plastoquinone, NAD(P)H deshydrogenase, tiorredoxin, plastocianin and ferredoxin (Meier et al, 2011). Moreover, according to the induction of carotenogenic genes during de-etiolation, chlorophyll genes are also induced (Woitsh & Römer, 2003) and the inhibition of lycopene cyclase with 2-(4 chlorophenylthio-triethyl-amine (CPTA) leads to accumulation of non-photoactive protochlorophyllide *a* (La Rocca et al., 2007). Also, PIF1 has been shown to bind to the promoter of PORC gene encoding Pchilide oxidoreductase whose activity is to convert Pchlide into chlorophylls (Moon et al., 2008).

Chlorophyll and carotenoid biosynthesis are also regulated indirectly by light through the redox potential generated during photosynthesis. In this process, plastoquinone acts as a redox potential sensor responsible for the induction of carotenogenic genes, indicating that the biosynthesis of carotenoids is under photosynthetic redox control (Jöet et al., 2002; Steinbrenner & Linden, 2003; Woitsch & Römer, 2003).

Different experimental approaches were used to determine the regulatory mechanism in which carotenoid and photosynthetic components are involved to determine the chloroplast biogenesis. Arabidopsis *pds3* knockout mutant, or plants treated by norflurazon (NF) exert white tissues (photooxidized plastids) due to inactivation of PDS. The *immutans (im)* variegation mutant, that has a defect in plastoquinol terminal oxidase IMMUTANS (IM) termed PTOX that transfers electrons from the plastoquinone (PQ) pool to molecular oxygen, presents variegated leaves. Considering the PQ pool as a potent initiator of retrograde signaling, a plausible hypothesis is that PDS activity exerts considerable control on excitation pressure, especially during chloroplast biogenesis when the photosynthetic electron transport chain is not yet fully functional and electrons from the desaturation reactions of carotenogenesis cannot be transferred efficiently to acceptors downstream of the PQ pool (Foudree et al., 2010).

Several different types of electronic interactions between carotenoids and chlorophylls have been proposed to play a key role as dissipation valves for excess excitation energy.

In Arabidopsis, the carotenoids–chlorophyll interactions parameter correlates with the nonphotochemical quenching (NPQ), and the fluorescence quenching of isolated major light-harvesting complex of photosystem II (LHCII). During the regulation of photosynthesis, the carotenoids excitation occurs after selective chlorophylls excitation.

Furthermore, the new possibility to quantify the carotenoids–chlorophyll interactions in real time in intact plants will allow the identification of the exact site of these regulating interactions, using plant mutants in which specific chlorophyll and carotenoide binding sites are disrupted (Bode et al., 2009).

3.3 Regulation of carotenoid expression in photosynthetic organs

Light is a stimulus that activates a broad range of plant genes that participate in photosynthesis and photomorphogenesis. Carotenoids are required during photosynthesis in plants and algae and therefore, genes that direct the biosynthesis of carotenoids in these organisms are also regulated by light (von Lintig et al., 1997; Welsch et al., 2000; Simkin et al., 2003; Woitsch & Römer, 2003, Ohmiya et al., 2006; Briggs et al., 2007).

The process of de-etiolation of leaves has been used to compare the levels of carotenoids and gene expression in dark-grown plants versus plants that were transferred to light after being in darkness. During de-etiolation of *A. thaliana*, the expression of *ggpps* and *pds* genes are relatively constant, whereas expression of the single copy gene, *psy* and *hdr* are significantly enhanced (von Lintig et al., 1997; Welsch et al., 2000, Botella-Pavía et al., 2004). Evidence indicates that the transcriptional activation of *psy, dxs and dxr* is essential for the induction of carotenoid biosynthesis in green organs (Welsch et al., 2003; Toledo-Ortiz et al., 2010).

During de-etiolation of tobacco (*Nicotiana tabacum*) and pepper, xanthophyll biosynthesis genes are transcriptionally activated after 3 or 5 h of continuous white-light illumination (Simkin et al., 2003; Woitsch & Römer, 2003). In *A. thaliana* and tomato, *lcyβ* mRNA expression increases 5 times when seedlings are transferred from a low light to a high light environment (Hirschberg, 2001). With the onset of red, blue or white light illumination, significant induction of the expression of carotenogenic genes was documented in etiolated seedlings of tobacco, regardless of the light quality used (Woitsch & Römer, 2003). The expression level was dependent of phytochrome and cryptochrome activities. However, considerable differences in expression levels were observed with respect to the type of light used to irradiate the seedlings. For example, *psy* gene expression was significantly induced after continuous red and white light illumination, pointing to an involvement of different photoreceptors in the regulation of their expression (Woitsch & Römer, 2003). PHY is involved in mediating the up-regulation of *psy2* gene expression during maize (*Zea mays*) seedling photoinduction (Li et al, 2008). Also *Lcyβ*, *cβhx* and *vde* are induced upon red light illumination. However, *zep* shows similar transcriptional activation in the presence of red or blue light (Woitsch & Römer, 2003).

Compared to normal carotenogenic gene induction mediated by light, the contribution of photo-oxidation to the amount of carotenoids produced in leaves is also important. Carotenoids are synthesized during light exposure but when light intensity increases from 150 to 280 $\mu mol/m^2/s$, the rate of photo oxidation is higher than the rate of synthesis and carotenoids are destroyed, reaching a certain basal level (Simkin et al., 2003). The level of expression of some carotenogenic genes is also reduced following prolonged illumination at moderate light intensities (Woitsch & Römer, 2003). During darkness, when photo oxidation of carotenoids does not occur, biosynthesis of carotenoids in leaves is stopped due principally to the very low level of expression of carotenogenic genes. In *C. annum*, *psy*, *pds*, *zds* and *lcyβ* genes are down regulated in darkness (Simkin et al., 2003) while in *A. thaliana* the *psy* and *hdr* are active in darkness only at basal levels (Welsch et al., 2003, Botella-Pavía et al., *2004*).

3.4 Effect of light in non-photosyntetic organs

Light has not only been analysed in photosynthetic tissue as a regulatory agent. In actual fact, light effect on carotenogenic pathway has been report in a number of species during physiological processes like fruit ripening and flower development (Zhu et al., 2003; Giovanonni, 2004; Adams-Phillips et al., 2004; Ohmiya et al., 2006).

In tomato, normal pigmentation of the fruits requires phytochrome-mediated light signal transduction, a process that does not affect other ripening characteristics, such as flavor (Alba et al., 2000). During tomato fruit ripening, carotenoid concentration increases 10 to 14 times, due mainly to accumulation of lycopene (Fraser et al., 1994). An increase in the synthesis of carotenoids is required during the transition from mature green to orange in tomato fruits. During this process, a coordinated upregulation of *dxs*, *hdr*, *pds* and *psy1* is observed, whilst at the same time the expression of *lcyβ*, *cycβ* and *lcyε* decreased (Fraser et al., 1994; Pecker et al., 1996; Ronen et al., 1999; Lois et al., 2000; Botella-Pavía et al., 2004). Two *lcyβ* genes have been identified in tomato, *cycβ* and *lcyβ*. The first is responsible for carotenoid biosynthesis in chromoplasts whereas *lcyβ* performs this role preferentially in chloroplasts (Ronen et al., 1999). The down regulation of *lcyβ* and *cycβ* in tomato during ripening leads to an accumulation of lycopene in chromoplasts of ripe fruits (Pecker et al., 1996; Ronen et al., 1999). In *C. annuum*, *lcyβ* is constitutively expressed during fruit ripening leading to an accumulation of β-carotene and the red-pigmented capsanthin (Hugueney et al., 1995). The *psy* gene also plays a considerable role in controlling carotenoid synthesis during fruit development and ripening (Fraser et al., 1999, Giuliano et al., 1993) and during flower development (Zhu et al., 2002, Zhu et al., 2003). In tomato, two distantly-related genes, *psy1* and *psy2* code for phytoene synthase, and the former was found to be transcriptionally activated only in petals and ripening tomato fruits after continuous blue and white-light illumination (Welsch et al., 2000; Schofield & Paliyath, 2005; Giorio et al., 2008). Transgenic tomato plants expressing an antisense fragment of *psy1* showed a 97% reduction in carotenoid levels in the fruit, while leaf carotenoids remained unaltered due to the expression of *psy2* (Fraser et al., 1999). *psy2* is expressed in all plant organs, preferentially in tomato leaves and petals (Giorio et al., 2008), but in green or ripe fruits it is only expressed at low levels (Bartley & Scolnik, 1993; Fraser et al., 1999; Giorio et al., 2008). *psy1* is also induced in the presence of ethylene, the major senescence hormone implicated in fruit ripening, indicating that PSY is a branch point in the regulation of carotenoid synthesis (Lois et al., 2000).

Evidence emphasizing the importance of light effectors during fruit ripening and carotenoid accumulation was obtained through post-transcriptionally silencing of negative regulators of light signal transduction such as HP1 and HP2, as described above (Mustilli et al., 1999, Liu et al., 2004, Giovannoni, 2004). These high–pigment tomato mutants (*hp1* and *hp2*) have increased total ripe fruit carotenoids and are hypersensitive to light, having little effect on other ripening characteristics, similar to transgenic tomato plants that overexpress CRY (Davuluri et al., 2004; Giliberto et al. 2005).

The up regulation of carotenoid gene expression during ripening has also been reported in other species. In Japanese apricot (*Prunus mume*) *psy*, *lcyβ*, *cβhx* and *zep* transcripts accumulate in parallel with the synthesis of carotenoids (Kita et al., 2007). In juice sacs of Satsuma mandarin (*Citrus reticulata*), Valencia orange (*C. sinensis*) and Lisbon lemon (*C. limon*) the expression of carotenoid biosynthetic genes such as CitPSY, CitPDS, CitZDS, CitLCYb, CitHYb, and CitZEP increases during fruit maturation, co-ordinately with the synthesis of carotenes and xanthophylls (Kato et al., 2004). In citrus of the "Star Ruby" cultivar, the high level of lycopene was correlated with a decrease in *CβHx* and *lcyb2* expression, genes associated to the synthesis of carotenoids in chromoplast (Alquezar et al., 2009). In *G. lutea* analysis of the expression of carotenogenic genes during flower development and in different plant organs indicated that *psy* was expressed in flowers concomitant with carotenoid synthesis but not in stems and leaves (Zhu et al., 2002).

Carotenoids are also present in amyloplasts of potato and cereal seeds such as maize and wheat (*Triticum aestivum*; Panfili et al., 2004, Howitt & Pogson 2006; Nesterenko & Sink, 2003). Both potatoes and cereals accumulate low levels of carotenoid in the dark (Nesterenko and Sink, 2003) in contrast to the highly pigmented modified root of carrots.

Daucus carota L. (carrot, 2n=18) is a biennal plant whose orange storage or modified root is consumed worldwide. Orange carrot contains high levels of α-carotene and β-carotene (8 mg/g dry weight, Fraser, 2004) that together constitute up to 95% of total carotenoids in the storage root Baranska et al., 2006). The kinetics of the transcript accumulation of some of the carotenogenic genes correlates with total carotenoid composition during the development of storage roots grown in the dark (Clotault et al., 2008).

We are focused in the study of carotenoid regulation in this novel plant model, taken in account that carotenoids in carrot are synthesized in leaves exposed to light, and also in the storage root that develops in darkness. All carotenogenic genes in carrot are expressed in both, leaves and roots during plant development, but the expression level is higher in leaves maybe due the faster exchange rate of carotenoids during photosynthesis (Beisel y et al., 2010). *Lcyb1* gene presents the higher increase in transcript level during leaves development and the paralogous genes, *psy1* and *psy2* are differentially expressed during development.

In roots, the expression of almost all carotenogenic genes are induced during storage root development and it correlates with carotenoid accumulation. In this organ carotenoids are stored in plastoglobuli in the chromoplasts, where they are more photo-stable than in chloroplasts (Merzlyak & Solovchenko, 2002). Therefore, photo-oxidation does not affect carotenoid content in these organs, even when they are exposed to light.

When roots were exposed to light, they did not develop normally and the expression of almost all genes differs from the pattern obtained in dark-grown roots during development (Figure 3A). In addition, the roots developed in the presence of light have the same carotenoid composition and amount as in leaves (Stange et al., 2008; Fuentes et al., 2011 in preparation). The thin non-orange carrot root also accumulates chloroplasts instead of chromoplasts, as leaves, and the carotenoid gene expression profile is almost the same as those expressed in the photosynthetic organ.

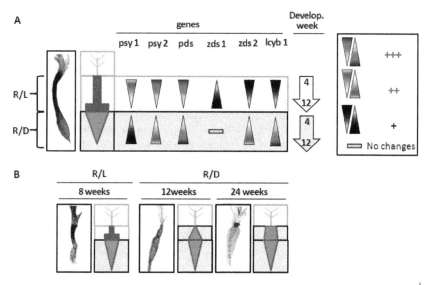

+++: high gene expression level, ++: middle gene expression level, +: low gene expression level. ⁄ :
expression increases during development, \ : expression decreases during development

Fig. 3. **Light affects morphology and carotenogenic gene expression in carrot roots A**; a comparison of carotenogenic gene expression in roots under light (R/L) and dark (R/D) conditions during the developmental process from 4 weeks to 12 weeks. Abbreviations: phytoene synthase 1 (psy 1), phytoene synthase 2 (psy 2), phytoene desaturase (pds), ζ-carotene desaturase 1 (zds1), ζ-carotene desaturase 2 (zds2), lycopene β cyclase 1 (lcyb1), Develoment (Develop). **B**; changes in the phenotype of a 8 weeks old carrot root grown in light (R/L) and then transferred to dark conditions (R/D) until 12 weeks and 24 weeks. The root normal development is inhibited by light in a reversible manner (Modified from Stange et al., 2008).

Also, when the carrot root of an 8 weeks old plant was transferred from light to darkness, the root started to develop (Figure 3B). Therefore, light alters the morphology and development of carrot modified roots in a reversible manner (Stange et al., 2008). Light inhibited storage root development, possibly because some transcriptional or growth factors are repressed, although more extensive studies are needed to investigate this phenomenon.

4. Conclusion

Light induces photomorphogenesis, chlorophyll and carotenoid biosynthesis through the signal transduction mediated by photoreceptors such as PHYA, PHYB and CRY in photosynthetic organs. At present, the principal components involved in the carotenogenic pathway have been described in many plant models, but fundamental knowledge regarding to the regulation is still necessary. In fact, *psy* gene may be the rate limiting step on carotenoid biosynthesis in leaves and also in chromoplasts accumulating organs. In addition, the highly regulated machinery on carotenoid biosynthesis can also be displayed through the organ specificity associated with carotenogenic gene function and their correlation with chlorophyll biosynthesis.

New strategies aimed to elucidate the regulation of carotenoid pathway could be associated with transcriptome analysis which could provide insights into regulatory branch points of the pathway. Conventional studies focused on the identification and characterization of carotenogenic gene promoters could also help to understand the regulation of the expression of the genes in photosynthetic and in non-photosynthetic organs. In fact, light responsive elements (LRE) in such promoters could be associated with transcription factors involved in carotenogenic and chlorophyll gene expression. On the other hand, research focused in the adjustment of the light- mediated signal transduction machinery would also be an effective metabolic approach for modulating chlorophyll and fruit carotenoid composition in economically valuable plants.

5. Acknowledgements

Acknowledgements to the Chilean Grant Fondecyt 11080066

6. References

Adams-Philips, L.; Barry, C. & Giovannoni, J. (2004). Signal transduction system regulating fruit ripening. *Trends Plant Sci*, Vol.9, No.7, (July 2004), pp. 331-338.

Ahmad, M.; Jarillo, JA.; Smirnova O. & Cashmore, AR. (1998). The CRY1 blue light photoreceptor of Arabidopsis interacts with phytochrome A in vitro. *Mol Cell*, Vol.1, No.7, (June 1998), pp. 939–948.

Alba, R.; Cordonnier-Pratt MM. & Pratt LH. (2000). Fruit localized phytochromes regulate lycopene accumulation independently of ethylene production in tomato. *Plant Physiol*, Vol.123, No.1, (May 2000), pp. 363-370.

Alquézar, B.; Zacarías, L. & Rodrigo, MJ. (2009). Molecular and functional characterization of a novel chromoplast-specific lycopene β-cyclase from *Citrus* and its relation to lycopene accumulation. *Journal of Experimental Botany*, Vol.60, No.6, (March 2009), pp.1783-97.

Aluru, M.; Xu, Y.; Guo, R.; Wang, Z.; Li, S.; White, W. & Rodermel, S. (2008). Generation of transgenic maize with enhanced provitamin A content. *J Exp Bot*, Vol.59, No.13, (Agust 2008), pp.3551-62.

Ang, LH.; Chattopadhyay, S.; Wei, N.; Oyama, T.; Okada, K.; Batschauer, A. & Deng, XW. (1998). Molecular interaction between COP1 and HY5 defines a regulatory switch for light control of Arabidopsis development. *Mol Cell*, Vol.1, No.2, (January), pp.213-222.

Apel, R.; Rock, R. (2009). Enhancement of Carotenoid Biosynthesis in Transplastomic Tomatoes by Induced Lycopene-to-Provitamin A Conversion. *Plant Physiol*, Vol.151, No.1, (September 2009), pp.59-66.

Averina, NG. (1998). Mechanism of regulation and interplastid localization of chlorophyll biosynthesis. *Membr. Cell Biol.*, Vol.12, No.5, pp.627-643.

Bae, G. & Choi, G. (2008). Decoding of light signals by plant phytochromes and their interacting proteins. *Annu Rev Plant Biol*, Vol.59, (June 2008), pp.281-311.

Ballesteros, ML.; Bolle, C.; Lois, LM.; Moore, JM.; Vielle-Calzada, JP.; Grossniklaus, U. & Chua, N. (2001). LAF1, a MYB transcription activator for phytochrome A signaling. *Genes Dev.* , Vol.15, No.19, (October), pp.2613-25.

Baranska M, Baranski R, Schulz H, Nothnagel T. (2006). Tissue-specific accumulation of carotenoids in carrot roots. *Planta*, Vol.224, No.5, (October 2006), pp. 1028-37.

Baroli, I. & Nigoyi, KK. (2000). Molecular genetics of xanthophylls-dependent photoprotection in green alge and plants. *Philos Trans R Soc Lond B Biol Sci.* Vol.355, No.1402, (October 2000), pp.1385–94.

Bartley, G. & Scolnik, P. (1993). cDNA cloning expression during fruit development and genome mapping of PSY2, a second tomato gene encoding phytoene synthase. *J Biol Chem*, Vol.268, No.34, (December 1993), pp.25718-21.

Bartley, G. & Scolnik, P. (1995). Plant carotenoids: pigments for photoprotection, visual attraction and human health. *Plant Cell*, Vol.7, No.7, (July 1995), pp.1032.

Beisel, KG.; Jahnke, S.; Hofmann, D.; Köppchen, S.; Schurr, U. & Matsubara, S. (2010). Continuous turnover of carotenes and chlorophyll a in mature leaves of Arabidopsis revealed by 14CO2 pulse-chase labeling. *Plant Physiol*, Vol.152, No.4, (April 2010), pp.2188 - 99.

Bode, S.; Quentmeier, C.; Liao, P.; Hafi, N.; Barros, T.; Wilk, L,; Bittner, F.; & Walla, PJ. (2009). On the regulation of photosynthesis by excitonic interactions between carotenoids and chlorophylls. *Proc Natl Acad Sci U S A*, Vol.106, No.30, (June 2009), pp. 12311–12316.

Botella-Pavía, P.; Besumbes, O.; Phillips, M.; Carretero-Paulet, L.; Boronat, A. & Rodríguez-Concepción, M. (2004). Regulation of carotenoid biosynthesis in plants: evidence for a key role of hydroxymethylbutenyl diphosphate reductase in controlling the supply of plastidial isoprenoid precursors. *Plant Cell*, Vol.40, No.2, (October 2004), pp.188–199.

Breimer, L. (1990). Molecular mechanisms of oxygen radical carcinogenesis and mutagenesis: the role of DNA base damage. *Mol. Carcinog.*, Vol.3, No.4, pp.188-197.

Briggs, W.; Tseng, T.S.; Cho, H-T.; Swartz, T.; Sullivan, S.; Bogomolni, R.; Kaiserli, E. & Christie, J. (2007). Phototropins and their LOV domains: versatile plant blue-light receptors. *J Integr Plant Biol*, Vol.49, No.1, (January 2007), pp.4-10.

Briggs, W. & Olney, M. (2001). Photoreceptors in plant photomorphogenesis to date. Five phytochromes, two Cryptochromes, one phototropin, and one superchrom. *Plant Physiol.*, Vol.125, No.1, (January 2001), pp.85-88.

Briggs, W. (1999). Blue-light photoreceptors in higher plants. *Annu Rev Cell Dev Biol*, Vol.15, (November 1999), pp.33-62.

Britton, G. (1995). Regulation of carotenoid formation during tomato fruit ripening and development. *J Exp Bot*, Vol.53, No.377, (October 1995), pp.2107-2113.

Britton G. (Ed.), Liaaen-Jensen S. (Ed.), Pfander H. (Ed.) (1998). *Carotenoids: Biosynthesis and Metabolism* (1 edition), Birkhauser Basel, ISBN-10: 3764358297, Switzerland.

Casal, JJ. (2000). Phytochromes, Cryptochromes, phototropin: Photoreceptor interaction in plants. *Photochem Photobiol*, Vol.71, No.1, (May 2000), pp.1–11.

Cazzonelli, CI. & Pogson, BJ. (2010). Source to sink: regulation of carotenoid biosynthesis in plants. *Trends Plant Sci*, Vol.15, No.5, (May 2010), pp. 266 - 274.

Chen, Y.; Li, F. & Wurtzel, E. (2010) Isolation and Characterization of the *S-ISO* Gene Encoding a Missing Component of Carotenoid Biosynthesis in Plants. *Plant Physiol*, Vol. 153, (May 2010) pp. 66-79.

Clotault J.; Peltier, D.; Berruyer, R.; Thomas, M.; Briard, M. & Geoffriau, E. (2008). Expression of carotenoid biosynthesis genes during carrot root development. *J Exp Bot*, Vol.59, No.13, (Agust 2008), pp.3563-73

Crozier, A.; Kamiya, Y.; Bishop, G. & Yolota, T. (2000). Biosynthesis of hormone and elicitor molecules. Pages 865-872. In: B. Buchanan, W. Gruissem, & R. Jones (eds.), Biochemistry and Molecular Biology of Plants. American Society of Plant Physiologist.

Cunningham, FX. & Gantt, E. (1998). Genes and enzymes of carotenoid biosynthesis in plants. *Annu Rev Plant Physiol Plant Mol Biol*, Vol.49, (June 1998), pp.557-583.

Cunningham, FX. (2002). Regulation of carotenoid synthesis and accumulation in plants. *Pure Appli Chem*, Vol.74, No., (8), pp.1409-17.

Davuluri, GR.; Van Tuinen, A.; Mustilli, AC.; Manfredonia, A.; Newman, R.; Burgess, D.; Brummell, DA.; King, SK.; Palys, J.; Uhlig, J.; Pennings, HM. & Bowler, C. (2004). Manipulation of DET1 expression in tomato results in photomorphogenic phenotypes caused by post-transcriptional gene silencing. *Plant J*, Vol.40, No.3, (November 2004), pp.344-354.

Esterbauer, H.; Gebiki, J.; Puhl, H. & Jurgens, G. (1992). The role of lipid peroxidation and antioxidants in oxidative modification of LDL. *Free Radic Biol Med*, Vol.13, No.4, (October 1992), pp.341-391.

Flügge, UI. & Gao, W. (2005). Transport of isoprenoid intermediates across chloroplast envelope membranes. *Plant Biol*, Vol.7, No.1, (January 2005), pp. 97-97.

Franklin, KA.; Larner, VS. & Whitelam, GC. (2005). The signal transducing photoreceptor of plants. *Int. J. Dev. Biol.*, Vol.49, No.5-6, pp.653-664.

Franklin, K.; T. Allen, & G. Whitelam. (2007). Phytochrome A is an irradiance-dependent red light sensor. *Plant J*, Vol.50, No.1, (April 2007), pp.108-117.

Fraser, PD.; Truesdale, MR.; Bird, CR.; Schuch, W. & Bramley PM. (1994). Carotenoid biosynthesis during tomato fruit development. *Plant Physiol.*, Vol.105, No.1, (May 1994), pp.405-413.

Fuentes, P.; Pizarro, L.; Handford, M.; Rodriguez-Concepción, M.& Stange C (2011). Light-dependent changes in plastid differentiation influence carotenoid gene expression and accumulation in carrot roots.*Plant Mol. Biol*, under evaluation, July 2011

Foudree,   A.;   Aluru,   M.;   &   Rodermel,   S.   (2010).   PDS activity acts as a rheostat of retrograde signaling during early chloroplast biogenesis. *Plant Signal Behav*, Vol.5, No.12, (December 2010), pp. 1629–1632

Giliberto, L.; Perrotta, G.; Pallara, P.; Weller, JL.; Fraser, PD.; Bramley, PM.; Fiore, A.; Tavazza, M. & Giuliano, G. (2005). Manipulation of the blue light photoreceptor cryptochrome 2 in tomato affects vegetative development, flowering time, and fruit antioxidant content. *Plant Physiol.*, Vol.137, No.1, (January 2005), pp.199-208.

Giorio, G.; Stigliani, AL. & D'ambrosio, C. (2008). Phytoene synthase genes in tomato (Solanum lycopersicum L.) – new data on the structures, the deduced amino acid sequences and the expression patterns. *FEBS J.*, Vol.275, No.3, (February 2008), pp.527–535.

Giovannoni, JJ. (2004). Genetic regulation of fruit development and ripening. *Plant Cell*, Vol.16, Suppl.1, (June 2004), pp.S170-S180.

Giuliano, G.; Bartley, GE. & Scolnik, PA. (1993). Regulation of carotenoid biosynthesis during tomato development. *Plant Cell*, Vol.5, No.4, (April 1993), pp.379-387.

Grotewold, E. (2006). The genetics and biochemistry of floral pigments. *Annu Rev Plant Biol*, Vol.57, (June 2006), pp.761-780.

Guo, H.; Duong, H.; Ma, N. & Lin, C. (1999). The Arabidopsis blue light receptor cryptochrome 2 is a nuclear protein regulated by a blue light-dependent posttranscriptional mechanism. *Plant J*, Vol.19, No.3, (August 1999), pp.279-287.

Hirschberg, J. (2001). Carotenoids biosynthesis in flowering plants. *Curr Opin Plant Biol*, Vol.4, No.3, (June 2001), pp.210-218.

Holm, M.; Ma, LG.; Qu, LJ. & Deng, XW. (2002). Two interacting bZIP proteins are direct targets of COP1-mediated control of light dependent gene expression in Arabidopsis. *Genes Dev.* , Vol.16, No.10, (May 2002), pp.1247–59.

Howitt, CA.; & Pogson, BJ. (2006). Carotenoid accumulation and function in seeds and non-green tissues. Plant Cell Environ., Vol.29, No.3, (March 2006), pp.435-445.

Hugueney P, Badillo A, Chen HC, Klein A, Hirschberg J, Camara B, Kuntz M. (1995). Metabolism of cyclic carotenoids: a model for the alteration of this biosynthetic pathway in Capsicum annuum chromoplasts. *Plant J*, Vol.8, No.3, (September 1995), pp. 417-424.

Huq, E.; Al-sady, B. & Quail, PH. (2003). Nuclear translocation of the photoreceptor phytochrome B is necessary for its biological function in seedling photomorphogenesis. *Plant J*, Vol.35, No.5, (September 2003), pp.660–664.

Isaacson, T.; Ohad, I.; Beyer, P. & Hirschberg, J. (2004). Analysis in vitro of the enzyme CRTISO establishes a poly-cis-carotenoid pathway in plants. *Plant Physiol.*, Vol.136, No.4, (December 2004), pp.4246-4255.

Joët, T.; Genty, B.; Josse, EM.; Kuntz, M.; Cuornac, L. & Peltier, G. (2002). Involvement of a plastid terminal oxidase in plastoquinone oxidation as evidenced by expression of the Arabidopsis thaliana enzyme in tobacco. *J Biol Chem*, Vol.277, No.35, (August 2002), pp.31623-31630.

Joyard, J.; Ferro, M.; Masselon, C.; Seigneurin-Berny, D.; Salvi, D.; Garin, J. & Rolland, N. (2009) Chloroplast Proteomics and the Compartmentation of Plastidial Isoprenoid Biosynthetic Pathways. *Molec Plant*, Vol. 2, No. 6, (November 2009), pp. 1154-80

Jung, IC.; Yang, JY.; Seo, HS. & Chua, NH. (2005). HFRA is target by COP1 E3 ligase for post-transcriptional proteolysis during phytochrome A signaling. *Genes Develop*, Vol.19, No.5, (March 2005), pp.593-602.

Kato, M.; Ikoma, Y.; Matsumoto, H.; Sugiura, M.; Hyodo, H. & Yano, M. (2004). Accumulation of carotenoids and expression of carotenoid biosynthetic genes during maturation in citrus fruit. *Plant Physiol.*, Vol.134, No.2, (February 2004),

Ohmiya A, Kishimoto S, Aida R, Yoshioka S, Sumitomo K. (2006). Carotenoid cleavage dioxygenase (CmCCD4a) contributes to white color formation in chrysanthemum petals. *Plant Physiol*, Vol 142, No.3, (November 2006), pp. 1193 -201.

Krinsky, NI.; Wang, XD.; Tang, G. & Russell, RM. (1994). Cleavage of β-carotene to retinoid. In book: in: Retinoids: From Basic Science to Clinical Applications (Livrea, MA & Vidali, G, Eds.) pp. 21-28, Birkhaüser, Basel , Alemania. ISBN 3-7643-2812-6

La Rocca N, Rascio N, Oster U, Rüdiger W. (2007). Inhibition of lycopene cyclase results in accumulation of chlorophyll precursors. *Planta*, Vol.255, No.4, (March 2007), pp.1019-29.

Lange, BM. & Ghassemian, M. (2003). Genome organization in Arabidopsis thaliana: a survey for genes involved in isoprenoid and chlorophyll metabolism. *Plant Mol Biol*, Vol.51, No.6, (April 2003), pp.925-948.

Leivar P., & Quail, PH. (2011) PIFs: pivotal components in a cellular signaling hub. Trends Plant Sci, Vol.16, No.1, (January 2011), pp.19-28.

Leivar, P.; Tepperman, J.M.; Monte, E.; Calderon, R.H.; Liu, T.L.; Quail, P.H. (2009) Definition of Early Transcriptional Circuitry Involved in Light-Induced Reversal of PIF-Imposed Repression of Photomorphogenesis in Young Arabidopsis Seedlings. *Plant Cell*, Vol.21, No.11, (November 2009), pp.3535-53.

Li, F.; Vallabhaneni, R. & Wurtzel, L. (2008). PSY3, a new member of the phytoene synthase gene family conserved in the poaceae and regulator of abiotic stress-induced root carotenogenesis. *Plant Physiol*, Vol.146, No.3, (March 2008), pp.1333-45.

Lin, C. & Shalitin, D. (2003). Cryptochrome structure and signal transduction. *Annu Rev Plant Biol*, Vol.54, (June 2003), pp.469-96.

Liu, Y.; Roof, S.; Ye, Z., Barry, C.; van Tuinen, A.; Vrebalov, J.; Bowler, C. & Giovannoni, J. (2004). Manipulation of light signal transduction as a means of modifying fruit nutritional quality in tomato. *Proc Natl Acad Sci USA*, Vol.101, No.26, (June 2004), pp.9897-9902

Lois, L.; Rodriguez, C.; Gallego, F.; Campos, N. & Boronat, A. (2000). Carotenoid biosynthesis during tomato fruit development: regulatory role of 1-deoxy-D-xylulose 5-phosphate synthase. *Plant J*, Vol.22, No.6, (June 2000), pp.503-513.

Meier, S.; Tzfadia, O.; Vallabhaneni, R.; Gehring, C. & Wurtzel, ET. (2011). A transcriptional analysis of carotenoid, chlorophyll and plastidial isoprenoid biosynthesis genes during development and osmotic stress responses in Arabidopsis thaliana. *BMC Syst Biol*, Vol. 5, No.77, (May 2011), pp. 1-19.

Merzlyak, MN. & Solovchenko, AE. (2002). Photostability of pigments in ripening apple fruit: a possible photoprotective role of carotenoids during plant senescence. *Plant Sci*, Vol.163, No.4, (October 2002), pp.881-888.

Monte, E.; Alonso, JM.; Ecker, JR.; Zhang, Y.; Li, X.; Young, J.; Austin-Phillips, S. & Quail, PH. (2003). Isolation and characterization of phyC mutants in Arabidopsis reveals complex crosstalk between phytochrome signaling pathways. *Plant Cell*, Vol.15, No.9, (September 2003), pp.1962-80.

Moon, J.; Zhu, L.; Shen, H., & Huq, E. (2008) PIF1 directly and indirectly regulates chlorophyll biosynthesis to optimize the greening process in Arabidopsis. *Proc Natl Acad Sci U S A*, Vol.105, No.27, (Lujy 2008), pp.9433-38.

Mustilli, A.; Fenzi, F.; Ciliento, R.; Alfano, F. & Bowler, C. (1999). Phenotype of tomato high pigment-2 mutants is caused for a mutation in the tomato homolog of Deetiolated1. *Plant Cell*, Vol.11, No.2, (February 1999), pp.145-157.

Nagatani, A. (2004). Light-regulated nuclear localization of phytochromes. *Curr Opin Plant Biol*, Vol.7, No.6, (December 2004), pp.708-711.

Nesterenko, S. & Sink, KC. (2003). Carotenoid profiles of potato breeding lines and selected cultivars. *HortScience*, Vol.38, No.6, (October 2003), pp.1173-77.

Osterlund, MT.; Hardtke, CS.; Wei, N. & Deng, XW. (2000). Targeted destabilization of HY5 during light-regulated development of Arabidopsis. *Nature*, Vol.405, No.6785, (May 2000), pp.462–466.

Panfili, G.; Fratianni, A. & Irano, M. (2004). Improved normal-phase high-performance liquid chromatography procedure for the determination of carotenoids in cereals. *J Agric Food Chem*, Vol.52, No.21, (October 2004), pp.6373-6377.

Pecker, I.; Gubbay R.; Cunningham, FX. & Hirshberg, J. (1996). Cloning and characterization of the cDNA for lycopene beta-cyclase from tomato reveal a decrease in its expression during tomato ripening. *Plant Mol Biol*, Vol.30, No.4, (February 1996), pp.806-819.

Rao, AV. & Rao, LG.. (2007). Carotenoids and human health. *Pharmacological Res*, Vol.55, No., (March 2007), pp.207-216.

Römer, S & Fraser, PD. (2005). Recent advances in carotenoid biosynthesis, regulation and manipulation. *Planta*, Vol.221, No., (June 2005), pp.305-308.

Ronen, G.; Cohen, M.; Zamir, D. & Hirshberg, J. (1999). Regulation of carotenoid biosynthesis during tomato fruit development: expression of gene for lycopene epsilon cyclase is down regulated during ripening and is elevated in the mutant delta. *Plant J*, Vol.17, No.4, (February 1999), pp.341-351.

Ruckle, ME.; DeMarco, SM. Larkin RM. (2007). Plastid Signals Remodel Light Signaling Networks and Are Essential for Efficient Chloroplast Biogenesis in Arabidopsis. *The Plant Cell*, Vol.19, (December 2007), pp.3944-60.

Schepens, I.; Duek, P. & Fankhauser, C. (2004). Phytochrome-mediated light signaling in Arabidopsis. *Curr Opin Plant Biol*, Vol.7, No.5, (October 2004), pp.564–569.

Schofield, A. & Paliyath, G. (2005). Modulation of carotenoid biosynthesis during tomato fruit ripening through phytochrome regulation of phytoene synthase activity. *Plant Physiol Biochem*, Vol.43, No.12, (December 2005), pp.1052-1060.

Schmid, VH. (2008). Light-harvesting complexes of vascular plants. *Cell Mol Life Sci*, Vol.65, No.22, (November 2008), pp.3619-3639

Seo, HS.; Yang, JY.; Ishikawa, M.; Bolle, C.; Ballesteros, ML. & Chua NH. (2003). LAF1 ubiquitination by COP1 controls photomorphogenesis and is stimulated by SPA1. *Nature*, Vol.423, No.423, (June 2003), pp.995–999.

Seo, HS.; Watanabe, E.; Tokutomi, S.; Nagatani, A. & Chua, NH. (2004). Photoreceptor ubiquitination by COP1 E3 ligase desensitizes phytochrome A signaling. *Genes Dev*, Vol.18, No.6, (March 2004), pp.617-622.

Shalitin, D.; Yang, H.; Mockler, TC.; Maymon, M.; Guo, H.; Guitelam, GC. & Lin, C. (2002). Regulation of Arabidopsis cryptochrome 2 by blue light-dependent phosphorylation. *Nature*, Vol.417, No.6890, (June 2002), pp.763–767.

Shalitin, D.; Yu, X.; Maymon, M.; Mockler, T. & Lin, C. (2003). Blue light-dependent in vivo and in vitro phosphorylation of Arabidopsis cryptochrome 1. *Plant Cell*, Vol.15, No.10, (October 2003), pp.2421–2429.

Sharrock, R. & Clack, T. (2004). Heterodimerization of type II phytochromes in Arabidopsis. *Proc Natl Acad Sci USA*, Vol.101, No.31, (August 2004), pp.11500-11505.

Shewmaker, CK.; Sheehy, JA.; Daley, M.; Colburn, S. & Ke, DY. (1999). Seed-specific overexpresion of phytoene synthase: increase in carotenoids and other metabolic effects. *Plant J*, Vol.20, No.4, (November 1999), pp.401-412.

Shin, J.; Kim, K.; Kang, H.; Zulfugarov, IS.; Bae, G.; Lee, CH.; Lee, D. & Choi, G. (2009) Phytochromes promote seedling light responses by inhibiting four negatively-acting phytochrome-interacting factors. *Proc NatlAcad Sci U S A*, Vol.106, No.18, (May 2009), pp.7660-65.

Simkin, AJ.; Zhu, C.; Kuntz, M. & Sandmann, G. (2003). Light-dark regulation of carotenoid biosynthesis en pepper (Capsicum annuum) leaves. *J Plant Physiol*, Vol.160, No.5, (May 2003), pp.439-443.

Stange, C.; Fuentes, P.; Handford, M. & Pizarro, L. (2008). Daucus carota as a novel model to evaluate the effect of light on carotenogenic gene expression. *Biol Res*, Vol.41, No.3, (April 2008), pp.289-301.

Steinbrenner, J. & Linden, H. (2003). Light induction of carotenoid biosynthesis genes in the green alga Haematococcus pluvialis: regulation by photosynthetic redox control. *Plant Mol Biol* 52, Vol., No.2, (May 2003), pp.343-356.

Talbott, L.; Nikolova, G.; Ortíz, A.; Shmayevich, I. & Zeiger, E. (2002). Green light reversal of blue-light-stimulated stomatal opening is found in a diversity of plant species. *Am J Bot*, Vol.89, No.2, (February 2002), pp.366-368.

Telfer, A. (2005). Too much light? How beta-carotene protects the photosystem II reaction centre. *Photochem Photobiol Sci*, Vol.4, No.12, (December 2005), pp.950-956.

Toledo-Ortiz, G.; Huq, E. & Rodrígurz-Concepción, M. (2010). Direct regulation of phytoene synthase gene expression and carotenoid biosynthesis by Phytochrome-Interacting Factors. Vol.107, No.25, (June 2010), pp. 11626-11631.

Vishnevetsky, M.; Ovadis, M. & Vainstein, A. (1999). Carotenoid sequestration in plants: the role of carotenoid associated proteins. *Trends Plant Sci*, Vol.4, No.6, (June 1999), pp. 232-235.

von Lintig J. (2010). Colors with functions: elucidating the biochemical and molecular basis of carotenoid metabolism. *Annu Rev Nutr*, Vol.30, (August 2010), pp. 35-56.

Von Lintig, J.; Welsch, R.; Bonk, M.; Giuliano, G.; Batschauer, A. & Kleinig, H. (1997). Light-dependent regulation of carotenoid biosynthesis occurs at the level of phytoene synthase expression and is mediated by phytochrome in Sinapsis alba and Arabidopsis thaliana seedlings. *Plant J*, Vol.12, No.3, (September 1977), pp. 625-634.

Walter, MH.; Floss, D. & Strack, D. (2010). Apocarotenoids: hormones, mycorrhizal metabolites and aroma volatiles. *Planta*, Vol.232, No.1, (April 2010), pp 1-17.

Wang, H.; Ma LG.; Li, JM.; Zhao, HY. & Deng, XW. (2001). Direct interaction of Arabidopsis cryptochromes with COP1 in light control development. *Science*, Vol.294, No.5540, (August 2001), pp.154–158.

Welsch, R.; Beyer, P.; Hugueney, P.; Kleinig, H. & von Lintig, J. (2000). Regulation and activation of phytoene synthase, a key enzyme in carotenoid biosynthesis, during photomorphogenesis. *Planta*, Vol.211, No.6, (November 2000), pp.846-854.

Welsch, R.; Medina, J.; Giuliano, G.; Beyer, P. & von Lintig, J. (2003). Structural and functional characterization of the phytoene synthase promoter from Arabidopsis thaliana. *Planta*, Vol.216, No.3, (January 2003), pp.523-534.

Woitsch, S. & Römer, S. (2003). Expression of xanthophyll biosynthetic genes during light-dependent chloroplast differentiation. *Plant Physiol*, Vol.132, No.3, (July 2003), pp.1508-1517.

Yadav, V.; Kundu, S.; Chattopadhyay, D.; Negi, P.; Wei, N.; Deng, XW. & Chattopadhyay, S. (2002). Light regulated modulation of Z-box containing promoters by photoreceptors and downstream regulatory components, COP1 and HY5, in Arabidopsis. *Plant J*, Vol.31, No.6, (September 2002), pp.741-753.

Yanawaga, J.; Sullivan, JA.; Komatsu, S.; Gusmaroli, G.; Suzuki, G.; Yin, J.; Ishibashi, T.; Saijo, Y. ; Rubio, V.; Kimura, S.; Wang, J. & Deng, XW. (2004). Arabidopsis COP10 forms a complex with DDB1 and DET1 in vivo and enhances the activity of ubiquitin conjugating enzymes. *Genes Dev*, Vol.18, No.17, (September 2004), pp.2172–2181.

Yang, HQ.; Tang, RH. & Cashmore, AR. (2001). The signalling mechanism of Arabidopsis CRY1 involves direct interaction with COP1. *Plant Cell*, Vol.13, No.12, (December 2001), pp.2573–2587.

Ye, X.; Al-Babili, S.; Klot, A.; Zhang, J.; Lucca, P.; Beyer, P. & Potrycus, I. (2000). Engineering the provitamin A (β-carotene) biosynthetic pathway into (carotenoid-free) rice endosperm. *Science*, Vol.287, No.5451, (January 2000), pp.303-305.

Zhu, C.; Yamamura, S.; Koiwa, H.; Nishihara, M. & Sandmann, G. (2002). cDNA cloning and expression of carotenogenic genes during flower development in Gentiana lutea. *Plant Mol Biol*, Vol.48, No.3, (February 2002), pp.277-285.

Zhu, C.; Yamamura, S.; Nishihara, M.; Koiwa, H. & Sandmann, G. (2003). cDNAs for the synthesis of cyclic carotenoids in petals of Gentiana lutea and their regulation during flower development. *Biochem et Biophys Acta*, Vol.1625, No.3, (February 2003), pp.305-308.

3

The Guiding Force of Photons

Kevin M. Folta

*Horticultural Sciences Department, Graduate Program in Plant Molecular
and Cellular Biology, University of Florida, Gainesville, FL
USA*

1. Introduction

In a book titled *Photosynthesis* it is easy to forget that light is not simply the energy driving plant metabolism. Light also is the central environmental factor that affects plant size, shape and development. In fact, light activation of photomorphogenic signaling pathways sets the stage for photosynthesis and ensures the maintenance of the apparatus. The effects of specific wavebands of light exert their influence on plant biology from the molecular level all the way up to the higher morphological level, and even contribute to the canopy form as a whole. The wide influence is based on the fact that the light wavelengths that optimally activate photosynthesis also strongly modulate mechanisms that control plant morphology, such as the length of internodes, expansion of leaves or even leaf position. The same light qualities also guide the development, activity and maintenance of the chloroplast, as the demands of light-driven autotrophy require specialized communication and coordination between the plastid and nucleus to ensure full function of the organelle. Contrastingly, the light qualities that lend relatively little power to photosynthesis provide important information about the ambient environment as well that lead to adaptive adjustments in physiology.

Clearly, integrating information from the light environment is an important prerequisite of photosynthesis, as it sets the stage for photosynthetic activity and later maintains the core apparatus. Precise regulatory mechanisms guide the non-photosynthetic plastid, the etioplast, toward photosynthetic competence. Sensing of the first photons of light sparks a rapid cascade of events that shift the role of the plastid from a warehouse of essential materials to a dynamic center of metabolism. The control of gene expression associated with the conversion of etioplast to chloroplast has been well described, and is a central theme of this chapter.

Nuclear genes are required for photosynthetic competence. Studies of Ribulose bisphosphosphate carboxylase-oxygenase small subunit (*Rbcs*) and Chlorophyll a/b binding protein (*cab*; synonymous with Light-harvesting, chlorophyll-binding protein or *Lhcb*) transcript accumulation have been models of photomorphogenic gene expression going back almost three decades. The current literature describes the transition in the parlance of genome-wide changes, and a complementary proteomics literature adds additional understanding of how the dark-state of the plastid matures rapidly into a light-harvesting sub-cellular machine.

The first major heading of this chapter will cover the qualities of light and the receptors that sense them. The approach will be more historical and provide an understanding of how each receptor system delivers a signal and some of the processes that are controlled. Development and competence of the plastid is the second area of emphasis, describing events that transform the etioplast to the chloroplast. The final portion of the chapter will discuss the communication between the chloroplast and the nucleus. These separate compartments must be in constant and precise communication to ensure coordinated gene expression that fulfills the requirements of the plastid for new proteins. While many of the proteins required for photosynthesis are encoded in the plastid itself, a subset of important genes reside in the nucleus, and their precise expression is required for normal chloroplast operation. Many of these are subunits of chloroplast protein complexes that are non-functional in the absence of nuclear encoded subunits. Careful communication between these compartments has been the subject of interest for decades, and recent findings have illuminated how antero- and retrograde mechanisms might mediate this critical network.

2. Connecting light to gene expression and development

2.1 Not all wavebands are created equal

Back in seventh grade I was introduced to the Spectronic 20, or "Spec 20" for short. For those readers that are unfamiliar, it is essentially a tan breadbox with two dials, an analog meter, a dial to determine the wavelength transmitted, and a chamber for introducing a sample in a test tube. The device was originally made by Bausch & Lomb back in 1954, and even modern iterations reflect the basic, industrial, sturdy simplicity of the original model.

Back in 1980 we were given the charge to determine which wavelengths chlorophyll absorbed best. I don't remember how we purified our sample, I just know that I zeroed the machine using a blank filled with water using the dial on the left, lowered the sample into the chamber, closed the lid, and then recorded the values for light absorption as we marched across the dial — from the UV to past the red wavelengths. From these readings we'd build a graph that would reflect an absorption spectrum for chlorophyll, absorbing light in the blue and red most efficiently, while offering little to no absorption in the green, yellow, orange and far-red regions of the spectrum.

Some basic hypotheses could have been constructed from these findings. Certainly the qualities of light that excite chlorophyll must provide information to the plant as well. In my third year of college I learned that this was so. I learned that red and blue light would trigger photomophogenic development. We discussed effects in a variety of plants, from peas to mung beans, to tobacco, to *Lemna* as well as studies of chloroplast orientation in the green alga *Mougeotia*. There were even studies in a strange plant called *Arabidopsis thaliana*. All showed developmental effects of light, but mostly blue and red wavelengths. The correlation between the wavelengths that stimulated development and drove light-regulated metabolism was no surprise. Some wavelengths impart valuable information to the plant that promotes growth, development and photosynthetic capacity. Other wavelengths, like far-red and green, are not so important for metabolism but they are not benign- they shape plant processes in other ways that optimize light capture and adaptation.

2.2 The light sensor collection

Plants interact with the ambient light environment through a series of light sensors. These specialized molecules capture photons of discrete wavelengths and initiate downstream signaling events that ultimately lead to changes in gene expression, development, and/or morphology. It is important to remember that plants rely on these environmental cues to drive, or in some cases constrain, their development. In the context of photosynthesis, light signaling is important to consider in two general contexts. The transition from etiolate to autotrophic growth is driven by light. As the developing seedling meanders through the soil, sensitive light receptors are in place and prepared to ignite a downstream flow of events upon capture of a photon. These signaling events in many cases prepare the plastid, shifting its function from that of an etioplast to the metabolic center of the chloroplast.

The sensory networks that transduce information starting with the capture of a photon into a suite of downstream responses are well known. Historically these pigments were postulated to control various aspects of plant growth and development, particularly germination, phototropic movements and the transition between vegetative and reproductive growth. Long before the genes and proteins were identified and characterized, a tremendous body of work produced evidence of their activities and effects on physiology. Light quality effects on germination were examined by Lewis H. Flint where he demonstrated the promotive effects of red light (Flint, 1936). These studies were expanded through collaboration with E.D. McAlister, a physicist that utilized a spectrograph to split the spectrum and illuminate seeds with discrete wavelengths. Together Flint and McAlister generated elegant action spectra that illustrated how red light promoted germination, while blue, green and far-red light were inhibitory (Flint and McAlister, 1937). Work E.S. Johnson followed earlier studies by Blaauw that implicated that shorter wavelengths of blue light were more effective in generating phototropic responses in oat coleoptiles (Johnson, 1937). Monochromatic light studies of this period are well documented in the book, *Pigment of the Imagination*, by Linda Sage (1992). The text documents the quest for higher fluence rates of pure monochromatic light, noting barriers like countless blown circuit breakers and generation of deadly gases, along with the use of all kinds of light sources from arc lamps and 200W incandescent filaments. One interesting passage describes the development of a large spectrograph, cobbled together from parts obtained from streetcars, movie theatres and other sundry sources, assembled in a windowless wine-racking room at the USDA laboratories in Beltsville, MD. This spectrograph projected a 14 m rainbow of light onto an adjacent wall powered by a 10,000 W cabon-arc lamp. This large spectral projection allowed great advances in understanding how specific light qualities affected discrete plant processes. Clearly different parts of the spectrum had unique abilities to spur developmental or morphological changes—and even plants placed outside of the visible spectrum exhibited treatment effects, indicating that plants responded to a wider span of wavelengths than the human eye.

In discussion of photoreceptor families it is important to define the nomenclature, as first presented in a Plant Cell Letter to the Editor (Quail et al., 1994). In this report the notation is as follows: wild-type gene: *PHY, PHYA, PHYD* mutant gene: *phy, phyA, phyD* apoprotein: PHY, PHYA, PHYD chromoprotein (apoprotein + chromophore) phy, phyA, phyD.

In the following pages the discussion on photosensory systems is broken down into sections on discovery, mechanism, and associated physiology.

2.2.1 The phytochromes

2.2.1.1 Discovery

The USDA spectrograph and tools like it led to the discovery that red and far-red light presented opposing effects on biological processes. Within a short time the red/far-red reversibility of Grand Rapids lettuce seed germination was described (Borthwick et al., 1952), and the floodgates of phytochrome research were open. Within several decades a series of light-sensing mutants were obtained from mutagenized *Arabidopsis thaliana* collections (Koornneef et al., 1980). Ultimately several of these would be shown to encode light-signaling components. The *hy3* mutation was shown to be a lesion in the phytochrome B receptor (Somers et al., 1991). Soon after, a separately-isolated mutant called *hy8* was shown to encode phytochrome A (Parks and Quail, 1993). These genetic studies now attached genes and their cognate proteins to processes controlled by red and far-red light. Additional phytochromes were isolated, a total of five in Arabidopsis, phyA-phyE (Clack et al., 1994). Phytochromes may be grouped by their stability in light. They Type I phytochromes are light-labile while the Type II's are light stable. In Arabidopsis phyA is the only Type I phytochrome, and it is also thought of as the dark phytochrome because of its abundance (Jordan et al., 1997). The other phytochromes are stable in light, where phyB makes up the majority of phytochrome in the cell (Chen et al., 2004; Franklin and Quail, 2010). The individual phytochromes form functional hetero- and homodimers (Sharrock and Clack, 2004; Clack et al., 2009).

2.2.1.2 Signaling mechanism

The hallmark photoreversiblity is achieved from switching phytochromes between two conformational states. In darkness, phytochromes exist in a form known as Pr. This form is biologically inactive and has an absorption peak of approximately 660 nm. When illuminated the Pr form converts to the Pfr form, which initiates biological activity (for review see Chen et al., 2004). The Pfr form may be photoconverted back to the Pr form immediately by illumination with far-red light, or over a longer period of time in darkness. Both conformations maintain some overlap in their spectral absorption profiles while maintaining their distinct sensitivities. When illuminated with light and an equilibrium is established between Pr and Pfr.

After conversion to Pfr phytochromes travel from the cytosol to the nucleus. The phyA receptor moves quickly to the nucleus. The phyA::GFP fusion proteins are detected only minutes after illumination, while the phyB receptor moves with different kinetics, showing up hours after light treatment (Kircher et al., 1999; Hisada et al., 2000). The timing of movement to the nucleus matches well with the earliest detected responses to phytochrome activation. Using high-resolution imaging and phytochrome mutants, Parks and Spalding (Parks et al., 1998) demonstrated that phyA and phyB activity control inhibition of stem elongation by red light with similar kinetics. The phyA receptor exerts a transient influence within minutes while phyB involvement is evident later, but persists longer. Here localization kinetics overlap impeccably with physiological events, suggesting that rate of nuclear localization is directly influencing plant growth and development. Later, regulated nuclear import of phyB using a steroid-inducible system demonstrated that both light and nuclear localization were required for phyB to induce its effects (Huq et al., 2003).

Once in the nucleus phytochromes interact with a suite of other proteins. Some of these were first identified in interaction screens using the C-terminal PHYB as bait (Ni et al., 1998).

Interactors were termed PHYTOCHROME INTERACTING FACTORS, or PIFs (for review Castillon et al., 2007; Leivar and Quail, 2011). Analysis of PIF function would prove complex. For instance, PIF3 (a bHLH protein) binds phyA or phyB upon illumination, yet through separate domains and with distinct affinities (Leivar and Quail, 2011). PIF3, PIF1 and PIF5 have been shown to be rapidly phosphorylated and degraded via a ubiquitin-dependent process, with half lives between 5-20 min. One interpretation is that PIFs repress photomorphognesis in darkness and they are degraded rapidly upon light exposure to initiate developmental responses (Leivar et al., 2008).

2.2.1.3 Associated physiology

Phytochromes are relevant to just about all aspects of light-mediated development because they absorb well in red, blue, far-red and UV portions of the spectrum. The sum of molecular and physiological processes controlled is too extreme to list here. The most relevant roles to applied phy biology are in regulation of plant stature. Phytochromes repress stem elongation, promote leaf expansion and alter plant body form in response to crowding. The phytochromes also contribute to flowering. In Arabidopsis phyA has a role in promoting flowering in response to far-red signals whereas phyB works against it after absorbing red (Valverde et al., 2004).

The role of phytochrome in establishing a platform for photosynthesis is clearly observed during photomorphogenic development. During this time there is a substantial contribution of phy to chloroplast developmental processes. Phytochromes regulate the accumulation of transcripts encoding CAB (LHCB) proteins required for anchoring the photosynthetic apparatus (Kaufman et al., 1985; Karlin-Neumann et al., 1988), as well as the small subunit of RUBISCO (Kaufman et al., 1984). Global analysis of gene expression shows that many transcripts encoding proteins destined for the plastid are induced within minutes to hours of phy activation (Tepperman et al., 2001; Tepperman et al., 2004). The major role of phy in plastid development is in de-repression of the PIF-mediated constraint of transcription and will be discussed later in this chapter. The effect is strong as developing seedlings treated with far-red light can actually be permanently disabled from greening and chloroplast development (Barnes et al., 1996).

2.2.2 The cryptochromes

2.2.2.1 Discovery

While the participation of phytochromes defined a mechanism for red light effects in many plant processes, there were clear effects of blue light that could not be easily ascribed to phytochromes. In fact, phototropic curvature, had been described as blue-favored by Charles Darwin (Darwin, 1897), complementing a battery of blue light responses characterized in the early part of the century (Briggs, 2006). Analysis of plant actions in response to red and/or blue light provided clear evidence that more than one photosensory pigment was involved in light responses. Throughout the 20th century there was considerable discussion about the nature of the pigment, fueled by experimentation in plants and fungi. Analysis of countless action spectra drove speculation that the receptor was based on a carotenoid, a flavin, or a pterin, since there were general peaks at 450, 475 and 420 nm that supported these possibilities, yet fine structure of action spectra left the absolute identification of the chromophore(s) ambiguous.

In 1979 Jonathan Gressel gave a name to the illusive photoreceptor controlling plant and cryptogam form and function in blue light, appropriately, *cryptochrome* (Gressel, 1979). In his review he notes that the name was "despised by many". Yet his moniker was quite accurate, as blue light responses would later be shown to be transduced by a series of receptors (including phytochromes) some requiring phytochrome co-activation. These ambiguities would hide the genetic nature of the cryptochrome gene for another fourteen years. Gressel also contended that the cryptochrome receptor would be the single blue-light receptor.

2.2.2.2 Molecular structure

The actual structure of the cryptochrome receptor was eventually elucidated in 1993, yet the path to its characterization was laid with a series of plants that grew long and tall under blue light in a 1980 report. A screen for light sensing mutants in a mutagenized *Arabidopsis thaliana* population revealed a seedling that failed to suppress elongation in light. This particular seedling, noted as *hy4* (the fourth of the hypocotyl elongation mutants), showed an especially strong presentation of the long-hypocotyl phenotype under blue light, moreso than red, green, far-red, or white light (Koornneef et al., 1980). These findings suggested a lesion in the blue-light sensing pathway. Later, several T-DNA mutants with long hypocotyls under blue light revealed the first sequence identity of the HY4 protein- a sequence that matched convincingly with the long-wavelength microbial DNA photolyases. DNA photolyases are chromophore-bound proteins that catalyze repair of pyrimidine dimers in DNA (REF). Later studies would show that the HY4 protein (later renamed to CRY1 for CRYPTOCHROME1) was the receptor controlling these blue light responses. The receptor maintains two chromophores — flavin adenine dinucleotide (FAD) and methenyltetrahydrofolate (MTHF), with the photon exciting MTHF and shuttling the excitation energy to FAD to initiate the signaling process (Cashmore et al., 1999).

2.2.2.3 Various types of cryptochromes

The cry proteins are distinguished by two domains that underlie its diverse functions (Lin and Shalitin, 2003). The first is an N-terminal photolyase related (PHR) domain. The other domain is a C-terminal extension. This latter domain is variable between the different cryptochromes and defines the function. While exhibiting variation in this extension there are short islands of conserved sequence. These motifs (from N to C) are DQXVP, an acidic region high in D and E, and a STAES sequence followed by GGXVP. Because their order and sequences are so highly conserved they are noted in the literature together as a DAS domain. The DAS organization has been conserved from the most rudimentary mosses to angiosperms, so cryptochromes date back approximately 400 million years. The DAS domain also dictates cry localization and interaction that defines how individual crys contribute to physiology. A comprehensive report on cry structure and function is presented by Lin and Shalitin (2003).

In Arabidopsis there are three cryptochromes. The CRY1 and CRY2 proteins are translated and then localized to the nucleus upon activation by light. The third cryptochrome is called cry3 or cry-DASH. This member is localized to the chloroplast and performs DNA repair, much like prototypical photolyases (Kleine et al., 2003). No other signaling role has been proposed.

A late flowering mutant (known at the time as *fha1*) would connect the cry2 receptor to control of the flowering transition (Guo et al., 1998; Mockler et al., 2003). The contribution to flowering time is probably the cryptochrome's most agriculturally relevant attribute. The transition is controlled by blue light activation of cry2, followed by its nuclear localization and enhanced stability (Valverde et al., 2004). The cry2 receptor contributes to seedling height under certain fluence rates (Lin et al., 1998), yet has potent effects on stem elongation during early development (Folta and Spalding, 2001). Perhaps the most well-studied output of cryptochrome activation is the modulation of gene expression. A number of reports have examined the role of cryptochromes using genomic-level analyses. Studies during blue light induced de-etiolation show that crys alter gene expression associated with the plant hormone gibberellic acid (Folta et al., 2003; Zhao et al., 2007), providing a means to connect light and cryptochromes to growth responses.

2.2.2.4 Signaling mechanism

What is the mechanism of cryptochrome action?   Great steps were made to pinpoint the transduction mechanism when the c-terminus of the CRY1 protein was overexpressed in transgenic plants (Yang et al., 2000). Such transgenic seedlings, known as CCT for CRY C-Terminus, exhibited *cop*-like phenotypes, meaning that that were presenting light-grown phenotypes even in darkness. This finding was exciting because it potentially linked the COP1 protein, a regulator known to repress the light response in darkness, to cry function. Two hybrid interactions and co-immunoprecipitation analysis in vitro would confirm the interaction between COP1 and the cryptochrome light sensors (Wang et al., 2001).

The mutant *hy5* locus was isolated as part of the original screen of Arabidopsis photomorphogenic mutants (Koornneef et al., 1980). The mutant exhibited light-insensitivity symptoms especially under blue light. Later it was observed that *HY5* exhibited epistatic interactions with *COP1* (Ang and Deng, 1994), a gene encoding a ring-finger E3 ubiquitin ligase that shows a constitutive photomorphogenic phenotype. *HY5* transcripts and proteins accumulated rapidly after illumination, presenting the hypothesis that they were causal to development. It was demonstrated that *HY5* was transcribed and translated in darkness, yet the protein did not accumulate (Osterlund et al., 2000). The lack of accumulation could be reversed with the application of proteosome inhibitors, indicating that HY5 was likely being degraded via a ubiquitin-mediated mechanism. Moreover, the protein also accumulated in the *cop1* mutant. The stage was set- a positive regulator of photomorphogenesis, HY5, was destabilized when it was not needed and mutation or pharmacological block of the degradation system caused hyperaccumulation. It was possible to infer a mechanism. Now how to connect it with the light sensor?

Studies soon after tested the possibility that the cry receptor itself interacted with the COP1 degradation system. Examination of COP1-cry interaction showed that the receptor did interact with COP1 through the CCT domain, and interaction between the receptor and COP1 would interrupt ubiquitin-mediated degradation of the positive regulator HY5 (Wang et al., 2001). The effect appears to take place predominantly in the nucleus.

While this mechanism is supported by many lines of evidence it is important to remember that crys also have effects outside of the nucleus. Constructs that exclude cry from the nucleus show physiological function (Wu and Spalding, 2007) and events at the depolarization events at the cell membrane seconds after illumination (Folta and Spalding, 2001) suggest that crys are indeed functional in other contexts.

The flavin chromophore of the cryptochromes, when activated, opens new absorption properties and signaling states for cryptochrome receptors. When treated with blue light the chromophore takes on a different oxidation state that absorbs in the green, yellow and into the red. Several lines of evidence show that the treatment of plants with green light can reverse cryptochrome mediated responses (Banerjee et al., 2007; Bouly et al., 2007), including anthocyanin accumulation, hypocotyl elongation and flowering. In this way the cry responses to blue light may be attenuated much like the red/far-red responses of phytochromes.

The cryptochromes are a stellar example of why examination of plant processes can have large-scale impacts. Cryptochromes were first identified in plants, yet since have been shown to have central positions in the animal circadian oscillator, and in magnetoperception that guides bird migration. Fungal cryptochromes have been identified, yet their precise functions remain elusive for the most part, and understood members bind DNA reminiscent of the cry3 (CRY-DASH) proteins of Arabidopsis. The cryptochrome receptors clearly control a great swath of responses relevant to all eukaryotes.

2.2.3 The phototropins

2.2.3.1 Discovery

Characterization of the cryptochromes gave plant science discrete receptors for red, far-red and blue light responses. A number of lines of evidence indicated that the effects of cryptochrome activation were distinct from those that regulated phototropism (Liscum et al., 1992; Liscum and Briggs, 1996; Lasceve et al., 1999), suggesting the existence of an additional blue light receptor class. A report in *Nature* showed that the *cry1cry2* double mutant was deficient in first-positive phototropism (Ahmad et al., 1998). Yet the results of this work did not bear out with further tests.

The pursuit for the blue-light photosensor controlling phototropism was heating up in concert with the characterization of the first cryptochrome receptor in the years leading up to 1993. Several independent research tracks were racing toward receptor identification that would ultimately converge in an Arabidopsis mutant with defects in the receptor. One approach was an attempt to identify the receptor genetically using the Arabidopsis system. The photophysiological characteristics of curvature were well understood in this species (Steinitz and Poff, 1986), and formed a sound basis for a mutant screen. Two non-complementing mutants with defects in phototropic curvature (JK224 and JK218) were isolated (Khurana and Poff, 1989). JK224 was defective in first-positive curvature, requiring substantially higher fluences to induce measurable change. The JK218 mutant also showed resistance to phototropic curvature, only bending after long treatments with unilateral blue light. Both mutants were perfectly gravitropic, suggesting that they were sensory mutants and not simply unable to respond to stimuli that induce differential growth (Khurana and Poff, 1989).

With a separate approach a team of scientists working under the direction of Winslow R. Briggs used biochemical methods to characterize the blue light sensor for phototropism. A 120 k-Da phosphorylated protein was identified in association with plasma membranes of pea epicotyls (Gallagher and Ellis, 1982; Short and Briggs, 1990; Short et al., 1993; Short et al., 1994). When biochemistry and physiology were compared through time and space, some important correlations were uncovered. The threshold and saturation in the phototropic fluence response (Baskin, 1986) mirrored the parameters of light induced phosphorylation (Short and Briggs, 1990; Short et al., 1992). The regions of the seedling that exhibit the strongest phototropic response show the highest degree of phosphorylation (Short and Briggs, 1990). Both responses obeyed the Bunsen-Roscoe Law of Reciprocity, and the phosphorylation reaction is complete just prior to the development of phototropic curvature. Such correlations were observed in other species as well (Palmer et al., 1993a; Palmer et al., 1993b).

The Arabidopsis mutants and the phosphorylation activity would become linked. Reymond et al. (1992) tested the diminutive Arabidopsis plants for the phosphorylation activity detected in the epicotyls of peas, zucchini, tomato and sunflower hypocotyls, and the coleoptiles of maize, barley and oat coleoptiles. The non-phototropic JK224 mutant exhibited low levels of phosphorylation upon illumination, suggesting that JK224 was in fact the receptor. On the other hand, JK218's levels were not significantly altered. This reported tied the autophosphorylation to phototropic curvature.

With a mutant genotype possessing defects in biochemistry and phenotype it would seem simple to move to the process of gene discovery. In the early 90's the Arabidopsis system was emerging as a tractable genetic system (Konieczny and Ausubel, 1993), it would be possible to map the gene if the phenotypes were robust. This was the problem. Poff's mutant phenotypes were solid, yet subtle, as first-positive phototropism would bend a seedling only 6-10 degrees. Furthermore, the defective seedlings did eventually bend over time. Because of the subtle differences, and the fact that these were not null mutants, it would have been extremely difficult to screen reliably in large populations suitable for genetic mapping.

Liscum and Briggs also performed a screen in the Arabidopsis system, yet they resorted to fast-neutron treated seeds in an attempt to find strong non-phototropic alleles (Liscum and Briggs, 1995, 1996). Four loci were identified. The nph1 mutant was allelic to JK224 and the nph3 mutant proved to correspond to JK218. The nph1 mutant had no detectable phosphorylation of the 120 kDa protein, leading to the hypothesis that it was locus encoding the receptor for phototropism. The gene was eventually cloned (Huala et al., 1997).

2.2.3.2 Phototropin structure

The NPH1 protein contains two highly similar domains reminiscent of LOV (Light, Oxygen and Voltage) PAS domains. These domains bind flavins (FAD) and perform a variety of functions relative to environmental sensing from bacterial aerotaxis to modulating K+ currents in Drosophila. NPH1 also possessed a serine-threonine kinase domain in the C-terminus. The NPH1 protein was shown to preferably bind FMN as a chromophore, and the absorption spectrum for the purified receptor mirrored that of phototropism (Christie et al., 1998). These findings prompted a functionalized name change from the locus NPH1 to the gene encoding the receptor, PHOT1.

Based on sequence homology the NPH-LIKE (NPL1) gene was soon identified (Jarillo et al., 1998). This sequence has a similar topology to that of NPH1, with the same conserved kinase region and LOV domains. The individual LOV domains (LOV1 and LOV2) have distinct roles in phototropin action. Mutation of Cys39 of the LOV domain in the LOV1 domain has no effect on phototropism, whereas this mutation in the LOV2 domain abolishes curvature. The LOV2 domain also is the critical domain for promoting leaf expansion (Cho et al., 2007). There is evidence that the role of the LOV1 domain is to attenuate LOV2 effects by acting as a site for dimerization.

Separating the LOV domains from the kinase domain is an alpha-helical hinge that holds LOV domains in proximity to the kinase domain. Upon activation with light, the protein opens around this hinge region, allowing the kinase to be phosphorylated (Tokutomi et al., 2008). This is the basis for the phototropin signaling mechanism.

2.2.3.3 Signaling mechanism

A tremendous wealth of information has arisen concerning the photocycle of the LOV domains and how it is translated into receptor function. The field of LOV domain receptors exploded from two labs in 2000 to at least 42 by 2004 (Letter to Plant Physiol. October 2010, Vol. 154, p. 1,). There are literally hundreds today.

The most progress has been made on understanding the light induced activation of the photosensor itself. The phot proteins associate with the plasma membrane. Here a photon of blue light is captured by the FMN chromophore bound to the LOV2 domain of the receptor. This excitation establishes a covalent bond between the FMN and the aforementioned Cys39 of the LOV domain. A conformational change occurs and an adjacent alpha helix (termed the J-domain) opens access to the kinase domain. The protein then autophosphorylates on multiple serine residues. These events are a simple sketch of how the receptor begins to excite the downstream events mediated by phototropins.

The mechanism that controls phototropism toward unilateral blue light relies on a simple starting point—the plant must transform a gradient of blue light into a chemical gradient capable of inducing differential growth. A framework for inter-molecular signaling in phototropic curvature was deduced from the members of the original genetic screen. The *nph1* (*phot1*), *nph3* and *nph4* mutants were all defective in phototropic curvature (Liscum and Briggs, 1996). As mentioned earlier, *NPH1* encodes the phototropin receptor. NPH3 is a phot1 interacting protein thought to be an adapter or scaffold protein (Motchoulski and Liscum, 1999). The action of NPH3 has remained unclear for the last decade, but recent studies show that members of a family that are likely involved in ubiquitination of substrates. The NPH3 protein is phosphorylated in darkness and upon light activation of phot1 it is dephosphorylated (Pedmale and Liscum, 2007).

Recent studies have shown that two of the PHYTOCHROME KINASE SUBSTRATE proteins, PKS1 and PKS2, co-immunoprecipitate with phot1 and phot2 (de Carbonnel et al., 2010). PKS1 binds to both phototropin1 and NPH3, and is required for phototropic curvature (Lariguet et al., 2006). These proteins have roles in the phot transduction to leaf position and flatness, but do not affect chloroplast relocation (de Carbonnel et al., 2010). A specific isoform of the 14-3-3 protein class also binds to phot1, but does not interact with phot2 (Sullivan et al., 2009). A growing list of proteins have been confirmed as interactors (reviewed in Inoue et al., 2010).

One confirmed interactor ties phototropin to redistribution of auxin. The auxin efflux carrier ABCB19 was shown to interact with phot1, and is a substrate for its kinase activity (Christie et al., 2011). The phosphorylated carrier fails to translocate auxin it accumulates in the cells, leading to lateral efflux by PIN3, developing the onset of phototropic curvature.

A variety of other downstream signaling events may be required for phot resposnes. Various reports have implicated calcium release from endomembranes or insolitol phosphates as potential links in phot signal transduction. These studies involved the use of pharmacological agents or reporters, so while coincident with phot signaling events, it is unclear how they precisely contribute to the processes.

2.2.3.4 Associated physiology in plants

Photosynthesis is constantly tuned on the biochemical and molecular level, yet many other adjustments happen at a level that one may witness simply with the naked eye and time. Phototropins dictate the position of plant organs and organelles to optimize light intercept.

They regulate guard cells that gate gas exchange may be restricted or opened to admit carbon dioxide. At the molecular level the transcripts of specific genes necessary for photosynthetic activity accumulate and decay in light-dependent ways. All of these diverse processes share phototropins as primary photoreceptors. All of these responses have the potential to affect on optimizing photosynthesis, which appears to be the major codifying theme for the phototropins.

The physiology regulated by phot1 and phot2 may be broken down into responses that have contrasting fluence thresholds, time courses and areas of action. The phot receptors have been implicated in phototropism (Christie et al., 1998), chloroplast relocation (Jarillo et al., 2001a; Kagawa et al., 2001), stomatal opening (Kinoshita et al., 2001), leaf expansion (Sakamoto and Briggs, 2002), control of stem elongation (Folta and Spalding, 2001), inflorescence, stem and petiole positioning (Kagawa et al., 2009), leaf positioning (Inoue et al., 2008), growth responses in low-light environments (Takemiya et al., 2005) and post-translational stability of transcripts encoding chloroplast-targeted proteins (Folta and Kaufman, 2003).

Its isolation as a genetic mutant proved the importance of phot1 to phototropism in Arabidopsis. The null *phot1* mutants show severe defects in phototropic curvature. There is some evidence of redundant function between the two phot receptors. For instance, while *phot1-5* is a null mutant, it eventually will bend toward unilateral light based on compensatory activity of phot2 (Sakai et al., 2001).

While the phot2 receptor has a clear role in phototropism in response to higher fluence blue light at lengthy time course, the receptor controls the predominance of other functions. The control of stomatal opening is also mediated by redundant function of the two phot receptors (Kinoshita et al., 2001). Both receptors are capable of modulating the response with similar light sensitivity and time course. The phot2 receptor also controls the accumulation of chloroplasts into a plane perpendicular to low fluence rate light. The chloroplasts move in the cell to orient themselves to optimize position for photosynthesis. This is known as the accumulation response. In times of low light the chloroplasts will align themselves to intercept incoming light. When light is extreme, the chloroplasts retreat to positions perpendicular to incoming light, shielding themselves essentially by hiding behind other chloroplasts. This is known as the avoidance response. It has been demonstrated that both phot1 and phot2 contribute to the accumulation response to low light, but the avoidance responses are controlled by phot2 (Jarillo et al., 2001a; Kagawa et al., 2001).

The phot1 receptor solely mediates the first phase of hypocotyl growth inhibition in response to blue light. Upon first illumination hypocotyl growth slows significantly within minutes. This primary, sensitive and early response is due to phot1 (Folta and Spalding, 2001) Sustained effects are cry dependent. The phot1 receptor also controls the stability of the *Lhcb* transcript in response to a short, single pulse of blue light. Whereas the accumulation from low fluence blue light requires the plant g-protein and the GCR1 receptor (Warpeha et al., 2007), the transcript is destabilized in a manner that requires phot1 (Folta and Kaufman, 2003). Phots have also been shown to control leaf expansion (Sakamoto and Briggs, 2002), leaf and petiole position (Kagawa et al., 2009), and will probably be shown to control solar tracking (Briggs and Christie, 2002). All of these responses, from molecular to macroscopic, utilize the phot system to optimize the position and content of the hardware for photosynthesis.

2.2.4 The other LOV domain photosensors

The central flavin-binding, photocycling domain of the phototropin receptor, the LOV domain, has emerged as a recurrent theme in many proteins spanning many species. Their direct connection to processes germane to photosynthesis is limited at this point, but their existence merits discussion. LOV domain proteins have been identified in non-vascular plants, several types of algae, in fungi and bacteria. They are found within transcription factors, kinases, phosphatases, and proteins with undefined function. Their function outside of plants is diverse, with LOV domain proteins regulating processes as ranging from plant light signaling to virulence in *Brucella* (Swartz et al., 2007), to transcriptional changes in fungi (Ballario et al., 1998).

In Arabidopsis three non-phot, LOV domain proteins reside in the genome. These same genes were identified in genetic screens for defects in the circadian clock and flowering time. These are ZEITLUPE/ADAGIO (Somers et al., 2000; Jarillo et al., 2001b), FKF1 (Nelson et al., 2000), and LKP2 (Schultz et al., 2001). All three undergo a photocycle that mirrors that of the phototropin LOV domains (Salomon et al., 2000). The three proteins share a common role in using light to coordinate the stability and accumulation of regulatory proteins.

The other main class of LOV domain proteins comes from studies in *Adiantum*. In these organisms phototropic curvature and chloroplast relocation, canonical blue light responses in plants, are induced by red light (Kawai et al., 2003). Genetic analysis of phototropic deficient mutants showed that the fern receptor is a hybrid between the red/far-red sensor phytochrome and the LOV-domain sensors. The receptor is a fusion between two receptor types that has adapted to exploitation of the understory.

2.2.5 A UV-B receptor

2.2.5.1 Discovery, structure and physiology

The light from the sun presents the plant with a double-edged problem. While necessary for photosynthetic growth, the mixture of light energies contain parcels of poison that could impart damage to DNA to the detriment of the organism. Plants being anchored to the earth by a root must therefore have means to detect ultraviolet (UV) light energies and tailor appropriate physiological and molecular countermeasures to combat the problems associated with UV exposure. Growing evidence to support this hypothesis has mounted for decades and recently resolved in the elucidation of a UV-B (280-320 nm) photosensor.

Observation of many plant physiological and molecular responses pointed to the existence of this receptor (for review, Ulm and Nagy, 2005; Jenkins, 2009). A suite of plant responses to UV-B were reported, including increases in intercellular calcium (Frohnmeyer et al., 1999), strong effects on hypocotyl growth inhibition (Shinkle et al., 2004; Shinkle et al., 2005), induction of genes associated with disease (Green and Fluhr, 1995), as well as patterns of global gene expression that differ from those observed from activation of cryptochromes or phytochromes (Ulm et al., 2004). Effects on stomatal opening have also been observed (Eisinger et al., 2003), and synergistic interactions with phytochromes have been long documented (Yatsuhashi and Hashimoto, 1985).

The quest for identification of the UV-B receptor followed a trail established from studies of other light sensors. As mentioned earlier, interaction between receptors and the ubiquitin E3 ligase COP1 is a regulatory node of light signaling. Additionally, photoreceptors have been shown to move to, or reside in, the nucleus upon illumination. The same patterns were observed for the protein UV RESISTANCE LOCUS 8 (UVR8; Kaiserli and Jenkins, 2007), and the results were UV-B specific (Favory et al., 2009). Mutations in UVR8 that abolished UV-B induced photomorphogenesis also impaired interaction with COP1, and interaction in yeast was UV-B dependent (Rizzini et al., 2011) presenting support for the hypothesis that UVR8 was the UV-B receptor. The mechanism of action was shown to be dependent on UVR8 dimers splitting to monomers when specific aromatic amino acids were activated by UV-B radiation. The UVR8 protein is constitutively expressed throughout the plant (Kaiserli and Jenkins, 2007; Favory et al., 2009), allowing all cells to maintain a system to respond to potentially damaging wavelengths.

2.2.6 Hypothetical green light receptors
A quick glance at the current receptor collection shows that the visible light spectrum is well blanketed with the absorption spectra of photosensors to receive it. As noted, the sensor collection extends plant signal perception clearly into the UV and far-red. Are there truly responses that cannot be account for by the current set of receptors? Are there likely to be more ways that a plant can sense the light environment? A series of green light responses that persist in the absence of known sensors suggest that there are additional players in the plant sensorium.

Green light can excite phytochrome, cryptochrome and phototropin responses, depending of course on fluence rate and time of illumination. Green wavebands can induce phytochrome-mediated germination (Shinomura et al., 1996), several effects via cryptochromes as discussed earlier (Banerjee et al., 2007; Bouly et al., 2007; Sellaro et al., 2011), and even phototropic curvature (Steinitz et al., 1985) that in retrospect must be phototropin dependent. Green light has also been shown to be transmitted efficiently within the plant body and efficiently drive photosynthesis in deeper layers of the leaf (Terashima et al., 2009).

However, examination of the literature presents a suite of green-light-dependent phenomena that cannot easily be described as the action of cryptochromes, phytochromes or LOV domain receptors. These actions are induced specifically by green wavebands (~500-540 nm) and tend to oppose those of red and blue light (for review, Folta and Maruhnich, 2007). Some of the first evidence was noted when plants were grown under white light, or the same light source with various parts of the spectrum filtered to skew the quality of illumination. In early studies Frits Went observed that tomato seedlings grown under white light (red, blue and green) had a lower dry mass than tomato plants grown under red and blue light alone (Went, 1957). The effect was observed across fluence rates, so it was not simply an effect of limiting photosynthetic capacity. It was as if the presence of green wavebands contradicted the effects of red and blue.

Later, similar "reversal" effects in plant growth and the performance of tissue cultures were observed (Klein et al., 1965; Klein and Edsall, 1967). A curious blue-green reversal of stomatal open was described in Arabidopsis (Frechilla et al., 2000), sunflower (Wang et al., 2011a), and other species (Talbott, 2002). During experiments testing Arabidopsis stem growth kinetics in response to blue and red light, effects of green illumination were observed that were quite unusual. Unlike the inhibition caused by other wavebands, green light caused an increase in stem elongation rate. This finding was surprising because the etiolated elongation rate was always presumed to be the most rapid. The response was analyzed for its photophysiological and genetic parameters (Folta, 2004) and the results indicated that the green light induced stimulation of hypocotyl growth rate was not likely mediated by known photosensors.

Based on the results of this study a microarray experiment assessed the state of the transcriptome in green light treated, etiolated seedlings. Surprisingly, a dim pulse of green light, far below "safelight" energies, excited large-scale changes in the transcriptome. The most conspicuous difference observed as the lower abundance of transcripts associated with the chloroplast, especially those playing a role in photosynthesis. Green light incuded reduction of steady-state transcripts encoding (among many others) the large subunit of RUBISCO (RbcL), psaA, and psbD was observed (Dhingra et al., 2006). This response was shown to be excited by low fluence pulses of light, occur within minutes, happen only in response to green light, and persist in the suite of photoreceptor mutants tested. These findings also indicated that a response to dim green light could drive a series of counterintuitive adaptive responses.

Additional observations now show that the addition of green wavebands to a background of red and blue light can attenuate light responses. Green light can induce shade avoidance phenotypes (Mullen et al., 2006; Zhang et al., 2011) and directly antagonize the effects of red light on stem growth inhibition, but not blue light (Y. Wang and K. Folta, unpublished). These effects point to the presence of a yet-to-be-characterized green light sensor that works in concert with other light sensing systems to optimize plant physiology in low-light environments.

3. The transition to photosynthetic competence

3.1 The plastic plastid

As mentioned previously, the plastid housed within the cells of the etiolate seedling is simply a structure poised to rapidly mature into a center of light-driven metabolic activity. In darkness the etiolated plastid, or etioplast, maintains a process known as skotomorphogenesis, or the developmental state occurring in the absence of light. The etioplast should not be considered simply a default, ground state. When we consider that the plastid has evolved from an endosymbiont that was photosynthetic (Margulis, 1970), the etioplast must be a derived state, a structure that provides a selective advantage for the emerging plant. This interpretation is supported by the observation that etioplasts are specialized. They feature a unique arrangement of thylakoid membrane precursors into a highly-ordered pro-lamellar body (Selstam and Widell-Wigge, 1993). This structure contains a storehouse of lipids and proteins (Selstam and Widell-Wigge, 1993; Kleffmann et al., 2007) required for the greening process. The prolamellar body's paracrystalline matrix also contains carotenoids, chlorophyll precursors and NADPH:protochlorophyllide oxidoreductase (Rosinski and Rosen, 1972 ; Selstam and Sandelius, 1984; Masuda and Takamiya, 2004). In angiosperms, POR is the central enzyme required for the production of chlorophyll via a light dependent reaction (Lebedev and Timko, 1998). In the etioplast POR complexes with protochlorophyllide which is immediately (within 2 ms) converted to chlorophyllide upon activation with light (Heyes and Hunter, 2005). POR is responsible for the majority of chlorophyll synthesis as the prolamellar body with unstacked prothylakoids transitions to the mature thylakoids of photosynthetically active chloroplasts (Solymosi et al., 2007). Not only is POR activity light dependent, but the expression of POR-encoding genes has also been shown to be driven by light. Here a handful of photons steers the competence of the developing chloroplast by generating chlorophyll. While necessary for photosynthesis, it is certainly not the sole entity that is required for the process. A cast of additional factors must be recruited to the rapidly developing plastid to facilitate photosynthetic functions. Their coordinated manufacture and assembly underlie maturation of the chloroplast during the transition to the light environment.

3.2 Molecular control of plastid development

Two of the phytochrome interacting factors (PIF1 and PIF3) have been generally shown to limit chloroplast development, primarily by interacting with the promoters of target genes. Their repression is lifted by activation of phytochrome as the PIFs are degraded by ubiquitin-mediated proteolysis

Its counterpart, PIF1, is generally regarded as a negative regulator of phytochrome activity. It also has been shown to be an active repressor of blue light response It also has been shown to repress chlorophyll biosynthesis (Huq et al., 2004) by binding to the G-box in genes associated with chlorophyll synthesis (Moon et al., 2008), as well as limit carotenoid biosynthesis by binding directly to the *PHYTOENE SYNTHASE* promoter (Toledo-Ortiz et al., 2010).

The phytochrome interacting bHLH protein PIF3 has been described to promote chloroplast development (Monte et al., 2004). Other reports examined early chloroplast development in *pif* mutants, unveiling clear roles as repressors of chloroplast development (Stephenson et al., 2009). The *pif1pif3* mutant exhibited a constitutively photomorphogenic phenotype in the dark. The plants accumulates protochlorophyllide, and showed more evidence of thylakoid stacking. Genes associated with chlorophyll and heme synthesis were also mis-regulated in the mutant, permitting accumulation of protochlorophyllide in darkness. The transition from darkness to light is phytochrome mediated, but the precise mechanisms that control the cross talk between compartments are now being elucidated. This is the subject of the next section of this chapter.

4. Biochemical communication between compartments

The hardware of photosynthesis is composed of many components that are encoded in the nucleus. Some of these components are labile, requiring a constant reloading of the plastid from parts transcribed in the nucleus, translated in the cytosol and located to the chloroplast. How do these two separate intracellular entities coordinate activities to ensure efficient interaction?

Earlier in this chapter there was a discussion of the etioplast and its light-driven transition to the chloroplast. This transition is critical, the timing is important, and the requirements of the plastid tax the cell as a whole. These organellar demands increase with the maintenance of autotrophy, as many proteins required for chloroplast function are encoded by genes in the nucleus. These two retrograde mechanisms have been referred to as developmental control and operational control, respectively (Pogson et al., 2008). To satisfy these demands, lines of careful biochemical coordination network the chloroplast and nucleus. This feedback between cellular genomes come as no surprise, as at some point (or probably many points) there was genetic exchange between the genes of the endosymbiont and the new resident cell. Evidence of this is rich, with islands of plastid genes present in the nucleus, likely benefiting from a finer control of gene expression, splicing and useful economic properties of the nuclear environment.

The cell requires the chloroplast to be in harmony with the nucleus- the two compartments working in concert with great precision. The chloroplast contains genome fewer than one-hundred open reading frames, yet function of the chloroplast requires over three thousand proteins. Regulatory steps that communicate demands to the nucleus to start construction of these proteins need to be precise. There are challenges to fluid exchange of signals between chloroplasts and the nucleus, in particular the presence of membranes that block passage of the vast majority of ions, peptides or other small molecules. The co-evolution between the endosymbiont and the plant cell had to involve a way to bypass these barriers.

There are many lines of evidence that show evidence of communication between chloroplast and nucleus. Pharmacological disruption of transcription in the plastid transcription or chlorophyll synthesis results in aberrant nuclear gene expression. Using tagetitoxin (a potent phytotoxin that inhibits select RNA polymerases including the plastidic one) to repress transcription in the chloroplast, Rapp and Mullet (1991) illustrated that the nuclear *cab* (now *Lhcb*) and *RbcS* transcripts failed to normally accumulate when the chloroplast was impaired. Other nuclear-encoded transcripts like actin responded normally, and the plants grew in a typical fashion. Disruption of plastid translation with lincomycin or disruption of carotenoid biosynthesis with norflurazon, an inhibitor of phytoene desaturase, also inhibits the normal accumulation of *Lhcb* and *RbcS* transcripts (Gray et al., 2003). The same treatments do not affect mitochondrial gene expression patterns. The chloroplast specificity was demonstrated through the use of erythromycin in peas, a compound that inhibits plastid translation but does not affect translation in the mitochondrion (Sullivan and Gray, 1999). The use of thujaplicin arrests the production of protochlorophyllide, causing a back-accumulation of Mg-ProtoIX and Mg-ProtoIXme. The treatment also hinders *Lhcb* accumulation (Oster et al., 1996), a result that is important to underscore, as chlorophyll biosynthetic mutants will later show similar effects when this step is interrupted.

While these pharmacological studies had utility, the introduction of nucleus-chloroplast signaling mutants brought new illumination to the processes of intra-compartmental feedback. The barley genotypes *albostrians* and *Saskatoon* fail to accumulate chlorophyll in various sectors. Analysis of nuclear gene expression indicated that *RbcS* and *Lhcb* gene expression levels were repressed in these regions (Hess et al., 1991; Hess et al., 1994). Early studies in other plants, including maize (Mayfield and Taylor, 1984) and mustard (Oelmüller et al., 1986), showed that plants deficient in carotenoid synthesis similar breakdowns in nuclear gene expression. There is also evidence that the redox state of photosynthetic electron transport affects the expression of these transcripts (Pfannschmidt et al., 2001; Masuda et al., 2003; Brautigam et al., 2009). Other evidence suggests that an accumulation of nuclear encoded proteins that fail to localize to the chloroplast is a retrograde signal (Kakizaki et al., 2009).

Together the observation that *Lhcb, RbcS,* and other nuclear genes are repressed when the chloroplast is not functioning correctly is significant because these transcripts accumulate rapidly in response to phytochrome activation. Active repression of these transcripts indicated that some factor reflecting the state of the chloroplast was overriding the normal response. The repression was selectively affecting specific nuclear genes, impairing activity of those required for chloroplast function. In times of internal dystrophy the plastid is instructing the nucleus that there is no need for various gene products.

The understanding of chloroplast-nuclear communication was accelerated with studies in *Arabidopsis thaliana*. By exploiting the powerful genetics of this system a series of mutants were isolated that affected retrograde signaling. It was well demonstrated that application of norflurazon treatment actively repressed *Lhcb* accumulation by disrupting chlorophyll synthesis. Therefore, mutagenized plants with lesions in the repressing pathway should be allow expression of a reporter from an *Lhcb* promoter in the presence of the norflurazon. Susek et al. (1993) utilized this approach, using a truncation of the Arabidopsis *CAB3* promoter to drive hygromycin resistance and the *uidA* (GUS) gene. Seedlings growing on hygromycin and norflurazon would be candidates for genotypes possibly deficient in the plastid to nuclear signal. Results could be confirmed using the colormetric detection. The results of this screen identified non-complementing alleles called *gun* (for genomes uncoupled) mutants — *gun1, gun2* and *gun3*. These mutants were deficient in *Lhcb* and *Rbcs* repression, indicating that nuclear gene expression could be uncoupled from the chloroplast, allowing light-mediated changes in gene expression while the chloroplast remained undeveloped (Susek et al., 1993). Three other loci, *gun4, gun5* (Mochizuki et al., 2001) and *gun6* (Woodson et al., 2011), were later isolated.

Analysis of *GUN1* shows it to encode a pentatricopeptide repeat protein localized to the chloroplast (Koussevitzky et al., 2007). Analysis of promoters affected by *gun1* (and also *gun5*) mutation presented a suite of genes that shared an abscisic acid response element in the promoter, suggesting a role for ABA in retrograde signaling. The *gun2* and *gun3* mutants were shown to possess lesions in heme oxygenase and biliverdin reducase, respectively. Later it was shown that *GUN4* encodes a protein required for normal Mg chelatase activity, while *GUN5* encodes a required subunit of the Mg chelatase enzyme (Mochizuki et al., 2001). The *gun2-gun5* loss-of-function mutants disrupt genes essential for tetrapyrrole metabolism. Genetic evidence shows that they participate in the same signaling pathway, supporting the hypothesis that accumulation of a precursor could be the retrograde signal. *GUN6*-1D is a gain-of-function mutant that overexpresses a plastidic ferrochelatase (Woodson et al., 2011). Its overexpression leads to the hyper-accumulation of heme that could serve as a retrograde signal. In addition to the *GUN* genes, the *GOLDEN2-LIKE* (*GLK*) genes also have been shown to control similar sets of genes relevant to chlorophyll synthesis and antenna binding (Waters et al., 2009), playing central roles in communication between plastid and nucleus (Fitter et al., 2002).

When considered together the mutants and pharmacological treatments demonstrate that blocks in tetrapyrrole and/or heme synthesis may cause accumulation of precursor compounds that would leave the plastid (or trigger another mobile signal) leading to repression of plastid-associated, nuclear-encoded genes. While attractive, several lines of evidence reject this hypothesis. Mainly, there is no observed difference in Mg-ProtoIX or Mg-ProtoIXme is detected in plants with disrupted signaling responses (Mochizuki et al., 2008). Using sensitive LC/MS methods to identify chlorophyll precursors in norflurazon treated plants, it was shown that there was no effect on MgProtoIX when chlorophyll synthesis was disrupted (Moulin et al., 2008).

There is an undeniable chemical communication link between the chloroplast and nucleus. Genetic and biochemical tools suggest that chlorophyll precursors and/or heme play a part in the process, yet clearly it is not as simple as over-accumulation of a compound like MgProtoIX. While many careers and high-profile publications frame this question, there are answers to be resolved before a complete picture of retrograde signaling is understood.

5. Conclusions

The last two decades have brought tremendous resolution about how the guiding force of photons shapes plant biology, especially processes germane to photosynthesis. The 1990's produced a wellspring of genetic tools that would define several major classes of photosensors and their contiguous protein transduction partners. The last decade brought the utility of genomics tools that would help define the mechanisms and targets of light signal transduction events. New methods in imaging and improved reporter genes have allowed researchers to monitor small changes in plant growth and development, as well as localization and interaction between proteins in vivo. The challenge of the next decade will be to apply these basic discoveries in meaningful ways that escape the models. Here the rules that integrate light signals, change gene expression, alter development, and shape plant form may be manipulated to improve the production of food with less environmental impact.

6. References

Ahmad, M., Jarillo, J.A., Smirnova, O., and Cashmore, A.R. (1998). Cryptochrome blue-light photoreceptors of Arabidopsis implicated in phototropism. Nature 392, 720-723.

Ang, L.H., and Deng, X.W. (1994). Regulatory Hierarchy of Photomorphogenic Loci: Allele-Specific and Light-Dependent Interaction between the HY5 and COP1 Loci. The Plant Cell Online 6, 613-628.

Ballario, P., Talora, C., Galli, D., Linden, H., and Macino, G. (1998). Roles in dimerization and blue light photoresponse of the PAS and LOV domains of Neurospora crassa white collar proteins. Molecular Microbiology 29, 719-729.

Banerjee, R., Schleicher, E., Meier, S., Munoz Viana, R., Pokorny, R., Ahmad, M., Bittl, R., and Batschauer, A. (2007). The signaling state of Arabidopsis cryptochrome 2 contains flavin semiquinone. J Biol Chem.

Barnes, S.A., Nishizawa, N.K., Quaggio, R.B., Whitelam, G.C., and Chua, N.H. (1996). Far-red light blocks greening of Arabidopsis seedlings via a phytochrome A-mediated change in plastid development. Plant Cell 8, 601-615.

Baskin, T.I. (1986). Redistribution of Growth during Phototropism and Nutation in the Pea Epicotyl. Planta 169, 406-414.

Borthwick, H., Hendricks, S., Parker, M., Toole, E., and Toole, V. (1952). A reversible photoreaction controlling seed germination. Proc Natl Acad Sci U S A 38:, 662-666.

Bouly, J.P., Schleicher, E., Dionisio-Sese, M., Vandenbussche, F., Van der Straeten, D., Bakrim, N., Meier, S., Batschauer, A., Galland, P., Bittl, R., and Ahmad, M. (2007). Cryptochrome blue-light photoreceptors are activated through interconversion of flavin redox states. J Biol Chem.

Brautigam, K., Dietzel, L., Kleine, T., Straher, E., Wormuth, D., Dietz, K.-J., Radke, D., Wirtz, M., Hell, R., Darmann, P., Nunes-Nesi, A., Schauer, N., Fernie, A.R., Oliver, S.N., Geigenberger, P., Leister, D., and Pfannschmidt, T. (2009). Dynamic Plastid Redox Signals Integrate Gene Expression and Metabolism to Induce Distinct Metabolic States in Photosynthetic Acclimation in Arabidopsis. The Plant Cell Online 21, 2715-2732.

Briggs, W.R. (2006). Blue/UV-A receptors: Historical overview. In Photomorphogenesis in Plants and Bacteria, S.E.a.N. F, ed (Springer), pp. 171-219.

Briggs, W.R., and Christie, J.M. (2002). Phototropins 1 and 2: versatile plant blue-light receptors. Trends Plant Sci 7, 204-210.

Cashmore, A.R., Jarillo, J.A., Wu, Y.J., and Liu, D. (1999). Cryptochromes: blue light receptors for plants and animals. Science 284, 760-765.

Castillon, A., Shen, H., and Huq, E. (2007). Phytochrome Interacting Factors: central players in phytochrome-mediated light signaling networks. Trends in Plant Science 12, 514-521.

Chen, M., Chory, J., and Fankhauser, C. (2004). Light signal transduction in higher plants. Annu Rev Genet 38, 87-117.

Cho, H.Y., Tseng, T.S., Kaiserli, E., Sullivan, S., Christie, J.M., and Briggs, W.R. (2007). Physiological roles of the light, oxygen, or voltage domains of phototropin 1 and phototropin 2 in Arabidopsis. Plant Physiol 143, 517-529.

Christie, J.M., Reymond, P., Powell, G.K., Bernasconi, P., Raibekas, A.A., Liscum, E., and Briggs, W.R. (1998). Arabidopsis NPH1: a flavoprotein with the properties of a photoreceptor for phototropism. Science 282, 1698-1701.

Christie, J.M., Yang, H., Richter, G.L., Sullivan, S., Thomson, C.E., Lin, J., Titapiwatanakun, B., Ennis, M., Kaiserli, E., Lee, O.R., Adamec, J., Peer, W.A., and Murphy, A.S. (2011). phot1 Inhibition of ABCB19 Primes Lateral Auxin Fluxes in the Shoot Apex Required For Phototropism. PLoS Biol 9, e1001076.

Clack, T., Mathews, S., and Sharrock, R.A. (1994). The phytochrome apoprotein family in Arabidopsis is encoded by five genes: the sequences and expression of PHYD and PHYE. Plant Mol Biol 25, 413-427.

Clack, T., Shokry, A., Moffet, M., Liu, P., Faul, M., and Sharrock, R.A. (2009). Obligate heterodimerization of Arabidopsis phytochromes C and E and interaction with the PIF3 basic helix-loop-helix transcription factor. Plant Cell 21, 786-799.

Darwin, C. (1897). Power of Movement in Plants. (New York: D. Appleton and Co.).

de Carbonnel, M., Davis, P., Roelfsema, M.R.G., Inoue, S.-i., Schepens, I., Lariguet, P., Geisler, M., Shimazaki, K.-i., Hangarter, R., and Fankhauser, C. (2010). The Arabidopsis PHYTOCHROME KINASE SUBSTRATE2 Protein Is a Phototropin Signaling Element That Regulates Leaf Flattening and Leaf Positioning. Plant Physiology 152, 1391-1405.

Dhingra, A., Bies, D.H., Lehner, K.R., and Folta, K.M. (2006). Green light adjusts the plastid transcriptome during early photomorphogenic development. Plant Physiol 142, 1256-1266.

Eisinger, W.R., Bogomolni, R.A., and Taiz, L. (2003). Interactions between a blue-green reversible photoreceptor and a separate UV-B receptor in stomatal guard cells. Am. J. Bot. 90, 1560-1566.

Favory, J.-J., Stec, A., Gruber, H., Rizzini, L., Oravecz, A., Funk, M., Albert, A., Cloix, C., Jenkins, G.I., Oakeley, E.J., Seidlitz, H.K., Nagy, F., and Ulm, R. (2009). Interaction of COP1 and UVR8 regulates UV-B-induced photomorphogenesis and stress acclimation in Arabidopsis. Embo J 28, 591-601.

Fitter, D.W., Martin, D.J., Copley, M.J., Scotland, R.W., and Langdale, J.A. (2002). GLK gene pairs regulate chloroplast development in diverse plant species. The Plant Journal 31, 713-727.

Flint, L.H. (1936). The action of radiation of specific wave-lengths in relation to the germination of light sensitive lettuce seed. . Proc. Int. Seed. Test. Assoc. 8, 1-4.

Flint, L.H., and McAlister, E.D. (1937). Wavelengths of radiation in the visible spectrum promoting the germination of light-sensitive lettuce seed. . Smithsonian Misc. Collect. 94, 1-11.

Folta, K.M. (2004). Green light stimulates early stem elongation, antagonizing light-mediated growth inhibition. Plant Physiol 135, 1407-1416.

Folta, K.M., and Spalding, E.P. (2001). Unexpected roles for cryptochrome 2 and phototropin revealed by high-resolution analysis of blue light-mediated hypocotyl growth inhibition. Plant J 26, 471-478.

Folta, K.M., and Kaufman, L.S. (2003). Phototropin 1 is required for high-fluence blue-light-mediated mRNA destabilization. Plant Mol Biol 51, 609-618.

Folta, K.M., and Maruhnich, S.A. (2007). Green light: a signal to slow down or stop. J Exp Bot 58, 3099-3111.

Folta, K.M., Pontin, M.A., Karlin-Neumann, G., Bottini, R., and Spalding, E.P. (2003). Genomic and physiological studies demonstrate roles for auxin and gibberellin in the early phase of cryptochrome 1 action in blue light. Plant J 36, 203-214.

Franklin, K.A., and Quail, P.H. (2010). Phytochrome functions in Arabidopsis development. J Exp Bot 61, 11-24.

Frechilla, S., Talbott, L.D., Bogomolni, R.A., and Zeiger, E. (2000). Reversal of blue light-stimulated stomatal opening by green light. Plant Cell Physiol 41, 171-176.

Frohnmeyer, H., Loyall, L., Blatt, M.R., and Grabov, A. (1999). Millisecond UV-B irradiation evokes prolonged elevation of cytosolic-free Ca2+ and stimulates gene expression in transgenic parsley cell cultures. Plant J 20, 109-117.

Gallagher, T.F., and Ellis, R.J. (1982). Light-stimulated transcription of genes for 2 chloroplast polypeptides in isolated pea leaf nuclei. Embo J 1, 1493-1498.

Gray, J.C., Sullivan, J.A., Wang, J.H., Jerome, C.A., and MacLean, D. (2003). Coordination of plastid and nuclear gene expression. Philos Trans R Soc Lond B Biol Sci 358, 135-144; discussion 144-135.

Green, R., and Fluhr, R. (1995). UV-B-Induced PR-1 Accumulation Is Mediated by Active Oxygen Species. Plant Cell 7, 203-212.

Gressel, J. (1979). BLUE-LIGHT PHOTORECEPTION. Photochem. Photobiol. 30, 749-754.

Guo, H., Yang, H., Mockler, T.C., and Lin, C. (1998). Regulation of flowering time by Arabidopsis photoreceptors. Science 279, 1360-1363.

Hess, W.R., Schendel, R.B., Borner, T.R., and Rudiger, W. (1991). Reduction of mRNA level for two nuclear encoded light regulated genes in the barley mutant albostrians is not correlated with phytochrome content and activity. . J Plant Physiol 138, 292-298.

Hess, W.R., Muller, A., Nagy, F., and Borner, T. (1994). Ribosome-deficient plastids affect transcription of light-induced nuclear genes: genetic evidence for a plastid-derived signal. Mol Gen Genet 242, 305-312.

Heyes, D.J., and Hunter, C.N. (2005). Making light work of enzyme catalysis: protochlorophyllide oxidoreductase. Trends Biochem.Sci. 30, 642-649.

Hisada, A., Hanzawa, H., Weller, J.L., Nagatani, A., Reid, J.B., and Furuya, M. (2000). Light-Induced Nuclear Translocation of Endogenous Pea Phytochrome A Visualized by Immunocytochemical Procedures. Plant Cell 12, 1063-1078.

Huala, E., Oeller, P.W., Liscum, E., Han, I.S., Larsen, E., and Briggs, W.R. (1997). Arabidopsis NPH1: a protein kinase with a putative redox-sensing domain. Science 278, 2120-2123.

Huq, E., Al-Sady, B., and Quail, P.H. (2003). Nuclear translocation of the photoreceptor phytochrome B is necessary for its biological function in seedling photomorphogenesis. Plant J 35, 660-664.

Huq, E., Al-Sady, B., Hudson, M., Kim, C., Apel, K., and Quail, P.H. (2004). PHYTOCHROME-INTERACTING FACTOR 1 Is a Critical bHLH Regulator of Chlorophyll Biosynthesis. Science 305, 1937-1941.

Inoue, S.-i., Takemiya, A., and Shimazaki, K.-i. (2010). Phototropin signaling and stomatal opening as a model case. Current Opinion in Plant Biology 13, 587-593.

Inoue, S.-i., Kinoshita, T., Takemiya, A., Doi, M., and Shimazaki, K.-i. (2008). Leaf Positioning of Arabidopsis in Response to Blue Light. Molecular Plant 1, 15-26.

Jarillo, J.A., Ahmad, M., and Cashmore, A.R. (1998). NPL1: A second member of the NPH1 serine/threonine kinase family of Arabidopsis (PGR 98–100) Plant Physiology 117, 719.

Jarillo, J.A., Gabrys, H., Capel, J., Alonso, J.M., Ecker, J.R., and Cashmore, A.R. (2001a). Phototropin-related NPL1 controls chloroplast relocation induced by blue light. Nature 410, 952-954.

Jarillo, J.A., Capel, J., Tang, R.H., Yang, H.Q., Alonso, J.M., Ecker, J.R., and Cashmore, A.R. (2001b). An Arabidopsis circadian clock component interacts with both CRY1 and phyB. Nature 410, 487-490.

Jenkins, G.I. (2009). Signal transduction in responses to UV-B radiation. Annu Rev Plant Biol 60, 407-431.

Johnson, E.S. (1937). Growth of Avena coleoptile and first internode in different wavebands of the visible spectrum. . Smithsonian Misc. Collect. 96, 1-19.

Jordan, E.T., Marita, J.M., Clough, R.C., and Vierstra, R.D. (1997). Characterization of regions within the N-terminal 6-kilodalton domain of phytochrome A that modulate its biological activity. Plant Physiol 115, 693-704.

Kagawa, T., Kimura, M., and Wada, M. (2009). Blue Light-Induced Phototropism of Inflorescence Stems and Petioles is Mediated by Phototropin Family Members phot1 and phot2. Plant and Cell Physiology 50, 1774-1785.

Kagawa, T., Sakai, T., Suetsugu, N., Oikawa, K., Ishiguro, S., Kato, T., Tabata, S., Okada, K., and Wada, M. (2001). Arabidopsis NPL1: a phototropin homolog controlling the chloroplast high-light avoidance response. Science 291, 2138-2141.

Kaiserli, E., and Jenkins, G.I. (2007). UV-B Promotes Rapid Nuclear Translocation of the Arabidopsis UV-Bâ€"Specific Signaling Component UVR8 and Activates Its Function in the Nucleus. The Plant Cell Online 19, 2662-2673.

Kakizaki, T., Matsumura, H., Nakayama, K., Che, F.S., Terauchi, R., and Inaba, T. (2009). Coordination of plastid protein import and nuclear gene expression by plastid-to-nucleus retrograde signaling. Plant Physiol 151, 1339-1353.

Karlin-Neumann, G.A., Sun, L., and Tobin, E.M. (1988). Expression of Light-Harvesting Chlorophyll a/B-Protein Genes Is Phytochrome-Regulated in Etiolated Arabidopsis-Thaliana Seedlings. Plant Physiology 88, 1323-1331.

Kaufman, L.S., Thompson, W.F., and Briggs, W.R. (1984). Different Red Light Requirements for Phytochrome-Induced Accumulation of cab RNA and rbcS RNA. Science 226, 1447-1449.

Kaufman, L.S., Briggs, W.R., and Thompson, W.F. (1985). Phytochrome Control of Specific mRNA Levels in Developing Pea Buds : The Presence of Both Very Low Fluence and Low Fluence Responses. Plant Physiol 78, 388-393.

Kawai, H., Kanegae, T., Christensen, S., Kiyosue, T., Sato, Y., Imaizumi, T., Kadota, A., and Wada, M. (2003). Responses of ferns to red light are mediated by an unconventional photoreceptor. Nature 421, 287-290.

Khurana, J.P., and Poff, K.L. (1989). Mutants of Arabidopsis thaliana with altered phototropism. Planta 178, 400-406.

Kinoshita, T., Doi, M., Suetsugu, N., Kagawa, T., Wada, M., and Shimazaki, K. (2001). Phot1 and phot2 mediate blue light regulation of stomatal opening. Nature 414, 656-660.

Kircher, S., Kozma-Bognar, L., Kim, L., Adam, E., Harter, K., Schafer, E., and Nagy, F. (1999). Light quality-dependent nuclear import of the plant photoreceptors phytochrome A and B. Plant Cell 11, 1445-1456.

Kleffmann, T., von Zychlinski, A., Russenberger, D., Hirsch-Hoffmann, M., Gehrig, P., Gruissem, W., and Baginsky, S. (2007). Proteome dynamics during plastid differentiation in rice. Plant Physiol 143, 912-923.

Klein, R.M., and Edsall, P.C. (1967). Interference by near ultraviolet and green light with growth of animal and plant cell cultures. Photochem Photobiol 6, 841-850.

Klein, R.M., Edsall, P.C., and Gentile, A.C. (1965). Effects of near ultraviolet and green radiations on plant growth. Plant Physiol 40, 903-906.

Kleine, T., Lockhart, P., and Batschauer, A. (2003). An Arabidopsis protein closely related to Synechocystis cryptochrome is targeted to organelles. Plant J 35, 93-103.

Konieczny, A., and Ausubel, F.M. (1993). A procedure for mapping Arabidopsis mutations using co-dominant ecotype-specific PCR-based markers. The Plant Journal 4, 403-410.

Koornneef, M., Rolff, E., and Spruit, C. (1980). Genetic control of light-inhibited hypocotyl elongation in Arabidopsis thaliana (L.) Heynh. Z. Pflanzenphysiol. 100:, 147-160.

Koussevitzky, S., Nott, A., Mockler, T.C., Hong, F., Sachetto-Martins, G., Surpin, M., Lim, J., Mittler, R., and Chory, J. (2007). Signals from Chloroplasts Converge to Regulate Nuclear Gene Expression. Science 316, 715-719.

Lariguet, P., Schepens, I., Hodgson, D., Pedmale, U.V., Trevisan, M., Kami, C., de Carbonnel, M., Alonso, J.M., Ecker, J.R., Liscum, E., and Fankhauser, C. (2006). PHYTOCHROME KINASE SUBSTRATE 1 is a phototropin 1 binding protein required for phototropism. Proc Natl Acad Sci U S A 103, 10134-10139.

Lasceve, G., Leymarie, J., Olney, M.A., Liscum, E., Christie, J.M., Vavasseur, A., and Briggs, W.R. (1999). Arabidopsis Contains at Least Four Independent Blue-Light-Activated Signal Transduction Pathways. Plant Physiol. 120, 605-614.

Lebedev, N., and Timko, M.P. (1998). Protochlorophyllide photoreduction Photosynth Res 58, 5-23.

Leivar, P., and Quail, P.H. (2011). PIFs: pivotal components in a cellular signaling hub. Trends in Plant Science 16, 19-28.

Leivar, P., Monte, E., Oka, Y., Liu, T., Carle, C., Castillon, A., Huq, E., and Quail, P.H. (2008). Multiple Phytochrome-Interacting bHLH Transcription Factors Repress Premature Seedling Photomorphogenesis in Darkness. Current Biology 18, 1815-1823.

Lin, C., and Shalitin, D. (2003). CRYPTOCHROME STRUCTURE AND SIGNAL TRANSDUCTION. Annual Review of Plant Biology 54, 469-496.

Lin, C., Yang, H., Guo, H., Mockler, T., Chen, J., and Cashmore, A.R. (1998). Enhancement of blue-light sensitivity of Arabidopsis seedlings by a blue light receptor cryptochrome 2. Proc Natl Acad Sci U S A 95, 2686-2690.

Liscum, E., and Briggs, W.R. (1995). Mutations in the NPH1 locus of Arabidopsis disrupt the perception of phototropic stimuli. Plant Cell 7, 473-485.

Liscum, E., and Briggs, W.R. (1996). Mutations of Arabidopsis in potential transduction and response components of the phototropic signaling pathway. Plant Physiol 112, 291-296.

Liscum, E., Young, J.C., Poff, K.L., and Hangarter, R.P. (1992). Genetic separation of phototropism and blue light inhibition of stem elongation. Plant Physiol 100, 267-271.

Margulis, L. (1970). Recombination of non-chromosomal genes in Chlamydomonas: assortment of mitochondria and chloroplasts? J Theor Biol 26, 337-342.

Masuda, T., and Takamiya, K. (2004). Novel Insights into the Enzymology, Regulation and Physiological Functions of Light-dependent Protochlorophyllide Oxidoreductase in Angiosperms. Photosynth Res 81, 1-29.

Masuda, T., Tanaka, A., and Melis, A. (2003). Chlorophyll antenna size adjustments by irradiance in Dunaliella salina involve coordinate regulation of chlorophyll a oxygenase (CAO) and Lhcb gene expression. Plant Mol Biol 51, 757-771.

Mayfield, S.P., and Taylor, W.C. (1984). Carotenoid-deficient maize seedlings fail to accumulate light-harvesting chlorophyll a/b binding protein (LHCP) mRNA. Eur J Biochem 144, 79-84.

Mochizuki, N., Brusslan, J.A., Larkin, R., Nagatani, A., and Chory, J. (2001). Arabidopsis genomes uncoupled 5 (GUN5) mutant reveals the involvement of Mg-chelatase H subunit in plastid-to-nucleus signal transduction. Proceedings of the National Academy of Sciences 98, 2053-2058.

Mochizuki, N., Tanaka, R., Tanaka, A., Masuda, T., and Nagatani, A. (2008). The steady-state level of Mg-protoporphyrin IX is not a determinant of plastid-to-nucleus signaling in Arabidopsis. Proceedings of the National Academy of Sciences 105, 15184-15189.

Mockler, T., Yang, H., Yu, X., Parikh, D., Cheng, Y.C., Dolan, S., and Lin, C. (2003). Regulation of photoperiodic flowering by Arabidopsis photoreceptors. Proc Natl Acad Sci U S A 100, 2140-2145.

Monte, E., Tepperman, J.M., Al-Sady, B., Kaczorowski, K.A., Alonso, J.M., Ecker, J.R., Li, X., Zhang, Y., and Quail, P.H. (2004). The phytochrome-interacting transcription factor, PIF3, acts early, selectively, and positively in light-induced chloroplast development. Proc Natl Acad Sci U S A 101, 16091-16098.

Moon, J., Zhu, L., Shen, H., and Huq, E. (2008). PIF1 directly and indirectly regulates chlorophyll biosynthesis to optimize the greening process in Arabidopsis. Proceedings of the National Academy of Sciences 105, 9433-9438.

Motchoulski, A., and Liscum, E. (1999). Arabidopsis NPH3: A NPH1 photoreceptor-interacting protein essential for phototropism. Science 286, 961-964.

Moulin, M., McCormac, A.C., Terry, M.J., and Smith, A.G. (2008). Tetrapyrrole profiling in Arabidopsis seedlings reveals that retrograde plastid nuclear signaling is not due to Mg-protoporphyrin IX accumulation. Proceedings of the National Academy of Sciences 105, 15178-15183.

Mullen, J.L., Weinig, C., and Hangarter, R.P. (2006). Shade avoidance and the regulation of leaf inclination in Arabidopsis. Plant Cell and Environment 29, 1099-1106.

Nelson, D.C., Lasswell, J., Rogg, L.E., Cohen, M.A., and Bartel, B. (2000). FKF1, a clock-controlled gene that regulates the transition to flowering in Arabidopsis. Cell 101, 331-340.

Ni, M., Tepperman, J.M., and Quail, P.H. (1998). PIF3, a phytochrome-interacting factor necessary for normal photoinduced signal transduction, is a novel basic helix-loop-helix protein. Cell 95, 657-667.

Oelmüller, R., Levitan, I., Bergfeld, R., Rajasekhar, V.K., and Mohr, H. (1986). Expression of nuclear genes as affected by treatments acting on the plastids. Planta 168, 482-492.

Oster, U., Brunner, H., and Rudiger, W. (1996). The greening process in cress seedlings.V. Possible interference of chlorophyll precursors, accumulated after thujaplicin treatment, with light regulated expression of Lhc genes. J. Photochem. Photobiol. 36, 255-261.

Osterlund, M.T., Hardtke, C.S., Wei, N., and Deng, X.W. (2000). Targeted destabilization of HY5 during light-regulated development of Arabidopsis. Nature 405, 462-466.

Palmer, J.M., Short, T.W., and Briggs, W.R. (1993a). Correlation of Blue Light-Induced Phosphorylation to Phototropism in Zea mays L. Plant Physiol 102, 1219-1225.

Palmer, J.M., Short, T.W., Gallagher, S., and Briggs, W.R. (1993b). Blue Light-Induced Phosphorylation of a Plasma Membrane-Associated Protein in Zea mays L. Plant Physiol 102, 1211-1218.

Parks, B.M., and Quail, P.H. (1993). hy8, a new class of arabidopsis long hypocotyl mutants deficient in functional phytochrome A. Plant Cell 5, 39-48.

Parks, B.M., Cho, M.H., and Spalding, E.P. (1998). Two genetically separable phases of growth inhibition induced by blue light in Arabidopsis seedlings. Plant Physiol 118, 609-615.

Pedmale, U.V., and Liscum, E. (2007). Regulation of Phototropic Signaling in Arabidopsis via Phosphorylation State Changes in the Phototropin 1-interacting Protein NPH3. Journal of Biological Chemistry 282, 19992-20001.

Pfannschmidt, T., Schutze, K., Brost, M., and Oelmuller, R. (2001). A novel mechanism of nuclear photosynthesis gene regulation by redox signals from the chloroplast during photosystem stoichiometry adjustment. J Biol Chem 276, 36125-36130.

Pogson, B.J., Woo, N.S., Forster, B., and Small, I.D. (2008). Plastid signalling to the nucleus and beyond. Trends Plant Sci 13, 602-609.

Quail, P.H., Briggs, W., Chory, J., Hangarter, R.P., Harberd, N.P., Kendrick, C.I., Koornneef, M., Parks, B.M., Sharrock, R.A., Schafer, E., Thompson, W.E., and Whitelam, G. (1994). Spotlight on Phytochrome Nomenclature. Plant Cell 6, 468-471.

Rapp, J.C., and Mullet, J.E. (1991). Chloroplast transcription is required to express the nuclear genes rbcS and cab. Plastid DNA copy number is regulated independently. Plant Mol Biol 17, 813-823.

Reymond, P., Short, T.W., Briggs, W.R., and Poff, K.L. (1992). Light-induced phosphorylation of a membrane protein plays an early role in signal transduction for phototropism in Arabidopsis thaliana. Proc Natl Acad Sci U S A 89, 4718-4721.

Rizzini, L., Favory, J.-J., Cloix, C., Faggionato, D., O’Hara, A., Kaiserli, E., Baumeister, R., Schäfer, E., Nagy, F., Jenkins, G.I., and Ulm, R. (2011). Perception of UV-B by the Arabidopsis UVR8 Protein. Science 332, 103-106.

Rosinski, J., and Rosen, W.G. (1972). Chloroplast development: fine structure and chlorophyll synthesis. Q Rev Biol 47, 160-191.

Sage, L.C. (1992). Pigment of the Imagination. (San Diego, CA: Academic Press).

Sakai, T., Kagawa, T., Kasahara, M., Swartz, T.E., Christie, J.M., Briggs, W.R., Wada, M., and Okada, K. (2001). Arabidopsis nph1 and npl1: blue light receptors that mediate both phototropism and chloroplast relocation. Proc Natl Acad Sci U S A 98, 6969-6974.

Sakamoto, K., and Briggs, W.R. (2002). Cellular and subcellular localization of phototropin 1. Plant Cell 14, 1723-1735.

Salomon, M., Christie, J.M., Knieb, E., Lempert, U., and Briggs, W.R. (2000). Photochemical and mutational analysis of the FMN-binding domains of the plant blue light receptor, phototropin. Biochemistry 39, 9401-9410.

Schultz, T.F., Kiyosue, T., Yanovsky, M., Wada, M., and Kay, S.A. (2001). A Role for LKP2 in the Circadian Clock of Arabidopsis. The Plant Cell Online 13, 2659-2670.

Sellaro, R., Crepy, M., Trupkin, S.A., Karayekov, E., Buchovsky, A.S., Rossi, C., and Casal, J.J. (2011). Cryptochrome as a sensor of the blue/green ratio of natural radiation in Arabidopsis. Plant Physiol 154, 401-409.

Selstam, E., and Sandelius, A.S. (1984). A Comparison between Prolamellar Bodies and Prothylakoid Membranes of Etioplasts of Dark-Grown Wheat Concerning Lipid and Polypeptide Composition. Plant Physiol 76, 1036-1040.

Selstam, E., and Widell-Wigge, A. (1993). Chloroplast lipids and the assembly of membranes. In Pigment-protein complexes in plastids: synthesis and assembly., C. Sundqvist and M. Ryberg, eds (San Diego: Academic Press), pp. 241-277.

Sharrock, R.A., and Clack, T. (2004). Heterodimerization of type II phytochromes in Arabidopsis. Proc Natl Acad Sci U S A 101, 11500-11505.

Shinkle, J.R., Derickson, D.L., and Barnes, P.W. (2005). Comparative photobiology of growth responses to two UV-B wavebands and UV-C in dim-red-light- and white-light-grown cucumber (Cucumis sativus) seedlings: physiological evidence for photoreactivation. Photochem Photobiol 81, 1069-1074.

Shinkle, J.R., Atkins, A.K., Humphrey, E.E., Rodgers, C.W., Wheeler, S.L., and Barnes, P.W. (2004). Growth and morphological responses to different UV wavebands in cucumber (Cucumis sativum) and other dicotyledonous seedlings. Physiol Plant 120, 240-248.

Shinomura, T., Nagatani, A., Hanzawa, H., Kubota, M., Watanabe, M., and Furuya, M. (1996). Action spectra for phytochrome A- and B-specific photoinduction of seed germination in Arabidopsis thaliana. Proc Natl Acad Sci U S A 93, 8129-8133.

Short, T.W., and Briggs, W.R. (1990). Characterization of a Rapid, Blue Light-Mediated Change in Detectable Phosphorylation of a Plasma Membrane Protein from Etiolated Pea (Pisum sativum L.) Seedlings. Plant Physiol 92, 179-185.

Short, T.W., Porst, M., and Briggs, W.R. (1992). A Photoreceptor System Regulating Invivo and Invitro Phosphorylation of a Pea Plasma-Membrane Protein. Photochem. Photobiol. 55, 773-781.

Short, T.W., Reymond, P., and Briggs, W.R. (1993). A Pea Plasma Membrane Protein Exhibiting Blue Light-Induced Phosphorylation Retains Photosensitivity following Triton Solubilization. Plant Physiol 101, 647-655.

Short, T.W., Porst, M., Palmer, J., Fernbach, E., and Briggs, W.R. (1994). Blue Light Induces Phosphorylation at Seryl Residues on a Pea (Pisum sativum L.) Plasma Membrane Protein. Plant Physiol 104, 1317-1324.

Solymosi, K., Smeller, L., Ryberg, M., Sundqvist, C., Fidy, J., and Boddi, B. (2007). Molecular rearrangement in POR macrodomains as a reason for the blue shift of chlorophyllide fluorescence observed after phototransformation. Biochim. Biophys. Acta-Biomembr. 1768, 1650-1658.

Somers, D.E., Sharrock, R.A., Tepperman, J.M., and Quail, P.H. (1991). The hy3 Long Hypocotyl Mutant of Arabidopsis Is Deficient in Phytochrome B. Plant Cell 3, 1263-1274.

Somers, D.E., Schultz, T.F., Milnamow, M., and Kay, S.A. (2000). ZEITLUPE encodes a novel clock-associated PAS protein from Arabidopsis. Cell 101, 319-329.

Steinitz, B., and Poff, K.L. (1986). A Single Positive Phototropic Response Induced with Pulsed-Light in Hypocotyls of Arabidopsis-Thaliana Seedlings. Planta 168, 305-315.

Steinitz, B., Ren, Z.L., and Poff, K.L. (1985). Blue and green light-induced phototropism in Arabidopsis thaliana and Lactuca-sativa L seedlings. Plant Physiology 77, 248-251.

Stephenson, P.G., Fankhauser, C., and Terry, M.J. (2009). PIF3 is a repressor of chloroplast development. Proc Natl Acad Sci U S A 106, 7654-7659.

Sullivan, J.A., and Gray, J.C. (1999). Plastid translation is required for the expression of nuclear photosynthesis genes in the dark and in roots of the pea lip1 mutant. Plant Cell 11, 901-910.

Sullivan, S., Thomson, C.E., Kaiserli, E., and Christie, J.M. (2009). Interaction specificity of Arabidopsis 14-3-3 proteins with phototropin receptor kinases. FEBS Letters 583, 2187-2193.

Susek, R.E., Ausubel, F.M., and Chory, J. (1993). Signal transduction mutants of arabidopsis uncouple nuclear CAB and RBCS gene expression from chloroplast development. Cell 74, 787-799.

Swartz, T.E., Tseng, T.-S., Frederickson, M.A., Paris, G.n., Comerci, D.J., Rajashekara, G., Kim, J.-G., Mudgett, M.B., Splitter, G.A., Ugalde, R.A., Goldbaum, F.A., Briggs, W.R., and Bogomolni, R.A. (2007). Blue-Light-Activated Histidine Kinases: Two-Component Sensors in Bacteria. Science 317, 1090-1093.

Takemiya, A., Inoue, S.-i., Doi, M., Kinoshita, T., and Shimazaki, K.-i. (2005). Phototropins Promote Plant Growth in Response to Blue Light in Low Light Environments. The Plant Cell Online 17, 1120-1127.

Talbott, L.D., Nikolova, G., Ortiz, A., Shmayevich, I., Zeiger, E. (2002). Green light reversal of blue-light-stimulated stomatal opening is found in a diversity of plant species. American Journal of Botany 89, 366-368.

Tepperman, J.M., Zhu, T., Chang, H.S., Wang, X., and Quail, P.H. (2001). Multiple transcription-factor genes are early targets of phytochrome A signaling. Proc Natl Acad Sci U S A 98, 9437-9442.

Tepperman, J.M., Hudson, M.E., Khanna, R., Zhu, T., Chang, S.H., Wang, X., and Quail, P.H. (2004). Expression profiling of phyB mutant demonstrates substantial contribution of other phytochromes to red-light-regulated gene expression during seedling de-etiolation. Plant J 38, 725-739.

Terashima, I., Fujita, T., Inoue, T., Chow, W.S., and Oguchi, R. (2009). Green Light Drives Leaf Photosynthesis More Efficiently than Red Light in Strong White Light:

Revisiting the Enigmatic Question of Why Leaves are Green. Plant and Cell Physiology 50, 684-697.

Tokutomi, S., Matsuoka, D., and Zikihara, K. (2008). Molecular structure and regulation of phototropin kinase by blue light. Biochimica et Biophysica Acta (BBA) - Proteins & Proteomics 1784, 133-142.

Toledo-Ortiz, G., Huq, E., and Rodriguez-Concepcon, M. (2010). Direct regulation of phytoene synthase gene expression and carotenoid biosynthesis by phytochrome-interacting factors. Proceedings of the National Academy of Sciences 107, 11626-11631.

Ulm, R., and Nagy, F. (2005). Signaling and gene regulation in response to ultraviolet light. Curr Opin Plant Biol 8, 477-482.

Ulm, R., Baumann, A., Oravecz, A., Mate, Z., Adam, E., Oakeley, E.J., Schafer, E., and Nagy, F. (2004). Genome-wide analysis of gene expression reveals function of the bZIP transcription factor HY5 in the UV-B response of Arabidopsis. Proc Natl Acad Sci U S A 101, 1397-1402.

Valverde, F., Mouradov, A., Soppe, W., Ravenscroft, D., Samach, A., and Coupland, G. (2004). Photoreceptor regulation of CONSTANS protein in photoperiodic flowering. Science 303, 1003-1006.

Wang, H., Ma, L.G., Li, J.M., Zhao, H.Y., and Deng, X.W. (2001). Direct interaction of Arabidopsis cryptochromes with COP1 in light control development. Science 294, 154-158.

Wang, Y., Noguchi, K., and Terashima, I. (2011a). Photosynthesis-Dependent and - Independent Responses of Stomata to Blue, Red and Green Monochromatic Light: Differences Between the Normally Oriented and Inverted Leaves of Sunflower. Plant and Cell Physiology 52, 479-489.

Warpeha, K.M., Upadhyay, S., Yeh, J., Adamiak, J., Hawkins, S.I., Lapik, Y.R., Anderson, M.B., and Kaufman, L.S. (2007). The GCR1, GPA1, PRN1, NF-Y Signal Chain Mediates Both Blue Light and Abscisic Acid Responses in Arabidopsis. Plant Physiology 143, 1590-1600.

Waters, M.T., Wang, P., Korkaric, M., Capper, R.G., Saunders, N.J., and Langdale, J.A. (2009). GLK Transcription Factors Coordinate Expression of the Photosynthetic Apparatus in Arabidopsis. The Plant Cell Online 21, 1109-1128.

Went, F.W. (1957). The Experimental Control of Plant Growth. (Waltham, MA: Chronica Botanica).

Woodson, Jesse D., Perez-Ruiz, Juan M., and Chory, J. (2011). Heme Synthesis by Plastid Ferrochelatase I Regulates Nuclear Gene Expression in Plants. Current Biology 21, 897-903.

Wu, G., and Spalding, E.P. (2007). Separate functions for nuclear and cytoplasmic cryptochrome 1 during photomorphogenesis of Arabidopsis seedlings. Proceedings of the National Academy of Sciences 104, 18813-18818.

Yang, H.Q., Wu, Y.J., Tang, R.H., Liu, D., Liu, Y., and Cashmore, A.R. (2000). The C termini of Arabidopsis cryptochromes mediate a constitutive light response. Cell 103, 815-827.

Yatsuhashi, H., and Hashimoto, T. (1985). MULTIPLICATIVE ACTION OF A UV-B PHOTORECEPTOR and PHYTOCHROME IN ANTHOCYANIN SYNTHESIS. Photochem. Photobiol. 41, 673-680.

Zhang, T., Maruhnich, S.A., and Folta, K.M. (2011). Green light induces shade avoidance symptoms. Plant Physiology. In Press

Zhao, X., Yu, X., Foo, E., Symons, G.M., Lopez, J., Bendehakkalu, K.T., Xiang, J., Weller, J.L., Liu, X., Reid, J.B., and Lin, C. (2007). A Study of Gibberellin Homeostasis and Cryptochrome-Mediated Blue Light Inhibition of Hypocotyl Elongation. Plant Physiology 145, 106-118.

4

Energy Conversion in Purple Bacteria Photosynthesis

Felipe Caycedo-Soler[1,2], Ferney J. Rodríguez[2], Luis Quiroga[2],
Guannan Zhao[3] and Neil F. Johnson[3]

[1]*Ulm University, Institute of Theoretical Physics, Ulm*
[2]*Departamento de Física, Universidad de los Andes, Bogotá*
[3]*Department of Physics, University of Miami, Coral Gables, Miami, Florida*
[1]*Germany*
[2]*Colombia*
[3]*USA*

1. Introduction

The study of how photosynthetic organisms convert light offers insight not only into nature's evolutionary process, but may also give clues as to how best to design and manipulate artificial photosynthetic systems – and also how far we can drive natural photosynthetic systems beyond normal operating conditions, so that they can harvest energy for us under otherwise extreme conditions. In addition to its interest from a basic scientific perspective, therefore, the goal to develop a deep quantitative understanding of photosynthesis offers the potential payoff of enhancing our current arsenal of alternative energy sources for the future. In the following Chapter, we consider the excitation dynamics of photosynthetic membranes in *Rps. Photometricum* purple bacteria. The first studies on purple bacteria photosynthetic membranes were concerned with the complex underlying detailed structure (Jamieson & al, 2002; McDermott et al., 1995; Roszack & et al, 2003; Waltz et al., 1998). The interested reader might find helpful the first section of this chapter where we present a summary of structures and processes that allow photosynthesis in purple bacteria. As improved resolution became available for light-harvesting structures, so too the community's interest increased in understanding the details of the rapid femto- to picosecond timescales for the excitation transfer process within a given harvesting complex – even to the regime where quantum effects are expected. Indeed such quantum effects have recently been confirmed in, for instance, the Fenna-Matthews-Olson complex of green sulfur bacteria (Engel et al., 2007). However, the processes occurring at this level of detail in terms of both structure and properties of the excitation harvesting, have not yet been shown as being crucial to the performance of the full harvesting membrane, nor with the primary goal of any given photosynthetic organism: to fuel its metabolism. We focus on the transfer among different complexes, in particular the inter-complex excitation transfer. We consider first a system of few complexes in order to understand the consequences on the dynamics of the complexes' connectedness. We also consider the relative amounts of harvesting complexes (i.e. stoichiometry) on small sized networks in order to help establish our understanding of the behavior of complete chromatophore vesicles.

As a whole, chromatophore vesicles comprise sections of the purple bacteria cytoplasmic membrane where a large number of harvesting complexes accommodate. The current capabilities to dissect these vesicles on the nanoscale through Atomic Force Microscopy (AFM) has provided evidence of changes in the conformation of chromatophores as a result of different environmental conditions. According to Ref.(Scheuring & Sturgis, 2005) , membranes grown under Low Light Intensity (LLI, 10 Watt/m^2) present a relative amount of harvesting complexes which is different to the stoichiometry observed for bacteria grown under High Light Intensity (HLI, 100 Watt/m^2), thereby pinpointing the importance of global changes in the complete vesicles as an important means of fulfilling the bacteria's metabolic requirements. In order to study these global conformational changes – given the fact that the inter-complex transfer time-scale involves several picoseconds and the excitation delocalization length is not expected to be beyond a single harvesting complex – we choose a model of excitation dynamics which is based on a classical random walk. This random walk is coupled to the main processes leading to electron/chemical energy transformation, and hence the bacteria's metabolic demands.

This Chapter is organized as follows. Section 2 provides a review of the basic structures involved in excitation transfer, along with a summary of the processes required for electron/chemical energy conversion. A discussion of the dynamics of excitations in a few model architectures is explored in section 3, in order to understand the results from complete LLI and HLI adapted chromatophores in section 4 and to guide our development of an analytical model in section 5 for determining both the efficiency and power output of any given chromatophore vesicle under arbitrary light intensity regimes. Lastly, in section 6 we explore the effect of incident light with extreme photon arrival statistics, on the resulting electronic/chemical energy conversion, in order to heuristically provide a survivability margin beyond which terrestrial bacteria could not survive. This is motivated by the fact that one day, it may be necessary to send simple bacteria into deep space and/or cope with extreme photon conditions here on Earth as a result of a catastrophic solar change.

2. Important processes for solar/chemical energy conversion

Purple bacteria sustain their metabolism using photosynthesis in anaerobic conditions and under the dim light excitation proper of several meters deeps at ponds, lagoons and streams (Pfenning, 1978). As depicted in Fig.1, aerobic organisms are present near the surface of water reservoirs, collect blue and red spectral components of sun's light, leaving only the green and far red ($>$ 750nm) components, from where purple bacteria must fulfill their energy requirements.

The light energy absorption is accomplished through intracytoplasmic membranes where different pigment-protein complexes accommodate. Light Harvesting complexes (LHs) have the function of absorbing light and transfer it to Reaction Centers (RCs), where a charge separation process is initiated (Codgell et al., 2006). The unpaired charge reduces a quinone, which using a periplasmic hydrogen, converts to quinol ($Q_B H_2$). RCs neutrality is restablished thanks to cytochrome cyt charge carrier, which after undocking from the RC, must find the bc_1 complex to receive an electron and start its cycle all over again. The electron in bc_1 is given due to cytoplasmic Quinol delivery in bc_1. The proton gradient induced by the charge carriers cycling becomes the precursor of adenosyn triphosphate (ATP) synthesis: ADP+P$\rightarrow$ ATP+Energy, where ADP and P refer respectively, to adenosine diphosphate and phosphorous. The cycle is depicted in Fig.2.

Fig. 1. Representation of a lake containing both aerobic and anaerobic phototrophic organisms. Note that purple bacterial photosynthesis is restricted to the lower anaerobic layer and so they only receive solar energy that has been filtered, mainly by chlorophylls belonging to algae, cyanobacteria and plants.

2.1 Harvesting function

Light absorption occurs through organic molecules, known as chromophores, inserted in protein complexes. Bacteriochlorophyls (BChl) and carotenoids (Car) chromophores are the main absorbers in purple bacteria photosynthesis, principally in the far red and green, respectively. The light absorption process occurs through chromophore's Qy electronic transition excitation. Several chromophores are embedded in protein helices, named α and β apoproteins, inside complexes, classified by their absorption spectral maximum. Light Harvesting complex 2 (LH2) reveals two concentric subunits that according to their absorption maxima are called B800 or B850, composed of nine pairs of apoproteins (McDermott et al., 1995) comprising an inner α-helix and an outer β-helix both crossing the harvesting membrane from periplasm to cytoplasm, in an $\alpha\beta$ unit that serves to anchor a highly interacting B850 dimer and one B800 chromophore. Hence, the B800 is composed of nine chromophores, while B850 include eighteen BChla chromophores, having dipole moments parallel and nearly perpendicular to the membrane plane, respectively. Raman spectra using different excitation wavelength (Gall et al., 2006) and stoichiometry analysis (Arellano et al., 1998), indicated that one carotenoid (Car) is present per each $\alpha\beta$ unit. Light Harvesting complex 1 (LH1) absorbs maximally at 883 nm, and contain 32 BChls, arranged in 16 bi-chromophore $\alpha\beta$ units, surrounding an RC (Karrasch et al., 1995) in the same geometrical arrangement as B850 chromophores. An RC presents a highly interacting dimer, the special pair (P), that is ionized due to the electronic excitation transferred from the surrounding LH1.

The complexes' photon absorption cross section has been calculated for LH1 and LH2 complexes, where all absorbing molecules and extinction coefficients (Francke & Amesz, 1995) have been taken into account. A photon of wavelentht λ, is part of the power spectrum of a

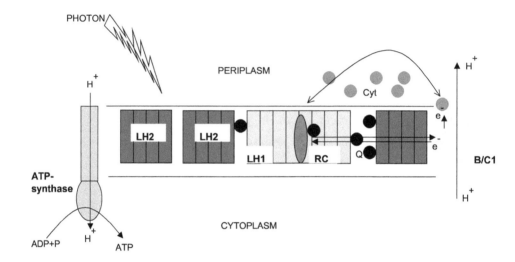

Fig. 2. Schematic representation of the photosynthetic apparatus in the intracytoplasmic membrane of purple bacteria. The reaction center (RC, red) is surrounded by the light-harvesting complex 1 (LH1, green) to form the LH1+RC complex, which is surrounded by multiple light-harvesting complexes (LH2s) (blue), forming altogether the photosynthetic unit (PSU). Photons are absorbed by LHs and excitation is transferred to the RC initiating a charge (electron-hole) separation. Electrons are shuttled back by the cytochrome c_2 charge carrier (blue) from the ubiquinone-cytochrome bc_1 complex (yellow) to the RC. The electron transfer across the membrane produces a proton gradient that drives the synthesis of ATP from ADP by the ATPase (orange). Electron e^- is represented in blue, and quinones Q_B, likely confined to the intramembrane space, in black.

source with occupation numbers $n(\lambda)$. Normalized to 18 W/m^2 intensity, the rate of photon absorption for circular LH1 complexes in *Rb. sphaeroides* (Geyer & Heims, 2006):

$$\gamma_1^A = \int n(\lambda)\sigma_{LH1}(\lambda)d\lambda = 18s^{-1}. \tag{1}$$

The same procedure applied to LH2 complexes, yields a photon capture rate of $\gamma_2^A = 10s^{-1}$. Since these rates are normalized to 18 W/m^2, the extension to arbitrary light intensity I is straightforward. The rate of photon absorption normalized to 1 W/m^2 intensity, will be $\gamma_{1(2)} = 1(0.55)s^{-1}$. From now on, subindexes 1 and 2 relate to quantities of LH1 and LH2 complexes, respectively. Vesicles containing several hundreds of complexes will have an absorption rate:

$$\gamma_A = I(\gamma_1 N_1 + \gamma_2 N_2) \tag{2}$$

where $N_{1(2)}$ is the number of LH1 (LH2) complexes in the vesicle.

2.2 Excitation transfer

Excitation transfer happens through Coulomb interaction of electrons, excited to the Q_y electronic transition in chromophores. The interaction energy can be formally written (van Amerongen et al., 2000):

$$V_{ij} = \frac{1}{2} \sum_{m,n,p,q} \sum_{\sigma,\sigma'} \langle \phi_m \phi_n | V | \phi_p \phi_q \rangle c_{m\sigma}^+ c_{p\sigma'}^+ c_{q\sigma} c_{n\sigma}, \tag{3}$$

where $c_{m\sigma}^+$, $c_{n\sigma'}$ are fermion creation and annihilation operators of electrons with spin σ and σ', in the mutually orthogonal atomic orbitals ϕ_m and ϕ_n. The overlap $\langle \phi_m \phi_n | V | \phi_p \phi_q \rangle$ is the Coulomb integral:

$$\langle \phi_m \phi_p | V | \phi_n \phi_q \rangle = \int \int d\vec{r}_1 \, d\vec{r}_2 \, \phi_m^*(\vec{r}_1 - \vec{r}_i) \phi_p(\vec{r}_1 - \vec{r}_i) \frac{e^2}{|\vec{r}_1 - \vec{r}_2|} \phi_n^*(\vec{r}_2 - \vec{r}_j) \phi_q(\vec{r}_2 - \vec{r}_j) \tag{4}$$

that accounts on inter-molecular exchange contribution when donor and acceptor are at a distance comparable to the extent of the molecules, and the direct Coulomb contribution for an electron that makes a transition between ϕ_m and ϕ_p, having both a finite value near the position of the donnor $\vec{r}_1 \approx \vec{r}_D$, while another electron is excited between ϕ_n and ϕ_q at the acceptor coordinate $\vec{r}_2 \approx \vec{r}_A$. In this latter situation, a commonly used framework concerns a tight-binding Hamiltonian, where details in the specific molecular orbitals $|\phi_n\rangle$ involving mainly the Q_y orbitals is left aside, and emphasis relies on occupation with a single index labeling an electronic state $|i\rangle$ concerning occupation in a given chromophore. The Hamiltonian H in the chromophore site basis $|i\rangle$,

$$H = \epsilon \sum_i |i\rangle \langle i| + \sum_{i,j} V_{ij} |i\rangle \langle j| \tag{5}$$

has diagonal elements $\langle i|H|i\rangle = \epsilon$, concerning the energy of the excitation, usually measured from the ground electronic state. Neighboring chromophores are too close to neglect their charge distribution and its interaction is determined such that the effective Hamiltonian spectrum matches the spectrum of an extensive quantum chemistry calculation (Hu et al., 1997). In the LH2 complex, the B850 ring, with nearest neighbors coupling $\langle i|H|i+1\rangle = 806$ cm^{-1} or 377 cm^{-1} for chromophores respectively, within or in different neighboring $\alpha\beta$ units. For next-to-neighboring chromophores the dipole-dipole approximation is usually used,

$$\langle i|H|j\rangle = V_{ij} = C \left(\frac{\vec{\mu}_i \cdot \vec{\mu}_j}{r_{ij}^3} - \frac{3(\vec{r}_{ij} \cdot \vec{\mu}_i)(\vec{r}_{ij} \cdot \vec{\mu}_j)}{r_{ij}^5} \right)$$

$$\text{with } i \neq j, \, i = j \neq \pm 1 \tag{6}$$

where $\vec{\mu}_i$ is the dipole moment and r is the distance between the interacting dipoles with constants: $\epsilon = 13059$ cm^{-1} and $C = 519044^3$ cm^{-1}. The LH1 complex has a spectrum maximum near 875 nm, with inter-complex distances and nearest neighbor interactions equal to the ones provided in the B850 LH2 ring. The molecular nature of chromophores involve vibrational degrees of freedom that provide a manifold within each electronic states which should be accounted. However, given these parameters within a harvesting complex, intra-complex energy transfer involves sub-picosecond time-scales, that imply relevance of electronic quantum features over the influence of thermalization in the vibrational manifold. If decoherence sources are not important within the time-scale of excitation dynamics on donor

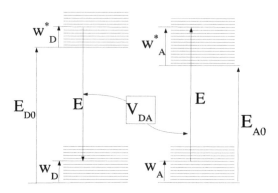

Fig. 3. Energy level scheme of donor and acceptor molecules. Although the zero phonon line might be different between both, energy conservation on the transfer applies due to vibrational levels.

or acceptor aggregates, they can be known from the eigenvector problem $H|\phi_k\rangle = E_k|\phi_k\rangle$ whose solution provides excitonic fully delocalized states

$$|\phi_k\rangle = \sum_n a_{k,i}|i\rangle \tag{7}$$

considered to best describe the B850 ring. A slightly less interaction strength allows thermalization through the vibrational manifold in comparable time-scales, degrading delocalization in the B800 ring over 2-3 pigments, however, able to improve robustness of B800→B850 energy transfer (Jang et al., 2004).

The fact that states involved in excitation transfer in molecules include the vibrational manifold, makes enormous the Hilbert space over which the sums of eq.(3) should be formally performed. When the transfer occurs with chromophores that belong to different complexes, say inter-complex energy transfer, rapid thermalization out stands direct Coulomb mechanism, and no defined phase relationship between donor and acceptor electronic states is expected. The hermitian nature of excitonic exchange is replaced due to decoherence, by a rate to describe electronic excitation transfer within thermalization in the vibrational manifold. According to Fermi's golden rule, adapted to the vibrational continuum, the rate of transfer is given by (Hu et al., 2002):

$$k_{DA} = \frac{2\pi}{\hbar} \int dE \int_{E_{A_0}} dw_A \int_{E_{D_0^*}} dw_D^* \frac{g_D^*(w_D^*)\exp(-w_D^*/k_B T)}{Z_D^*} \frac{g_A(w_A)\exp(-w_A/k_B T)}{Z_A} |\tilde{U}_{DA}|^2 \tag{8}$$

interpreted as a the sum of Coulomb contributions from electrons at donor (D) and acceptor (A) aggregates in ground or excited states (these latter dennoted by *), $\tilde{U}_{DA} = \langle \Psi_{D^*}\Psi_A|V_{DA}|\Psi_D\Psi_{A^*}\rangle$, weighted by Boltzmann factor (k_B is Botlzmann constant) and vibrational manifold multiplicity ($g_{D(A)}(w_{D(A)})$ for donor (acceptor)) at the electronic energies $w_{D(A)}$ measured from the donor (acceptor) zero phonon lines $E_{D(A)_0}$. With the Born-Oppenheimer approximation (Hu et al., 2002; van Amerongen et al., 2000) $|\Psi\rangle$ is assumed as products of electronic $|\phi\rangle$ and vibrational $|\chi\rangle$ molecular states:

$$\hat{U}_{DA} \approx \langle\phi_{D^*}\phi_A|V_{DA}|\phi_D\phi_{A^*}\rangle \times \langle\chi(w_D^*)|\chi(w_D)\rangle\langle\chi(w_A)|\chi(w_A^*)\rangle$$
$$\approx U_{DA}\langle\chi(w_D^*)|\chi(w_D)\rangle\langle\chi(w_A)|\chi(w_A^*)\rangle \tag{9}$$

where $U_{DA} = \langle\phi_{D^*}\psi_A|V_{DA}|\psi_D\psi_{A^*}\rangle$. Using the above approximation, the expression (8) now including the overlap between vibrational levels can be cast in a more illustrative form:

$$k_{DA} = \frac{2\pi}{\hbar}|U_{DA}|^2 \int dE\, G_D(E)G_A(E) \tag{10}$$

Here, $G_D(E)$ and $G_A(E)$ are often called the Franck-Condon weighted and thermally averaged combined density of states. Explicitly:

$$G_D(E) = \int_{E_{D_0^*}} dw_D^* \frac{g_D^*(w_D^*)\exp(-w_D^*/k_BT)|\langle\chi(w_D^*)|\chi(w_D)\rangle|^2}{Z_D^*}$$

with equal expression for acceptor molecule by replacement $D \rightarrow A$. Förster showed (Förster, 1965) that these distributions are related to extinction coefficient $\epsilon(E)$ and fluorescence spectrum $f_D(E)$ of direct experimental verification:

$$\epsilon(E) = \frac{2\pi N_0}{3\ln 10\hbar^2\, nc}|\mu_A|^2 E\, G_A(E), \quad f_D(E) = \frac{3\hbar^4 c^3\tau_0}{4n}|\mu_D|^2 E^3 G_D(E) \tag{11}$$

where $N_0 = 6.022 \times 10^{20}$ is the number of molecules per mol per cm^3, n is the refractive index of the molecule sample, c the speed of light, and τ_0 the mean fluorescence time of the donnor excited state. For normalized spectra

$$\hat{\epsilon}_A(E) = \frac{\epsilon_A(E)}{\int dE\, \epsilon_A(E)/E}, \quad \hat{f}_D(E) = \frac{f_D(E)}{\int dE\, f_D(E)/E^3} \tag{12}$$

and from the relations $w_D = E_{D_0^*} + w_D^* - E$ and $w_A^* = -E_{A_0} + w_A + E$ schematically presented in Fig.3, the Förster rate is cast

$$k_{DA} = \frac{2\pi}{\hbar}|U_{DA}|^2 \int dE \frac{\hat{\epsilon}_A(E)\hat{f}_D(E)}{E^4}. \tag{13}$$

Therefore, whenever fluorescence and absorption spectra are available, an estimate for the excitation transfer rate can be calculated.

Thermalization occurs firstly in the vibrational manifold of the electronic states involved. Due to the greater energy gap of electronic transitions compared with the one of vibrational nature, on a longer time-scale thermalization also occurs in excitonic states. Accordingly, the calculation in eq.(9) involves an statistical thermal mixture, explicitly:

$$\rho = \frac{1}{Tr\{\cdot\}}\sum_k \exp(-E_k/k_BT)|\phi_k\rangle\langle\phi_k| \tag{14}$$

where $Tr\{\cdot\}$ is trace of the numerator operator, used to normalize the state. Hence, in a straightforward fashion

$$U_{DA} = Tr\{\rho V_{DA}\} = \frac{1}{\sum_k \exp(-E_k/k_BT)}\sum_k\sum_p \exp(-E_k/k_BT)\langle\phi_k|V_{DA}|\phi_p\rangle$$

$$= \frac{1}{\sum_k \exp(-E_k/k_BT)}\sum_{k,p,i,j} a_{k,i}a_{p,j}^* V_{ij} \tag{15}$$

where the element $\langle \phi_k | V_{DA} | \phi_p \rangle$ are the elements of interaction among exciton states in molecules on different complexes and use is made of the individual contributions of excitonic states in chromophore's site basis eq.(7).

Summarizing, excitation transfer occurs through induced dipole transfer, among BChls transitions. The common inter-complex BChl distances 20-100 (Bahatyrova et al., 2004; Scheuring & Sturgis, 2005) cause excitation transfer to arise through the Coulomb interaction on the picosecond time-scale (Hu et al., 2002), while vibrational dephasing destroys coherences within a few hundred femtoseconds (Lee et al., 2007; Panitchayangkoona et al., 2010). As noted, the Coulomb interaction as dephasing occurs, makes the donor and acceptor phase become uncorrelated pointing into a classical rate behavior. Transfer rate measures from pump-probe experiments agree with the just outlined generalized Förster calculated rates (Hu et al., 2002), assuming intra-complex delocalization along thermodynamical equilibrium. LH2→LH2 transfer has not been measured experimentally, although an estimate of $t_{22} = 10$ ps has been calculated (Hu et al., 2002). LH2→ LH1 transfer has been measured for $R.$ $Sphaeroides$ as $t_{21} = 3.3$ps (Hess et al., 1995). Due to formation of excitonic states (Jang et al., 2007), back-transfer LH1→ LH2 is enhanced as compared to the canonical equilibrium rate for a two-level system, up to a value of $t_{12} = 15.5$ps. The LH1→LH1 mean transfer time t_{11} has not been measured, but the just mentioned generalized Förster calculation (Ritz et al., 2001) has reported an estimated mean time t_{11} of 20 ps. LH1→ RC transfer occurs due to ring symmetry breaking through optically forbidden (within ring symmetry) second and third lowest exciton lying states (Hu et al., 1997), as suggested by agreement with the experimental transfer time of 35-37 ps at 77 K (Visscher et al., 1989; 1991). Increased spectral overlap at room temperature improves the transfer time to $t_{1,RC} = 25$ ps (van Grondelle et al., 1994). A photo-protective design makes the back-transfer from an RC's fully populated lowest exciton state to higher-lying LH1 states occur in a calculated time of $t_{RC,1} = 8.1$ ps (Hu et al., 1997), close to the experimentally measured 7-9 ps estimated from decay kinetics after RC excitation (Timpmann et al., 1993).

Table 1 shows the results of mean transfer times presented in Ref.(Ritz et al., 2001) through the above mentioned calculation, compared with the experimental evidence restricted to different complex kind from the spectral resolution requirement of pump-probe spectroscopy. Since LH1↔ LH1 and LH2↔ LH2 transfer steps involve equal energy transitions, no experimental evidence is available regarding the rate at which these transitions occur. The experimentally determined B800→ B850 rate was 1/700fs (Shreve et al., 1991). The inter-complex transfer rate between LH2→ LH1 have been determined experimentally to be 1/3.3ps (Hess et al., 1995). Experimentally, LH1↔ RC forward transfer rate ranges between 1/50ps and 1/35ps, while back-transfer rate ranges between 1/12ps and 1/8ps (Timpmann et al., 1995; 1993; Visscher et al., 1989). It is interesting to note that exists a two fold difference in the experimental and theoretical determined LH2→LH1, ascribed to BChl$a$ Q_y dipole moment underestimation. It is assumed for theoretical calculation a value of 6.3 Debye, while a greater BChl$a$ Q_y dipole moment in Prostecochloris aestuarii (not a purple bacterium) of 7.7 Debye has been determined (Ritz et al., 2001). On the other hand, LH1→RC theoretical calculation gives a greater value for tranfer rate, thought to arise due to an overestimate of LH1 exciton delocalization (Ritz et al., 2001). This rate decreases when delocalization is assumed over fewer BChl's, therefore, further research is needed to understand the effect of decoherence sources (static inhomogeneities and dynamical disorder due to thermal fluctuations) on the delocalization length.

from \ to	LH1	LH2	RC
LH1	20.0/N.A.	15.5/N.A.	15.8/30-50
LH2	7.7/3.3	10.0/N.A.	N.A
RC	8.1/8	N.A	N.A

Table 1. Theoretical estimation/experimental evidence of inter-complex transfer times in picoseconds. N.A are not available data.

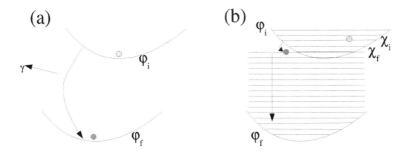

Fig. 4. Dissipation mechanisms. In (a), the electron de-excites due to its interaction with the quantized electromagnetic vacuum field through a fluorescent photon. In (b) internal conversion mechanism, where the vibrational levels overlap induces a transition between electronic excited and ground states. Dissipation overcomes when thermal equilibrium is reached in the vibrational manifold of electronic ground state.

2.3 Dissipation

Excitation in chromophores might be dissipated by two main mechanisms. The first is fluorescence, where the electronic excited state has a finite lifetime on the nanosecond time-scale, due to its interaction with the electromagnetic vacuum (Scully & Zubairy, 1997). The second is internal conversion, where the electronic energy is transferred to vibrational degrees of freedom.

Within the Born-Oppenheimer approximation, the molecular state Ψ, can be decomposed into purely electronic ϕ and (nuclear) vibrational χ states. The transition probability between initial state Ψ_i and final state Ψ_f, is proportional to $\langle \Psi_i | H | \Psi_f \rangle \propto \langle \chi_i | \chi_f \rangle$. Note that χ_i (χ_f) are vibrational levels in the ground (excited) electronic state manifold (see Fig.4). If the energy difference is small, and the overlap between vibrational levels of different electronic states is appreciable, the excitation can be transferred from the excited electronic state, to an excited vibrational level in the ground electronic state. This overlap increases with decreasing energy difference between electronic states. As higher electronic levels have smaller energy difference among their zero phonon lines, internal conversion process is more probable the higher energy electronic states have. Fluorescence and internal conversion between first excited singlet and ground electronic states, induce dissipation in a range of hundreds of picoseconds and a few nanoseconds. Numerical simulations are performed with a dissipation time including both fluorescence and internal conversion of $1/\gamma_D = 1$ ns, also used in (Ritz et al., 2001).

2.4 Special pair (SP) ionization

From the LH1 complex excitation reaches the RC, specifically the special pair (SP) dimer. The excitation can be transferred back to its surrounding LH1, or initiate a chain of ionizations along the A branch, probably, due to a tyrosine residue strategically positioned instead of a phenylalanine present in the B branch (Lia et al., 1993). Once the special pair is excited, it has been determined experimentally (Fleming et al., 1988) that takes 3-4 ps for the special pair to ionize and produce a reduced bacteriopheophytin, H_A^-, in a reaction $SP^* \rightarrow SP^+ H_A^-$. This reaction initiates an electron hop, to a quinone Q_A in about 200 ps, and to a second quinone, Q_B if available. Initially, the ionized quinol Q_B^+ captures an introcytoplasmic proton and produces hydroxiquinol $Q_B H$, which after a second ionization that produces $Q_B H^+$ to form quinol $Q_B H_2$. After any SP ionization a neutrality restablishment is required, provided by the cytochrome cyt charge carrier. After SP ionization, the cytochrome diffuses from the bc1 complex to a RC in order to replenish its neutrality $SP^+ \rightarrow SP$, within several microseconds (Milano et al., 2003). The first electron transfer step $P^* \rightarrow P^+$ occurs in the RC within $t_+ = 3$ ps, used for quinol ($Q_B H_2$) production (Hu et al., 2002).

2.5 Quinone-quinol cycling

The RC cycling dynamics also involves undocking of $Q_B H_2$ from the RC due to lower affinity among RC and this new product. Quinol starts a migration to the $bc1$ complex where enables the ionization of the cytochrome cyt charge carrier, while a new quinone Q_B molecule docks into the RC. The time before quinol unbinds, and a new Q_B is available, has been reported within milliseconds (Osváth & Maróti, 1997) to highlight quinol removal o as the rate limiting step (Osváth & Maróti, 1997) if compared to special pair restablishment.

Even though it has been reported that excitation dynamics change as a function of the RCs state (Borisov et al., 1985; Comayras et al., 2005), at a first glance the several orders of magnitude difference among the picosecond transfer, the nanosecond dissipation and the millisecond RC cycling, seems to disregard important effects due to these mechanisms' interplay. However, the quinol-quinone dynamics leaves the RC unable to promote further quinol production and eventually enhances the influence of dissipation of a wandering excitation, evident when none RC is available and the unique fate of any excitation is to be dissipated.

Interestingly, the quinone-quinol mechanism has been well established and thought to be of priority on adaptations of bacteria, that seem to respond to its dynamics. For instance, an observed trend for membranes to form clusters of same complex type (Scheuring, Rigaud & Sturgis, 2004) seems to affect diffusion of quinones, enhanced when, due to higher mobility of LH1s, left void spaces help travel quinones to the periplasm. Negligible mobility of LH2s in their domains, would restrict metabolically active quinones to LH1 domains (Scheuring & Sturgis, 2006). Easier diffusion of quinones, quinol and cytochromes promotes higher availability of charge carriers in RC domains under LLI conditions, increasing the rate at which RCs can cycle. The RC cycling dynamics and its connection to the membranes performance has been accounted in (Caycedo-Soler et al., 2010a;b) in a quantitative calculation to understand the effect of core clustering and stoichiometry variation in the RC supply or in the efficiency of the membranes from experimentally obtained Atomic Force Microscopy images, to be presented in this chapter.

3. Exciton kinetics

Figure 5 summarizes the relevant biomolecular complexes in purple bacteria *Rsp. Photometricum* (Scheuring, Rigaud & Sturgis, 2004), together with experimental– theoretical if the former are not available– timescales governing the excitation kinetics: absorption and transfer; and reaction center dynamics: quinol removal.

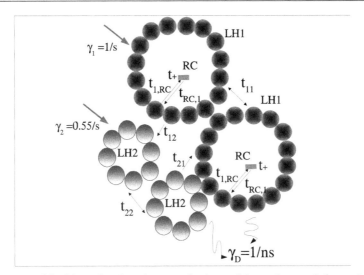

Fig. 5. Schematic of the biomolecular photosynthetic machinery in purple bacteria, together with relevant inter-complex mean transfer times t_{ij}, dissipation rate γ_D, and normalized light intensity rate $\gamma_{1(2)}$

3.1 Model

The theoretical framework used to describe the excitation transfer must be built around the experimental (if available) and theoretical parameters just outlined. Remind that the thermalization process occurs faster than inter-complex energy transfer, and provides the support to rely in a classical hopping process, since phase information is lost well within the time frame implied by direct Coulomb coupling. Accordingly, we base our analysis on a classical random walk for excitation dynamics along the full vesicle, by considering a collective state with $N = N_2 + 2N_1$ sites – resulting from N_2 LH2s, N_1 LH1s and hence N_1 RC complexes in the vesicle. The state vector $\vec{\rho} = (\rho_1, \rho_2, ..., \rho_M)$ has in each element the probability of occupation of a collective state comprising several excitations. If a single excitation is allowed in each complex, both excited and ground states of any complex should be accounted and the state space size is $M = \underbrace{2 \times 2 \times 2...}_{N} = 2^N$. On the other hand, if only one excitation that wanders in the whole network of complexes is allowed, a site basis can be used where each element of the state vector gives the probability of residence in the respective complex, and reduces the state vector size to $M = N$. In either case the state vector time evolution obeys a master equation

$$\partial_t \rho_i(t) = \sum_{j=1}^{M} G_{i,j} \rho_j(t). \tag{16}$$

where $G_{i,j}$ is the transition rate from a collective state or site i– whether many or a single excitation are accounted, respectively – to another collective state or site j. Since the transfer rates do not depend on time, this yields a formal solution $\vec{\rho}(t) = e^{\tilde{G}t} \vec{\rho}(0)$. However, the required framework depends on exciton abundance within the whole chromatophore at the regime of interest.

For instance, purple bacteria ecosystem concerns several meters depths, and should be reminded as a low light intensity environment. Within a typical range of 10-100 W/m^2 and a commonly sized chromatophore having ≈ 400 LH complexes, eq.(2) leads to an absorption rate $\gamma_A \approx$ 100-1000 s^{-1}, which compared with the dissipation mechanisms (rates of $\approx 10^9$ s^{-1}) imply that an absorption event occurs and then the excitation will be trapped by a RC or become dissipated within a nanosecond, and other excitation will visit the membrane not before some milliseconds have elapsed. However, it is important to remind the nature of thermal light where the possibility of having bunched small or long inter-photon times is greater than evenly spread, with greater deviations from poissonian statistics the grater its mean intensity is. Therefore, regardless of such deviations, under the biological light intensity conditions, the event of two excitations present simultaneously along the membrane will rarely occur and a single excitation model is accurate.

3.2 Small architectures
Small absorption rates lead to single excitation dynamics in the whole membrane, reducing the size of $\vec{\rho}(t)$ to the total number of sites N. The probability to have one excitation at a given complex initially, is proportional to its absorption cross section, and can be written as $\vec{\rho}(0) = \frac{1}{\gamma_A}(\underbrace{\gamma_1, ...,}_{N_1} \underbrace{\gamma_2, ...,}_{N_2} \underbrace{0, ..}_{N_1})$, where subsets correspond to the N_1 LH1s, the N_2 LH2s and the N_1 RCs respectively.

3.2.1 Complexes arrangement: architecture
To gain physical insight on the global behavior of the harvesting membrane, our interest lies in the probability to have an excitation at a given complex kind $k \in$ LH1,LH2 or RC, namely $\hat{p}_k$, given that at least one excitation resides in the network:

$$\hat{p}_k(t) = \frac{\rho_k(t)}{\sum_{i=1}^{N} \rho_i(t)} \ . \tag{17}$$

The effects that network architecture might have on the model's dynamics, are studied with different arrangements of complexes in small model networks, focusing on architectures which have the same amount of LH1, LH2 and RCs as shown in the top panel of Fig.6(a), (b) and (c). The bottom panel Fig.6 (d)-(e)-(f) shows that $\hat{p}_k$ values for RC, LH1 and LH2 complexes, respectively. First, it is important to notice that excitations trend is to stay within LH1 complexes, and not in the RC. Fig.6(d) shows that the highest RC population is obtained in configuration (c), followed by configuration (a) and (b) whose ordering relies in the connectedness of LH1s to antenna complexes. Clustering of LH1s will limit the number of links to LH2 complexes, and reduce the probability of RC ionization. For completeness, the probability of occupation in LH1 and LH2 complexes (Figs.6(e) and (f), respectively), shows that increased RC occupation benefits from population imbalance between LH1 enhancement and LH2 reduction. As connections among antenna complexes become more favored, the probability of finding an excitation on antenna complexes will become smaller, while the probability of finding excitations in RCs is enhanced. This preliminary result, illustrates that if the apparent requirement to funnel excitations down to RCs in bacterium were of primary importance, the greatest connectedness of LH1-LH2 complexes should occur in nature as a consequence of millions of years evolution. However, as will be presented, the real trend to form LH1 clusters, reduces its connectedness to antenna LH2 complexes and

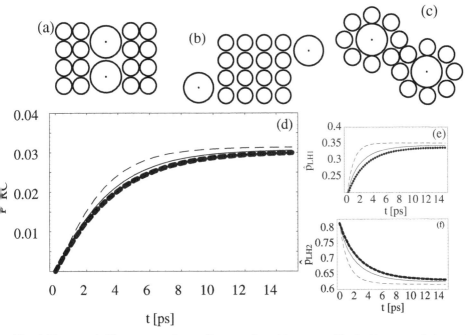

Fig. 6. Top panel: Three example small network architectures. The bottom panel shows the normalized probabilities for finding an excitation at an RC (see (d)), an LH1 (see (e)), or an LH2 (see (f)). In panels (d)-(f), we represent these architectures as follows: (a) is a continuous line; (b) is a dotted line; (c) is a dashed line.

somehow pinpoints other mechanisms as the rulers of harvesting membranes conformation and architecture.

3.2.2 Relative amount of complexes: Stoichiometry

We can also address with use of small architectures the effect of variation in the relative amount of LH1/LH2 complexes, able to change the population of the available states. Fig.7 shows small networks of LH-RC nodes, where the relative amount of LH2 and LH1 complexes quantified by stoichiometry $s = N_2/N_1$ is varied, in order to study the exciton dynamics. In Fig.8(a) the population ratio at stationary state of LHs demonstrate that as stoichiometry s becomes greater, the population of LH1s, becomes smaller, since their amount is reduced. It is apparent that RC population is quite small, and although their abundance increases the exciton trend to be found in any RC (Fig.8(b)), generally, excitations will be found in harvesting complexes. The population of LHs should be dependent on the ratio of complexes type. As verified in Fig.8(b), RCs have almost no population, and for the discussion below, they will not be taken into account. Populations can be written as:

$$\hat{p}_1(t \to \infty) = f_1(s)\frac{N_1}{N_1 + N_2} = \frac{f_1(s)}{1 + s} \tag{18}$$

$$\hat{p}_2(t \to \infty) = f_2(s)\frac{N_2}{N_1 + N_2} = \frac{s f_2(s)}{1 + s} \tag{19}$$

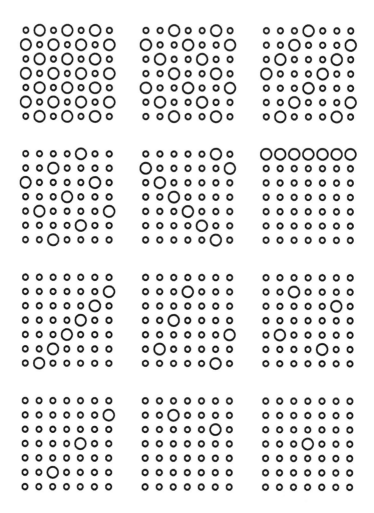

Fig. 7. Networks with different stoichiometries, from left to right, top to bottom, $s=\{1.04,$ 2.06, 3.08, 4.44, 5.125, 6, 7.16, 8.8, 11.25, 15.33, 23.5, 48\}, and equal number of harvesting complexes.

where the dependence on the amount of complexes is made explicit with the ratio $\frac{N_k}{N_1+N_2}$, and where $f_1(s)$ and $f_2(s)$ are enhancement factors. This factor provides information on how the population on individual complexes changes, beyond the features arising from their relative abundance. With use of eqs.(18-19), $f_1(s)$ and $f_2(s)$ can be numerically calculated provided that $\hat{p}_k(t \to \infty)$ can be known from the master equation, while s is a parameter given for each network. The results for enhancement factors are presented in Fig.8(c). The enhancement factor $f_2(s)$ for LH2 seems to saturate at values below one, as a consequence of the trend of excitations to remain in LH1s. This means that increasing further the number of LH2s will not enhance further the individual LH2 populations. On the other hand $f_1(s)$

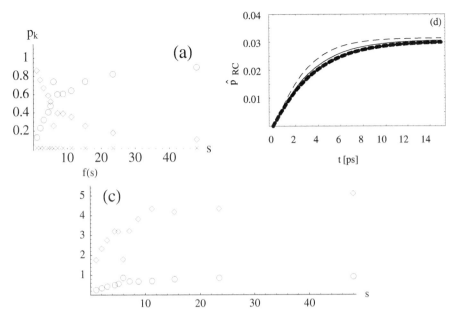

Fig. 8. In (a) stationary state populations for LH2s (circles), LH1s (diamonds) and RCs (crosses), as a function of the stoichiometry of membranes presented in Fig.7. In (b) a zoom of RC populations is made, and in (c) the enhancement factors $f_1(s)$ (diamonds) and $f_2(s)$ (circles) are presented.

has a broader range, and increases with s. This result reflects the fact that population of individual LH1s will become greater as more LH2 complexes surround a given LH1. An unconventional architecture (third column, second row in Fig.7) has an outermost line of LH1 complexes, whose connectedness to LH2s is compromised. In all the results in Fig. 8 (sixth point), this architecture does not follow the trends just pointed out, as LH1 and RC population, and enhancement factors, are clearly reduced. The population of LH1 complexes depends on their neighborhood and connectedness. Whenever connectedness of LH1 complexes is lowered, their population will also be reduced. Hence, deviations from populations trend with variation of stoichiometry, are a consequence of different degrees of connectedness of LH1s.

Up to this point, the master equation approach has helped us understand generally the effect of stoichiometry and architecture in small networks. Two conclusions can be made:

1. Connectedness of LH2 complexes to LH1s, facilitates transfer to RCs

2. The relative amount of LH2/LH1 complexes, namely, stoichiometry $s = N_2/N_1$, when augmented, induces smaller population on LH1-RC complexes. On the other hand, smaller s tends to increase the connectedness of LH1s to LH2s and hence, the population of individual LH1 complexes.

3.2.3 Special pair ionization

Another basic process involved in the solar energy conversion is the ionization of the special pair in the RC, and eventual quinol $Q_B H_2$ formation. Remind that once quinol is formed,

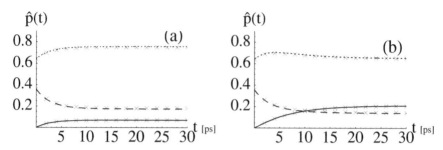

Fig. 9. Normalized probabilities $\hat{p}_k$ for finding the excitation at an LH2 (dashed), LH1 (dotted) or at an RC (continuous), for (a) $t_+ = 3$ps, and (b) $t_+ \to \infty$. Crosses are the results from the Monte Carlo simulation.

the special pair is unable to use further incoming excitations before quinol undocks and a new quinone replaces it. Even though the RC neutrality-diffusion process is propelled by complicated dynamics and involved mechanisms, in an easy approach, let us assume that the RC dynamics will proceed through a dichotomic process of "open" and "closed" RC states. In the open state, special pair oxidation is possible, while when closed, special pair oxidation to form quinol never happens, hence $t_+ \to \infty$

The effect of open and closed RC states changes the exciton kinetics. We start with a minimal configuration corresponding to a basic photosynthetic unit: one LH2, one LH1 and its RC. Figure 9(a) shows that if the RC is open, excitations will mostly be found in the LH1 complex, followed by occurrences at the LH2 and lastly at the RC. On the other hand, Figure 9(b)) shows clearly the different excitation kinetics which arise when the RC is initially unable to start the electron transfer $P^* \to P^+$, and then after ≈ 15ps the RC population becomes greater with respect to the LH2's. This confirms that a faithful description of the actual photosynthesis mechanism, even at the level of the minimal unit, must resort into RC cycling, given that its effects are by no means negligible. Moreover, comparison among Figs.6(d) and 9 also presents a feature that is usually undermined when small architectures are used to straightforward interpret its results as truth for greater, real biological vesicles. Energy funneling becomes smaller with the number of antenna LH2 complexes, thereby, in architectures with many harvesting antenna complexes, excitation will find it more difficult to arrive to any of the relatively spread RCs. Besides, although LH2→LH1 transfer rate is five-fold the back-transfer rate, the amount of smaller sized LH2s neighboring a given LH1 will increase the net back-transfer rate due to site availability. Hence, the funneling concept might be valid for small networks (Hu et al., 2002; Ritz et al., 2001), however, in natural scenarios involving entire chromatophores with many complexes, energy funneling might not be priority due to increased number of available states, provided from all LH2s surrounding a core complex, and globally, from the relative low RC abundance within a real vesicle.

It is important to mention that results for master equation calculations require several minutes in a standard computer to yield the results shown in Fig.8, and that these networks have an amount of nodes an order of magnitude smaller than the actual chromatophore vesicles. Dynamics concerning the RC cycling have not been described yet, fact that would increase further the dimension of possible membrane's states. To circumvent this problem, further analysis will proceed from *stochastic simulations*, and observables will be obtained from ensemble averages.

3.3 Full vesicles

A real vesicle involves several hundreds of harvesting complexes. Given the large state-space needed to describe such amount of complexes and our interest to inquire on a variety of incoming light statistics in the sections ahead, our subsequent model analysis will be based on a discrete-time random walk for excitation hopping between neighboring complexes.

3.3.1 Simulation algorithm

In particular, we use a Monte Carlo method to simulate the events of excitation transfer, the photon absorption, the dissipation, and the RC electron transfer. We have checked that our Monte Carlo simulations accurately reproduce the results of the population-based calculations described above, as can be seen from Figs.9(a) and (b). The Monte Carlo simulations proceed as follows. In general, any distribution of light might be used with the restriction of having a mean inter-photon time of γ_A^{-1} from eq.(2). Accordingly, a first photon is captured by the membrane and the time for the next absorption is set by inverting the cumulative distribution function from a [0,1] uniformly distributed (Unit Uniformly Distributed, UUD) random number. This inversing procedure is used for any transfer, dissipation or quinol removal event as well. The chosen absorbing complex is randomly selected first among LH1 or LH2 by a second UUD number compared to the probability of absorption in such complex kind, say $N_{1(2)}\gamma_{1(2)}/\gamma_A$ for LH1 (LH2), and a third UUD random number to specifically select any of the given complexes, with probability $1/N_{1(2)}$. Once the excitation is within a given complex, the conditional master equation given that full knowledge of the excitation residing in site i, only involves transfers outside such site, say $\partial_t\rho_i = -(\sum_j 1/t_{i,j} + \gamma_D)\rho_i$, whose solution is straightforward to provide the survival probability and its inverse, of use to choose the time t^* for the next event according to eq.(16) from a UUD number r: $-\log r/(\sum_j 1/t_{i,j} + \gamma_D) = t^*$. Once t^* is found, a particular event is chosen: transfer to a given neighboring complex j with probability $(1/t_{i,j})/(\sum_j 1/t_{i,j} + \gamma_D)$ or dissipation with probability $\gamma_D/(\sum_j 1/t_{i,j} + \gamma_D)$, which are assigned a proportional segment within [0,1] and compared with another UUD number to pinpoint the particular event. If the chosen event is a transfer step, then the excitation jumps to the chosen complex and the transfer-dissipation algorithm starts again. If dissipation occurs, the absorption algorithm is called to initiate a new excitation history. In a RC, the channel of quinol ionization is present with a rate $1/t_+$ in an event that if chosen, produces the same effect as dissipation. Nonetheless, the number of excitations that become SP ionizations are counted on each RC, such that when two excitations ionize a given RC and produce quinol, it becomes closed by temporally setting $1/t_+ = 0$ at such RC. Quinol unbinding will set "open" the RC, not before the RC-cycling time with mean τ, has elapsed, chosen according to a poissonian distribution. The algorithm can be summarized as follows:

1. Create the network: Obtain coordinates and type of LHs, and label complexes, for instance, by solely numerating them along its type, say complex 132 is of type 2 (we use 1 for LH1, 2 for LH2 and 3 for RC). Choose the j neighbors of complex i according to a maximum center to center distance less than $r_1 + r_2 + \delta$, $r_2 + r_2 + \delta$ and $r_1 + r_1 + \delta$ for respective complexes. We use $\delta = 20$, chosen such that only nearest neighbors are accounted and further increase of δ makes no difference on the amount of nearest neighbor connections, although further increase may include non-physical next to near-neighbors. In practice, the network creation was done by three arrays, one, say $neigh(i,j)$ with size $M \times S$, with M complexes as described above, and S as the maximum number of neighboring complexes among all the sites, hence requiring several attempts to be determined. Minimally $j \leq 1$ for an LH2, concerning the dissipation channel, $j \leq 2$ for LH1 including both dissipation and

transfer to its RC, and $j \leq 3$ for a RC accounting on dissipation, RC ionization an transfer to its surrounding LH1. The other arrays are built, say $size(i)$, with M positions, that keep on each the number of neighbors of the respective i labelled complex, and $rates(i, j)$ where at each position the inter-complex rate $i \rightarrow j$ is saved. For instance, $rates(i, 1)$ of any RC will be the ionization rate $1/t_+$.

2. Send photons to the network: On a time $t^* = -\log(r)/\gamma_A$ according to eq.(2), with r being an UUD number. Choose an LH2 or an LH1, according to the probability of absorption from the cross section of complex type $N_{1(2)}\gamma_{1(2)}/\gamma_A$. Add one excitation to the network, say $n = n + 1$, and assign the initial position $pos(n) = i$ of the excitation according to another UUD that selects an specific labelled i complex. Remind that n is bounded by the maximum amount of excitations allowed to be at the same time within the membrane, usually being one.

3. If the ith complex is excited, the construction of the above mentioned arrays make the cycle of excitation dynamics straightforward since the network is created only once, and dynamics only require to save the complex i where the excitation is, and then go through cycles of size $size(n, i)$ to acknowledge the stochastically generated next time for a given event. Excitation can be transferred to the available neighbors, become dissipated or a RC ionization event. Order all times for next events in order to know which will be the next in the array, say $listimes(p)$ with $p \leq n$, where $t_{min} = listemp(1)$. In parallel, update an organized array that saves the next process with the number of the neighbor to which hopping occurs, or say a negative number for RC ioinization and another negative number for dissipation.

4. Jump to next event: By cycling over the n present excitations, increase time up to the next event t_{min}. If RC cycling is accounted, check which time among t_{min} and the next opening RC time t_{RC} (its algorithm is to be discussed in the following) is the closest, and jump to it.

5. Change state of excitations or that of RCs: Update the current site of the excitation n, or whether it becomes a dissipation or a RC ionization. If the latter process occurs, keep in an array, say $rcstate(k)$ whose size equals the total amount of RCs, the number of excitations that have become ionizations from the last time the kth RC was opened.If $rcstate(k) = 2$ then the kth RC is closed by redefinition of $rate(i, 1) = 0$ and a poissonian stochastically generated opening time with mean τ is generated. This time interval is kept in an array $rctimes(k)$. Now, introduce this time interval into an ascending ordered list among all closed RCs opening times such that the minimum t_{RC} is obtained. If $t_{RC} < t_{min}$ then jump to that time and open the kth RC by letting $rcstate(k) = 0$ and $rates(i, 1) = 1/t_+$.

6. Look which is minimum among t^*, t_{RC} and t_{min} and jump to steps 2 or 4 according to whether $t^* < (t_{RC}, t_{min})$ or $(t_{RC}, t_{min}) < t^*$, respectively.

7. If the maximum amount of excitations chosen from the initialization, have been sent to the membrane, finish all processes and write external files.

The language used to program this algorithm was FORTRAN77, just to point out that these calculations do not require any high-level language.

3.3.2 Excitation dynamics trends in many node-complexes networks

In order to understand at a qualitative degree the excitation dynamics trends involved in full network chromatophores, a few toy architectures have been studied, shown in Fig.10.

In this preliminary study it is of interest to understand the excitation kinetics in complete chromatophores. In particular, it is useful to understand if any important feature arises according to nature's found tendency of forming clusters of the same complex type. In AFM images (Scheuring & Sturgis, 2005) it has been found that there is an apparent trend to form clusters of LH1 complexes with simultaneous formation of LH2 para-chrystalline domains. The reason that has been argued for this trend (Scheuring & Sturgis, 2005) involves the RC cycling dynamics and can be explained as follows.

The charge carrier quinone-quinol require diffusing through the intracytoplasmic membrane within the void spaces left among harvesting complexes, in order to reach the bc1 complex and complete the electric charge cycle with cytochorme that however, diffuses through the periplasm as schematically shown in Fig.2. The closeness among LH2 complexes in these para-chrystalline domains restricts the void spaces required for diffusion of quinone-quinol to the LH1 domains, where charge separation is taking place. Then, such aggregation indeed helps to improve the time it takes to quinone perform the whole RC-bc1-RC cycle, by restricting its presence to RC domains. However, an advantage concerning excitation dynamics has been heuristically proposed, where the path an excitation has to travel to reach an open RC once it encounters a closed RC is reduced due to LH1 clustering. In this section we investigate this latter possibility from a more quantitative point of view and address the former in the next section.

In order to understand exciton kinetics, for instance, let us fix our attention on the probability of a RC ionization when excitations start in a given LH2 on the configurations shown in Fig.10. To that end, let us introduce first the number n_{RCij} of excitations that are absorbed at site i and become ionization of the special pair at a given RC, labeled j. Also let $\hat{r}_{ij}$ be the unitary vector pointing in the direction from LH2 site i to RC site j. As an analogous to a force field, let vector field $\mathbf{v}$ correspond to a weighted sum of the directions on which RC ionizations occur, starting for a given LH2 complex

$$\mathbf{v}_i = \frac{\sum_j n_{RC_{ij}} \hat{r}_{ij}}{\sqrt{\sum_j n^2_{RC_{ij}}}}. \tag{20}$$

of help as a purely numerical calculation of graphical interpretation. The normalization gives a basis to compare only the directions for RC ionizations independent whether excitations are dissipated or not. Note that $\hat{r}_{ij}$ has both positive and negative signs, hence the vector $\mathbf{v}_i$ will have greater magnitude as RC ionizations occur more frequently at a given direction, or less magnitude as ionizations occur in more spread out directions.

Several features can be understood from this vector field. Figure 10 shows the vector fields attracted to the LH1 most abundant regions. The arrows illustrate the favored statistical direction in which excitations will become RC ionization, and they do not imply that an excitation will deterministically move in that direction if it were absorbed in such place. Firstly, in all cases, a tendency to point on clustered LH1s stands from distant LH2 complexes, a fact that emphasizes the very rapid transfer rates that lead to excitation's rapid global membrane sampling. Secondly, it is evident that better excitation attractors are made with increasing the core centers cluster size, easily observed in the top panel. Hence, according to this result one may naively state that a funneling effect is apparent and strongly dependent on cluster size. However, on the middle panel, it can be seen that this funneling effect is vanishingly small at sites where some degree of symmetry is present concerning the distribution of neighboring RCs, even though if these latter display some clustering degree. Third, a trend is displayed where the spreading of the flux becomes smaller with a lesser

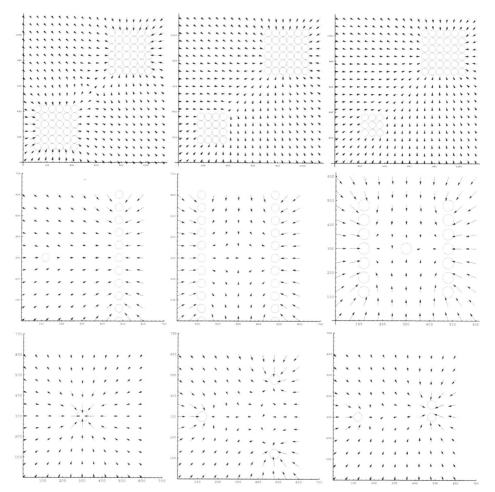

Fig. 10. Vector field **v** for different configurations, where the effect of core centers clustering is investigated (100000 runs starting in each site).

amount of RCs. However it is important to be aware that in more detail and in a general fashion, one also sees that even the LH2s closest to a given RC, "feel" the presence of further RCs. This implies that a significant amount of excitations that reach LH2 complexes neighboring a given LH1, will better prefer hopping to neighboring LH2 complexes to eventually reach a distant RC: no "funneling", as usually understood.

In a more realistic situation, the clustering of LH1 complexes is just a trend, and in average any cluster is formed by a few 3-4 core complexes in HLI situations and somehow less for LLI membranes as shown in Fig.11, where empirical architectures are presented in complete accordance with experimental data, taken from Ref.(Scheuring, Rigaud & Sturgis, 2004). Along these figures, the vector field calculated for the HLI and LLI empirical membranes is made, to highlight no clear trend of excitations to be transferred immediately to neighboring

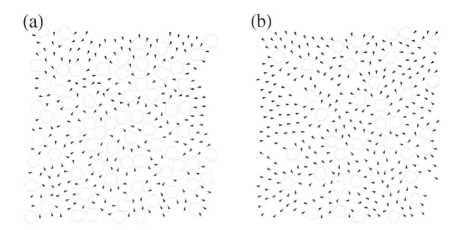

Fig. 11. Vector field **v** for HLI (a) and LLI (b) membranes (100000 runs each site).

LH1 complexes, and disregard a funneling LH2→LH1 as the unique direction of excitation flux when many complexes are accounted.

In detail, from the reasons just provided, in Fig.11(a) it is apparent that in HLI membranes, individual LH1 complexes are not efficient attractors of excitations. Here, not even the LH2s closest to a core center complex are notably attracted to such cores. On the other hand, a more uniform flux seems to be shared on clusters of LH2s in the LLI membrane (Fig.11(b)), that points to locations where more core centers are displayed. Hence, it is more evident the directionality of the flux to the most LH1 abundant locations that gives a statistical preference to ionize the closest RCs due their low global abundance in LLI membranes. The greater amount of LH1 complexes in the HLI membrane induces no significant preference for excitations to hop to the closest lying RCs and provides evidence that the excitation process proceeds without a clear funneling effect. If a single excitation can become an ionization in a single RC, the flux is clear, but if there are many and spread, the flux becomes random, as would be expected from any random walk. In summary, these results allow a mental picture of the excitation transfer process: Excitations become absorbed and start wandering along the membrane in a random walk regardless of a funneling effect, up to the moment where they reach a RC and become ionized, in a process where statistical preference is present in LLI membranes due to the reduced amount of RC, and presenting more clearly the random wandering on HLI due to higher spread out possibilities of ionization. This fact can also be quantitatively investigated.

Suppose that a given LH1 is surrounded by a few LH2 complexes. If an excitation is absorbed into one of these neighboring LH2s, it will have a survival probability decaying with rate $1/t_{21} + n/t_{22}$ where n wold be 3-4 corresponding to the number of LH2s neighboring such absorbing LH2. Given the presented transfer times (remind $t_{21} = 3.3$ps and $t_{22} = 10$ps) both terms in the survival decay rate become almost equal for $n = 3$, and hence no preference will occur on the excitation to be transferred to the closest LH1. If LH2→LH1 transfer would actually happen, the same exercise can be done for the survival probability within the LH1 complex, decaying with a rate $m/t_{12} + 1/t_{1,RC}$ (remind $t_{12} = 15.5$ps and $t_{1,RC} \approx 30$ps) which assuming a single surrounding LH2 complex ($m = 1$) would give a preference for excitations to return to antenna complexes. Hence, if the availability of neighbors is accounted,

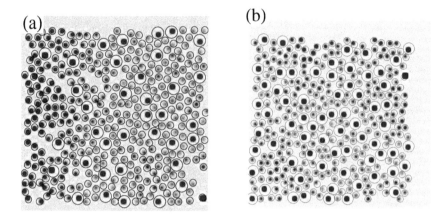

Fig. 12. Contour plots for dissipation in LLI (a) and HLI (b) membranes. Greater contrast means higher values for dissipation. The simulation is shown after 10^6 excitations were absorbed by the membrane with rate γ_A.

the funneling LH2→LH1 is lost, which along the enhanced back-transfer rate LH1→LH2 as compared to LH1→RC, provides the basis for understanding complete chromatophores as networks where actual performance depends upon the RC availability, since the event of ionization only depends on the probability to get to RC sites having no preference to be visited within the network. Excitations sample great portions of the whole membrane in its hopping, hence become able to reach RCs far away from the absorption sites before dissipation overcomes.

Dissipation itself also provides interesting features. The dissipation d_i measures the probability for excitations to dissipate at site i from the the total amount of excitations being absorbed in the membrane n_A:

$$d_i = \frac{n_{D_i}}{n_A} \tag{21}$$

that has an straightforward relation with the global efficiency η of the membrane, that accounts on the probability of any excitation to be used as an SP ionization. Given that excitations can only become RC ionizations or be dissipated, the sum over all complexes of dissipation probability will give the probability of any excitation to be dissipated, hence

$$\eta = 1 - \sum_i d_i \tag{22}$$

Figure 12 shows the dissipation on the membranes shinned with the respective illumination rate of their growth (LLI 10 W/m^2 and HLI 100 W/m^2). Figure 12(a) shows that the LLI membrane has highly dissipative clusters of LH2s, in contrast to the uniform dissipation in the HLI membrane (see Fig.12(b)).

This result addresses distinct features probably connected with the own requirements for bacteria when different light intensity is used during growth. Under LLI, bacterium might require use of all the available solar energy to promote its metabolism. In the numerical simulations (as expected from nature), the dissipation rate is set equal on any site, and therefore dissipation is only dependent on the probability for an excitation to be in a given complex. If more dissipation is found in some regions, it can there be supposed that excitations remain more time at such domains. Hence the tendency for excitations to dissipate implies they reside longer in LH2 complex domains and justify the view of LH2 clusters as excitation reservoirs. Although RC cycling has not been accounted yet, a given history might include an event where the excitation reaches closed clustered RCs and then jumps back to these LH2 domains waiting before the RCs become available again. On the other hand, HLI membranes display an evenly distributed dissipation. For instance, at HLI not all excitations might be required and dissipation can be used as a photo-protective mechanism. In this case, if dissipation were highly concentrated, vibrational recombination would overheat such highly dissipating domains.

Beyond these local details, an analysis regarding average values of dissipation D_k and residence p_{R_k} probabilities on a complex type k ($k = 1,2$ are respectively LH1, LH2 complexes) better supports the view of a completely random excitation hopping process. Table 2 shows the numerical results of p_{R_k}, that concerns the probability to find an excitation in a given complex type and is calculated from the sum of the residence times of excitations at a given complex type. It is straightforward to see that p_{R_k} is closely related to the probability of dissipation at such complex type, therefore, dissipation can correctly measure where excitations are to be found in general. The randomness of excitation dynamics is illustrated by realizing that dissipation in a given complex type depends primarily on its relative abundance, since $\frac{D_k}{D_j} \approx \frac{N_k}{N_j}$, and justifies that apart from LH2 clusters local variations, in the mean, all sites behave equivalently, and no dynamical feature arises to set a difference among LH1s and LH2s able to argue on the stoichiometry adaptation experimentally found. Lastly, as expected from the dynamical equivalence of sites, along the result obtained with toy architectures of varying stoichiometry where RC population heavily rose due their abundance but relatively invariant to arrangement, the probability to reach a RC solely depends on its abundance, and therefore the global efficiency η is greater in HLI membranes.

Membrane	p_{R_2}	p_{R_1}	D_2	D_1	$\frac{D_2}{D_1}$	$s = \frac{N_2}{N_1}$	$\eta = 1 - \frac{n_D}{n_A}$
LLI	0.72	0.25	0.74	0.26	9.13	9.13	0.86
HLI	0.50	0.46	0.52	0.48	3.88	3.92	0.91

Table 2. Dissipation D_k, residence probability p_{R_k}, on $k = \{1,2\}$ corresponding to N_1 LH1 and N_2 LH2 complexes respectively. Stoichiometry s and efficiency η are also shown.

Hence, for the present discussion, the most important finding from these numerical simulations is that the adaptation of purple bacteria does *not* lie in the single excitation kinetics. In particular, LLI membranes are seen to reduce their efficiency globally at the point where photons are becoming scarcer – hence the answer to adaptation must lie in some more fundamental trade-off (as we will later show explicitly). Due to the dissimilar timescales between millisecond absorption from eq.(2) and nanosecond dissipation, multiple excitation dynamics are also unlikely to occur within a membrane. However, multiple excitation dynamics cannot be regarded *a priori* not to be a reason for purple bacteria adaptation. A numerical study in Ref.(Caycedo-Soler et al., 2010a) involving blockade in which two excitations can not occupy the same site and annihilation, where two excitations annihilate due to vibrational recombination occurring due to significant Frank-Condon overlap on higher exciton states, shows that these mechanisms decrease the efficiency of membranes equally on LLI and HLI membranes, therefore, keeping the best performance to HLI.

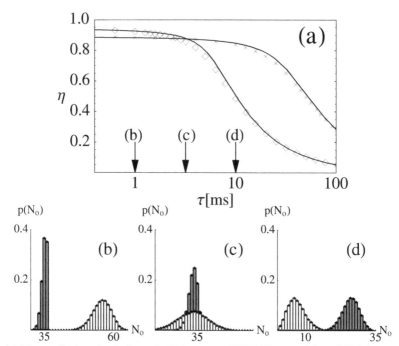

Fig. 13. (a) Monte Carlo calculation of efficiency η of HLI (diamonds) and LLI (crosses) grown membranes, as a function of the RC-cycling time τ. Continuous lines give the result of the analytical model. (b), (c) and (d) show the distributions $p(N_o)$ of the number of open RCs for the times shown with arrows in the main plot for HLI (filled bars) and LLI (white bars).

Nevertheless a more comprehensive study to be presented in the next section taking into account the coupled exciton-RC cycling is able to justify nature's choice of proceeding with the observed bacterial adaptations.

4. Complete chromatophores: Exciton and RC cycling coupled dynamics

It has been here repeatedly commented that RCs perform a cycle that provides the required exciton-chemical energy conversion. We now explain that the answer as to how adaptation can prefer the empirically observed HLI and LLI structures under different illumination conditions, lies in the *interplay* between the excitation kinetics and RC cycling dynamics. By virtue of quinones-quinol and cytochrome charge carriers, the RC dynamics features a 'dead' (or equivalently 'busy') time interval during which quinol is produced, removed and then a new quinone becomes available (Milano et al., 2003; Osváth & Maróti, 1997). Once quinol is produced, it leaves the RC and a new quinone becomes attached. These dynamics are introduced into the simulation algorithm as presented in section 3.3.1, by closing an RC for a (random) poissonian distributed time with mean τ after two excitations form quinol. The cycle can be sketched as follows: open RC$\rightarrow$ 2 ionizing excitations form quinol $\rightarrow$ closed RC in a time with mean $\tau\rightarrow$ open RC. This RC cycling time τ implies that at any given time, not all RCs are available for turning the electronic excitation into a useful charge separation. Therefore, the number of useful RCs decreases with increasing τ. Too many excitations will rapidly close RCs, implying that any subsequently available nearby excitation will tend to wander along the membrane and eventually be dissipated - hence reducing η. For the configurations resembling the empirical architectures (Fig.12), this effect is shown as a function of τ in Fig. 13(a) yielding a wide range of RC-cycling times at which LLI membrane is more efficient than HLI. Interestingly, this range corresponds to the measured time-scale for τ of milliseconds (Milano et al., 2003), and supports the suggestion that bacteria improve their performance in LLI conditions by enhancing quinone-quinol charge carrier dynamics as opposed to manipulating exciton transfer. As mentioned, a recent proposal (Scheuring & Sturgis, 2006) has shown numerically that the formation of LH2 para-crystalline domains produces a clustering trend of LH1 complexes with enhanced quinone availability – a fact that would reduce the RC cycling time.

However, the crossover of efficiency at $\tau \approx 3$ ms implies that even if no enhanced RC-cycling occurs, the HLI will be less efficient than the LLI membranes on the observed τ time-scale. The explanation is quantitatively related to the number N_0 of open RCs. Figs. 13(b), (c) and (d) present the distribution $p(N_0)$ of open RCs, for both HLI and LLI membranes and for the times shown with arrows in Fig.13(a). When the RC-cycling is of no importance (Fig. 13(b)) almost all RCs remain open, thereby making the HLI membrane more efficient than LLI since having more (open) RCs induces a higher probability for special pair oxidation. Near the crossover in Fig. 13, both membranes have distributions $p(N_0)$ centered around the same value (Fig. 13(c)), indicating that although more RCs are present in HLI vesicles, they are more frequently closed due to the ten fold light intensity difference, as compared to LLI conditions. Higher values of τ (Fig. 13(d)) present distributions where the LLI has more open RCs, in order to yield a better performance when photons are scarcer. Note that distributions become wider when RC cycling is increased, reflecting the mean-variance correspondence of Poissonian statistics used for simulation of τ. Therefore the trade-off between RC-cycling, the actual number of RCs and the light intensity, determines the number of open RCs and hence the performance of a given photosynthetic vesicle architecture (i.e. HLI versus LLI).

Hence, even though these adaptations show such distinct features in the experimentally relevant regimes for the RC-cycling time and illumination intensity magnitude (Milano et al., 2003; Osváth & Maróti, 1997; Scheuring, Sturgis, Prima, Bernadac & Rigaud, 2004), Figs.13(c) and (d) show that the distributions of open RCs actually overlap implying that despite such differences in growing environments, due to the adaptations arising, the resulting dynamics of the membranes become quite similar. Growth conditions generate adaptations that allow on LLI membrane to have a larger number of open RCs than the HLI adaptation and therefore LLI membrane will perform better than HLI with respect to RC ionization irrespective of any funneling dynamics. The inclusion of RC dynamics implies that the absorbed excitation will not find all RCs available, and somehow funneling would limit the chance of a necessary membrane sampling to explore further open RCs. Globally, a given amount of closed RCs will eventually alter the excitation's fate since probable states of oxidization are readily reduced. In a given lifetime, an excitation will find (depending on τ and current light intensity I) a number of available RCs – the *effective stoichiometry* – which is different from the actual number reported by Atomic Force Microscopy (Bahatyrova et al., 2004; Scheuring, Sturgis, Prima, Bernadac & Rigaud, 2004).

The effect of incident light intensity variations relative to the light intensity during growth with both membranes, presents a similar behavior. In Ref. (Caycedo-Soler et al., 2010a;b) such study is performed and it is concluded that: the LLI membrane performance starts to diminish well beyond the growth light intensity, while the HLI adaptation starts diminishing just above its growth intensity due to rapid RC closing that induce increased dissipation. Hence, in LLI

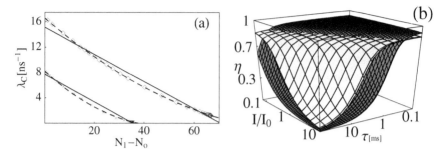

Fig. 14. (a) Numerical results showing the rate of ionization $\lambda_C(N_o)$ of an RC for HLI (diamonds) and LLI (crosses) membranes, together with a quadratic (dashed line) and linear (continuous) dependence on the number of closed RCs ($N_1 - N_o$). The fitting parameters for $a + bN_o$ are $a = \{15.16, 7.72\}$ns^{-1}, $b = \{-0.21, -0.21\}$ ns^{-1}; and for $a + bN_o + cN_o^2$, $a = \{16.61, 8.21\}$ns^{-1}, $b = \{-0.35, -0.33\}$, and $c = \{3.6, 1.5\}\mu$s^{-1}, for HLI and LLI membranes respectively. (b) η as function of τ and $\alpha = I/I_0$, obtained from the complete analytical solution for LLI (white) and HLI (grey) membranes

membranes excess photons are readily used for bacterial metabolism, and HLI membranes exploit dissipation in order to limit the number of processed excitations. In the same work, it is found that the effect of the arrangement itself is lost due to RC dynamics, since the effective stoichiometry with spread out open RCs becomes alike among different membranes sharing the same amount of RCs, equal cycling time τ and incident light intensity.

To summarize so far, the arrangement of complexes changes slightly the efficiency of the membranes when no RC dynamics is included – but with RC dynamics, the most important feature is the number of open RCs. Although the longer RC closing times make membranes more prone to dissipation and decreased efficiency, it also makes the architecture less relevant for the overall dynamics. The relevant network architecture instead becomes the dynamical one including sparse open RCs, not the static geometrical one involving the actual clustered RCs.

5. Analytical model

Within a typical fluorescence lifetime of 1 ns, a single excitation has travelled hundreds of sites and explored the available RCs globally. The actual arrangement or architecture of the complexes seems not to influence the excitation's fate, since the light intensity and RC cycling determine the number of open RCs and the availability for P oxidation.

5.1 Excitation transfer-RC cycling rate model

Here, we present an alternative rate model which is inspired by the findings of the numerical simulations, but which (1) globally describes the excitation dynamics and RC cycling, (2) leads to analytical expressions for the efficiency of the membrane and the rate of quinol production, and (3) sheds light on the trade-off between RC-cycling and exciton dynamics (Caycedo-Soler et al., 2010a;b).

Shortly, N_E excitations area absorbed by the membrane at a rate γ_A, and will find its way to become RC ionizations with a rate per particle $\lambda_C(N_o)$ whose dependence on the number of open RCs is made explicit, or be dissipated at a rate γ_D. On the other hand, RCs have their own dynamycs, closing at a rate $\lambda_C(N_o) \times N_E$ and individually opening at a rate $1/\tau$. The dependance $\lambda_C(N_o)$ is numerically found aidded by the stochastic model and shown in Fig.14(a). Rate equations can therefore be written:

$$\frac{dN_E}{dt} = -(\lambda_C(N_o) + \gamma_D)N_E + \gamma_A \tag{23}$$

$$\frac{dN_o}{dt} = \frac{1}{\tau}(N_1 - N_o) - \frac{\lambda_C(N_o)}{2}N_E. \tag{24}$$

to be solved, and of use for the calculation of the steady-state efficiency $\eta = n_{RC}/n_A$:

$$\eta = \frac{\lambda_C(N_o)N_E}{\gamma_A}. \tag{25}$$

A linear fit for $\lambda_C(N_o)$ allows an analytical expression for η:

$$\eta(\tau, \gamma_A(I)) = \frac{1}{\gamma_A \lambda_C^0 \tau} \left\{ 2N_1(\lambda_C^0 + \gamma_D) + \gamma_A \lambda_C^0 \tau - \right. \tag{26}$$

$$\left. \sqrt{4N_1^2(\lambda_C^0 + \gamma_D)^2 + 4N_1 \gamma_A \lambda_C^0(\gamma_D - \lambda_C^0)\tau + (\gamma_A \lambda_C^0 \tau)^2} \right\} \tag{27}$$

where λ_C^0 is the rate of RC ionization when no RC-cycling is accounted, dependent only on the amount of RCs present in the vesicle (Caycedo-Soler et al., 2010a). This analytical expression is shown in Fig.14(b) and illustrates that $\eta \geq 0.9$ if the transfer-P reduction time is less than a tenth of the dissipation time, not including RC cycling. As can be seen in Figs. 13(a), the analytical solution is in good quantitative agreement with the numerical stochastic simulation, and provides support for the assumptions made. Moreover, this model shows directly that the efficiency is driven by the interplay between the RC cycling time and light intensity. Figure 14(b) shows up an entire region of parameter space where LLI membranes are better than HLI in terms of causing P ionization, even though the actual number of RCs that they have is smaller. In view of these results, it is interesting to note how clever Nature has been in tinkering with the efficiency of LLI vesicles and the dissipative behavior of HLI adaptation, in order to meet the needs of bacteria subject to the illumination conditions of the growing environment.

5.2 Bacterial metabolic demands

Photosynthetic membranes must provide enough energy to fulfill the metabolic requirements of the living bacteria quantified by the quinol output or quinol rate

$$W = \frac{1}{2}\frac{dn_{RC}}{dt} \tag{28}$$

which depends directly on the excitations that ionize RCs n_{RC}. The factor $\frac{1}{2}$ accounts for the requirement of two ionizations to form a single quinol molecule. Although these membranes were grown under continuous illumination, the adaptations themselves are a product of millions of years of evolution. Using RC cycling times that preserve quinol rate in both adaptations, different behaviors emerge when the illumination intensity is varied (see Fig. 15(a). The increased illumination is readily used by the LLI adaptation, in order to profit from excess excitations in an otherwise low productivity regime. On the other hand, the HLI membrane maintains the quinone rate constant, thereby avoiding the risk of pH imbalance in the event that the light intensity suddenly increased. We stress that the number of RCs synthesized does not directly reflect the number of available states of ionization in the membrane. LLI synthesizes a small amount of RCs in order to enhance quinone diffusion, such that excess light intensity is utilized by the majority of special pairs. In HLI, the synthesis of more LH1-RC complexes slows down RC-cycling, which ensures that many of these RCs are unavailable and hence be advantageous of evenly distributed dissipation to steadily supply quinol independent of any excitation increase. The very good agreement between our analytic results and the stochastic simulations, yields additional physical insight concerning the stoichiometries found experimentally in *Rsp. Photometricum*.

A closed form expression regarding all dynamical parameters involved can be obtained (Caycedo-Soler et al., 2010a):

$$2W(s,I) = \frac{\gamma_A(s,I)}{2} + \frac{1}{B(s)}\left(1 + \frac{\gamma_D}{\lambda_c^0}\right) + \sqrt{[\frac{\gamma_A(s,I)}{2} + \frac{1}{B(s)}\left(1 + \frac{\gamma_D}{\lambda_c^0}\right)]^2 + \frac{\gamma_A(s,I)}{2B(s)}} \qquad (29)$$

where the dependence on stoichiometry is made explicit due to absoprtion cross section in γ_A and on $B(s) = \frac{\tau(s)(A_1 + sA_2)}{f(s)A_0}$, which is a parameter that depends on area $A_{1(2)}$ of individual LH1 (LH2) complexes and filling fraction $f(s)$. The filling chromatophore fraction dependence on s is available from experimental data of Ref.(Scheuring & Sturgis, 2006) and $\tau(s)$ is constructed from an interpolation scheme (Caycedo-Soler et al., 2010a).

As emphasized in Ref. (Scheuring & Sturgis, 2005), membranes with $s=6$ or $s=2$ were not observed, which is to be compared with the contour plots regarding constant quinol output, of s as a function of growing light intensity I_0, shown in Fig.15(b). These results support a dichotomic observation where $s \approx 4$ predominantly on a great range for growing light intensity. However, a prediction can be made for 30-40 W/m^2 where a great sensitivity of stoichiometry ratios rapidly build up the number of antenna LH2 complexes. Very recently (Liu et al., 2009), membranes were grown with 30W/m^2 and an experimental stoichiometry of 4.8 was found. The contour of 2200 s^{-1} predicts a value for stoichiometry of 4.72 at such growing light intensities. This agreement is quite remarkable, since a simple linear interpolation among the values $s \approx 4, I_0 = 100$ W/m^2 and $s \approx 8, I_0 = 10$ W/m^2 would wrongly predict $s = 7.1$ at 30 W/m^2. We encourage experimentalists to confirm the full range of predicted behaviors as a function of light-intensity and stoichiometry. Such exercise would without doubt confirm/expand the understanding on RC-exciton dynamics trade off, pinpointing a direction to pursue solar energy conversion research, provided by Nature itself.

6. Performance of photosynthetic membranes under extreme photon statistics

Photosynthetic (e.g purple) bacteria provide the crucial coupling between the Sun's energy and the production of food on Earth, and have adapted successfully to a variety of terrestrial conditions since the beginnings of life on Earth several billion years ago. In this section we explore whether terrestrial bacteria, which are the product of millions of years of evolutionary pressure on Earth, could survive if suddenly exposed to incident light with extreme statistics. We are therefore mimicking a scenario in which purple bacteria are either (i) suddenly transported to some unknown extreme solar environment elsewhere in the universe, or (ii) where our own Sun suddenly picks up extremal behavior in terms of the temporal statistics of its emitted photons, or (iii) the bacteria are subjected to extreme artificial light sources such as that in Ref. (Borlaug et al., 2009) involving stimulated Raman scattering, as well as in coherent

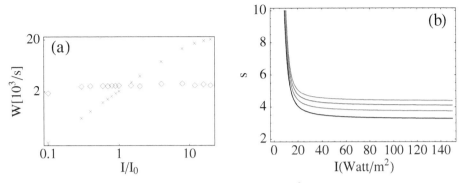

Fig. 15. (a) Quinol rate W in HLI (diamonds, $I_0 = 100W/m^2$) and LLI (crosses, $I_0 = 10W/m^2$) grown membranes, as a function of incident intensity I. (b) Quinol rate contours of $W = \{1900, 2000, 2100, 2200\}$ s^{-1} in black, blue, red and pink, respectively.

anti-Stokes Raman scattering in silicon, or (iv) the external membrane skin of the bacterium is modified in such a way that the absorption of photons takes on extreme statistical properties. Our discussion is qualitatively different from previous work looking at life in extreme conditions, since those discussions have tended to focus on *environmental* extremes affecting the biochemical metabolism or extremes in the incident light intensity. By contrast, our current discussion focuses on extremes in the incident photon statistics. Although the setting is largely hypothetical, our theoretical predictions are based on a realistic semi-empirical model which incorporates (i) high precision empirical AFM information about spatial locations of LH1 and LH2 biomolecular complexes in the membrane architecture of *Rsp. Photometricum* purple bacteria, and (ii) full-scale stochastic simulations of the excitation kinetics and reaction center dynamics within the empirical membrane.

As with any process involving events occurring in a stochastic way over time, the statistical properties of arriving photons may show deviations from a pure coin-toss process in two broad ways: burstiness and memory (Goh & Barabasi, 2008). First consider the simplest process in which the rate of arrival of a photon has a constant probability per unit time. It is well known that this so-called Poisson process produces a distribution of the waiting time for the next photon arrival which is exponential in form, given by $P_P(\tau) \sim \exp(-\tau/\tau_0)$. The extent to which the observed arrival time distribution $P(\tau)$ deviates from exponential, indicates how non-Poissonian the photon arrival is. Following Barabasi (Goh & Barabasi, 2008), we refer to this as 'burstiness' B and define it by its deviation from a purely Poisson process:

$$B \equiv \frac{(\sigma_\tau/m_\tau - 1)}{(\sigma_\tau/m_\tau + 1)} = \frac{(\sigma_\tau - m_\tau)}{(\sigma_\tau + m_\tau)}$$

where σ_τ and m_τ are the standard deviation and mean respectively of the empirical distribution $P(\tau)$. For a pure Poisson process, the mean and standard deviation of the arrival time distribution are equal and hence $B = 0$. The other property which can be noticeable for a non-Poisson process is the memory M between consecutive inter-arrival times which, following Barabasi, we define as:

$$M \equiv \frac{1}{n_\tau - 1} \sum_{i=1}^{n_\tau - 1} \frac{(\tau_i - m_1)(\tau_{i+1} - m_2)}{\sigma_1 \sigma_2}$$

where n_τ is the number of inter-arrival times measured from the signal and $m_1 (m_2)$ and $\sigma_1 (\sigma_2)$ are sample mean and sample standard deviation of $\tau_i (\tau_{i+1})$'s respectively ($i = 1, \ldots, n_\tau - 1$). For a pure Poisson process, $M = 0$. Both B and M range from -1 to $+1$, with the value $B = 0$ and $M = 0$ for a strict Poisson process. We will assume for simplicity that the arriving photons are all absorbed and hence one exciton created within the membrane by each photon. This can easily be generalized but at the expense of adding another layer of statistical analysis to connect the statistics of arriving photons to the statistics of the excitons being created within the LH2/LH1 membrane – indeed, taking a constant absorption probability less than unity would not change our main conclusions. We will also neglect the possibility that several photons arrive at exactly the same time. In principle, incident photons can be generated numerically with values of B and M which are arbitrarily close to any specific (M, B) value – however this is extremely time-consuming numerically. Instead, we focus here on specific processes where the B and M can be calculated analytically. Although this means that the entire (M, B) parameter space is not accessed, most of it can indeed be – and with the added advantage that analytic values for B and M are generated.

Figure 16 summarizes our findings in terms of the incident photons (i.e. excitation input) and metabolic output from the LH2/LH1 membrane, over the entire (M, B) parameter space. The subspace shown by the combination of the white and dotted regions (i.e. the region in the (M, B) space which is *not* diagonally shaded) comprises points with (M, B) values that can be generated by one of three relatively simple types of photon input: (a) step input, (b) bunched input and (c) power-law step input, as shown in the three panels respectively. Each point in the blank or dotted region denotes a time-series of initial excitations in the membrane with those particular burstiness and memory values (B and M). This train of initial excitations then migrates within the membrane of LH2/LH1 complexes, subject to the dynamical interplay of migration and trapping as discussed earlier in this paper, and gives rise to a given output time-series of 'food' (quinol) to the bacteria. Figure 16 shows explicitly a variety of initial (M, B) values (crosses and stars) and the trajectory represents the locus of resulting quinol outputs from the reaction centers (RC) for this particular (M, B) input. The trajectory is generated by varying the RC closing time within the range of physically reasonable values. The trajectory is finite since the range of physically reasonable RC closing times is also finite (20 to 1000Hz). As the time during which the RC is closed increases, the output becomes more Poisson-like, i.e. the B and M output values from the membrane after absorption at the RC, tend towards 0. Hence the trajectories head toward the center as shown.

The (M, B) values for natural sunlight on Earth lie near $M = 0$ and $B = 0$ since the incident photon arrival from the Sun is approximately a Poisson process. This produces a quinol output which is also a Poisson process ($M = 0$ and $B = 0$). Now suppose a terrestrial bacteria is suddenly subjected to an extreme incident light source with (M, B) values which lie at a general point in (M, B) parameter space. Can it survive? It is reasonable to expect that the bacteria will not survive if the resulting quinol output is very different from that on Earth (i.e. very different from $B = 0$ and $M = 0$) since the bacteria is well-tuned for Life on Earth only. Hence we say that the bacteria can only survive if the quinol output has (M, B) values within 0.01 of the terrestrial values of $B = 0$ and $M = 0$. The blank spaces denote (M, B) values for which the bacteria would survive, while the dotted spaces are where the bacteria would die. Remarkably, there are therefore substantial regions of extreme incident photon statistics (i.e. $B \neq 0$ and $M \neq 0$) where the bacteria would survive, producing a quinol output for the rest of the organism which resembles that on Earth (i.e. $|B| < \epsilon$ and $|M| < \epsilon$ where $\epsilon = 0.01$). The fact that these blank spaces have an irregular form is due to the highly nonlinear nature of the kinetic interplay between exciton migration and trapping, as discussed earlier in this chapter. For example, panel (c) shows that a bacterium which is only tuned to survive on Earth can have a maximally negative burstiness $B = -1$ and a maximal memory $M = 1$, meaning that the incident photons arrive like the regular ticks of a clock. The same applies to all the incident photon conditions demonstrated by the blank spaces. If we make the survival criterion more generous (i.e. $\epsilon > 0.01$) the blank spaces corresponding to survivability of the terrestrial bacterium increase in size until no dotted space remains (i.e. when $\epsilon = 1$). All the dotted areas in Fig. 16 become blank spaces and hence represent survivable incident photon conditions for the bacteria.

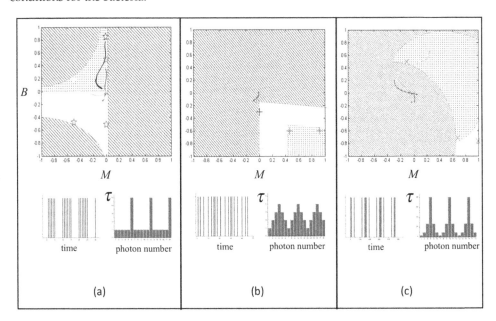

Fig. 16. The (M, B) phase diagram governing survivability of a terrestrial bacteria under a wide range of incident photon time-series with a given burstiness B and memory M. The sum of the blank and dotted areas in each diagram is the region accessible by each of the three photon processes shown. For each (M, B) value, a time-series of photons is created with these properties and used to generate excitations within the membrane. The resulting quinol output time-series is then calculated. Assuming that this output needs to be similar to that on Earth, the dotted region corresponds to photon (M, B) input values for which the bacteria would die, while the blank region corresponds to photon input for which the bacteria would survive. The trajectories represent the range of quinol outputs from the reaction centers (RC) as we span the range of physically reasonable RC closure times, given a particular input (shown as a star or cross of the same color). As the time during which the RC is closed increases, the B and M output values tend to move toward the origin (i.e. toward $B = 0$ and $M = 0$). Bottom: Photon inputs correspond to (a) step input, (b) bunched input and (c) power-law step input. Lower row in each case shows photon arrival process (left, black barcode) and waiting time τ between photon arrivals (right, red histogram).

7. Perspectives: Photosynthetic membranes of purple bacteria as a basis to develop stable energy conversion

This chapter has covered a comprehensive review of purple bacteria adaptation to light intensity conditions and has provided a basis to understand the chromatic adaptation of photosynthetic harvesting membranes as a consequence of the requirement imposed by nature on living organisms to develop a machinery capable of fuel production in a stable manner. It is worth to emulate, and certainly interesting to develop in artificial devices to surpass the usual bleaching on semiconductor panels, an analogous mechanism to the excitation kinetics and RC dynamics interplay, in order to shift the emphasis of requirements of fuel production and allow highly efficient energy transfer to available charge carriers (quinones) in low light intensity conditions, while consistently, a leveled fuel (quinol) production statistics is accomplished when photon arrival varies both in intensity and photon waiting time statistics.

8. Acknowledgments

The authors acknowledges financial support from Research Projects Facultad de Ciencias, Universidad de los Andes, Banco de la Republica and Fundacion Mazda.

9. References

Arellano, J. B., Raju, B. B., Nakvi, K. R. & Gilbro, T. (1998). Estimation of pigment stoichiometries in photosynthetic systems of purple bacteria: special reference to the (abscence of) second caroteniod in LH2, *Photochemistry and Photobiology* 68: 84.

Bahatyrova, S., Freese, R. N., Siebert, C. A., Olsen, J. D., van der Werf, K. A., van Grondelle, R., Niederman, R. A., Bullough, O. A., Otto, C. & Hunter, N. C. (2004). The native architecture of a photosynthetic membrane, *Nature* 430: 1058.

Borisov, A. Y., Freiberg, M., Godik, V. I., K., R. K. & Timpmann, K. E. (1985). Kinetics of picosecond bacteriochlorophyll luminescence in vivo as a function of the reaction center state, *Biochimica et Biophysica Acta* 807: 221.

Borlaug, D., Fathpour, S. & Jalali, B. (2009). Extreme value statistics in silicon photonics, *IEEE Photonics Journal* 1: 1.

Caycedo-Soler, F., Rodríguez, F., Quiroga, L. & Jonhson, N. (2010a). Interplay between excitation kinetics and reaction-center dynamics in purple bacteria, *New Journal of Physics* 12: 095008.

Caycedo-Soler, F., Rodríguez, F., Quiroga, L. & Jonhson, N. (2010b). Light-Harvesting Mechanism of Bacteria exploits a critical interplay between the dynamics of transfer and trapping, *Phys. Rev. Lett.* 104: 158302.

Codgell, R. J., Gall, A. & Kohler, J. (2006). The architecture and function of the light-harvesting apparatus of purple bacteria: From single molecule to in vivo membranes, *Quart. Rev. Biophys.* 39: 227–324.

Comayras, F., Jungas, C. & Lavergne, J. (2005). Functional consequences of the orgnization of the photosynthetic apparatus in rhodobacter sphaeroides. i. quinone domains and excitation transfer in chromatophores and reaction center antenna complexes, *J. Biol. Chem.* 280: 11203–11213.

Engel, G., Calhoun, T., Read, E., Ahn, T., Mancal, T., Cheng, Y., Blankenship, R. & Fleming, G. (2007). Evidence for wavelike energy transfer through quantum coherence in photosynthetic systems, *Nature* 446: 782–786.

Fleming, G. R., Martin, J. L. & Breton, J. (1988). Rates of primary electron transfer in photosynthetic reaction centres and their mechanistic implications, *Nature (London)* 333: 190.

Förster, T. (1965). *Delocalized excitation and excitation transfer*, Academic Press, New York.

Francke, C. & Amesz, J. (1995). The size of the photosynthetic unit in purple bacteria., *Photosynthetic Res.* 46: 347.

Gall, A., Gardiner, A. R., Gardiner, R. J. & Robert, B. (2006). Carotenoid stoichiometry in the LH2 crystal: no spectral evidence for the presence of the second molecule on the $\alpha\beta$-apoprotein dimer, *FEBS Letters* 580: 3841.

Geyer, T. & Heims, V. (2006). Reconstruction of a Kinetic model of the Chromophores Vesicles from Rhodobacter sphaeroides, *Biophyscal Journal* 91: 927.

Goh, K. & Barabasi, A. (2008). Burstiness and memory in complex systems, *Europhy. Lett.* 81.

Hess, S., Cachisvilis, M., Timpmann, K. E., Jones, M., Fowler, G., Hunter, N. C. & Sundstrom, V. (1995). Temporally and spectrally resolved subpicosecond energy transfer within the peripheral antenna complex (lh2) and from lh2 to the core antenna complex in photosynthetic purple bacteria, *Proceedings Natinal Academy of Sciences* 92: 12333.

Hu, X., Ritz, T., Damjanovic, A., Autenrie, F. & Schulten, K. (2002). Photosynthetic apparatus of purple bacteria., *Quart. Rev. of Biophys.* 35: 1–62.

Hu, X., Ritz, T., Damjanović, A. & Schulten, K. (1997). Pigment Organization and Transfer of Electronic Excitation in the Photosynthetic Unit of Purple Bacteria, *Journal of Physical Chemistry B* 101: 3854–3871.

Jamieson, S. J. & al, e. (2002). Projection structure of the photosynthetic reaction centre-antenna complex of *Rhodospirillum rubrum* at 8.5 resolution., *EMBO Journal* 21: 3927.

Jang, S., Newton, M. D. & Silbey, R. J. (2004). Multichromophoric Förster Resonance Energy Transfer, *Phys. Rev. Lett.* 92: 218301.

Jang, S., Newton, M. D. & Silbey, R. J. (2007). Multichromophoric fo1rster resonance energy transfer from b800 to b850 in the light harvesting complex 2: Evidence for subtle energetic optimization by purple bacteria, *J. Phys. Chem. B* 111: 6807–6814.

Karrasch, S., Bullough, P. A. & Ghosh, B. (1995). The 8.5 . projection map of the light harvesting complex I from *Rhodospirillum rubrum* reveals a ring composed of 16 subunits, *EMBO Journal* 14: 631.

Lee, H., Cheng, Y. & Fleming, G. (2007). Coherence dynamics in photosynthesis: Protein protection of excitonic coherence, *Science* 316: 1462.

Lia, Y., DiMagno, T. J. Chang, T. C., Wang, Z., Du, M., Hanson, D. K., Schiffer, M., Norris, J. R., R., F. G. & Popov, M. S. (1993). Inhomogeneous electron transfer kinetics in reaction centers of bacterial photosynthesis, *J. Phys. Chem.* 97: 13180.

Liu, L., Duquesne, K., Sturgis, J. & Scheuring, S. (2009). Quinone pathways in entire photosynthetic chromatophores of rhodospirillum photometricum, *J. Mol. Biol* 27: 393.

McDermott, G., Prince, S. M., Freer, A. A., Hawthornthwaite-Lawless, A. M., Papiz, M. Z. & Codgell, J. (1995). Crystal structure of an integral membrane light-harvesting complex from photosynthetic bacteria, *Nature* 374: 517.

Milano, F., A., A., Mavelli, F. & Trotta, M. (2003). Kinetics of the quinone binding reaction at the qb site of reaction centers from the purple bacteria rhodobacter sphaeroides reconstituted in liposomes, *Eur. Journ. Biochem* 270: 4595.

Osváth, S. & Maróti, P. (1997). Coupling of cytochrome and quinone turnovers in the photocycle of reaction centers from the photosynthetic bacterium rhodobacter sphaeroides, *Biophys. Journ* 73: 972–982.

Panitchayangkoona, G., Hayes D., Fransteda, K. A., Caram J. R., Harel, E., Wen, J., Blankenship, R. & Engel, G. S. (2010). Long-lived quantum coherence in photosynthetic complexes at physiological temperature, *PNAS* 107: 12766–12770.

Pfenning, N. (1978). *The Photosynthetic Bacteria*, Plenum Publishing Corporation, New York.

Ritz, T., Park, S. & Schulten, K. (2001). Kinetics of exciton migration and trapping in the photosynthetic unit of purple bacteria, *Journal of Physical Chemistry B* 105: 8259.

Roszack, A. W. & et al (2003). Crystal Structure the RC-LH1 core complex from Rhodopseudomonas palustris, *Science* 302: 1969.

Scheuring, S., Rigaud, J. L. & Sturgis, J. N. (2004). Variable lh2 stoichiometry and core clustering in native membranes of rhodospirillum photometricum, *EMBO* 23: 4127.

Scheuring, S. & Sturgis, J. (2005). Chromatic adaptation of purple bacteria, *Science* 309: 484.

Scheuring, S. & Sturgis, J. N. (2006). Dynamics and diffusion in photosynthetic membranes from rhodospirillum photometricum., *Biophys. J.* 91: 3707.

Scheuring, S., Sturgis, J. N., Prima, V., Bernadac, Levi, D. & Rigaud, J. L. (2004). Watching the photosytnhetic apparatus in native membranes, *Proceeding of National Academy of Sciences USA* 101: 1193.

Scully, M. O. & Zubairy, M. S. (1997). *Quantum Optics*, Cambrige University Press.

Shreve, A. P., Trautman, J. K., Franck, H. A., Owens, T. G. & Albrecht, A. C. (1991). Femtosecond energy-transfer processes in the b800-b850 light-harvesting complex of rhodobacter sphaeroides 2.4.1, *Biochimica et Biophysica Acta* 1058: 280–288.

Timpmann, K., Freiberg, A. & Sundström, V. (1995). Energy trapping and detrapping in the photosynthetic bacterium rhosopseudomonas viridis: transfer to trap limited dynamics, *Chemical Physics* 194: 275.

Timpmann, K., Zhang, F., Freiberg, A. & Sundström, V. (1993). Detrapping of excitation energy transfer from the reaction center in the photosynthetic purple bacterium rhodospirillum rubrum, *Biochemical et Biophysica Acta* 1183: 185.

van Amerongen, H., Valkunas, L. & R, G. (2000). *Photosynthetic Excitons*, World Scientific Publishing Co., Singapore.

van Grondelle, R. Dekker, J., Gillbro, T. & Sundstrom, V. (1994). Energy transfer and trapping in photosynthesis, *Biochim. et Biophys. Acta* 11087: 1–65.

Visscher, K., Bergström, H., Sündstrom, V., Hunter, C. & van Grondelle, R. (1989). Temperature dependence of energy transfer from the long wavelength antenna bchl-896 to the reaction center in rhodospirillum rubrum, rhodobacter sphaeroides (w.t. and m21 mutant) from 77 to 177k, studied by picosecond absorption spectroscopy, *Photosynthesys Research* 22: 211.

Visscher, K. J., Chang, M. C. van Mourik, F., Parkes-Loach, P. S., Heller, B. A., Loach, P. A. & van Grondelle, R. (1991). Fluorescence polarization and low-temperature absorption spectroscopy of a subunit form of light harvesting complex I from purple photosynthetic bacteria, *Biochemistry* 30: 5734.

Waltz, T., Jamieson, S. J., Bowers, C. M., Bullough, P. A. & Hunter, C. N. (1998). *Projection structure of three photosynthetic complexes from Rhodobacter sphaeroides: LH2 at 6 , LH1 and RC-LH1 at 25* 282: 883.

5

Mechanisms of Photoacclimation on Photosynthesis Level in Cyanobacteria

Sabina Jodłowska and Adam Latała

Department of Marine Ecosystems Functioning, Institute of Oceanography,
University of Gdańsk, Gdynia
Poland

1. Introduction

Cyanobacteria are oxygenic photoautotrophic prokaryotes, which develop in many aquatic environments, both freshwater and marine. They successfully grow in response to increasing eutrophication of water, but also because of shifts in the equilibrium of ecosystems (Stal et al., 2003). Cyanobacteria possess many unique adaptations allowing optimal growth and persistence, and the ability to out-compete algae during favorable conditions. For instance, many species are buoyant due to the possession of gas vesicles, some of them are capable of fixing N_2, and unlike algae, which require carbon dioxide gas for photosynthesis, most cyanobacteria can utilize other sources of carbon, like bicarbonate, which are more plentiful in alkaline or high pH environments. The cyanobacteria live in a dynamic environment and are exposed to diurnal fluctuations of light. Planktonic species experience differences in irradiance when mixed in the water column (Staal et al., 2002), whereas mat-forming cyanobacteria are exposed to changes in light intensity caused by sediment covering or sediment dispersion. Such rapidly changing environmental factors forced photoautotrophic organisms to develop many acclimation mechanisms to minimize stress due to low and high light intensities. High irradiance may damage photosynthetic apparatus by photooxidation of chlorophyll a molecules. Some carotenoid pigments may provide effective protection against such disadvantageous influence of light (Hirschberg & Chamovitz, 1994; Steiger et al., 1999; Lakatos et al., 2001; MacIntyre et al., 2002). Photosynthetic organisms respond to decreased light intensity by increasing the size or/and the number of photosynthetic units (PSU) whose changes, in turn, can be reflected in characteristic patterns of P-E curves (Platt et al., 1980; Prézelin, 1981; Ramus, 1981; Richardson et al., 1983; Henley, 1993; Dring, 1998; Mouget et al., 1999; MacIntyre et al., 2002; Jodłowska & Latała, 2010). Variation in α and P_m (expressed per biomass or per chlorophyll a unit) plays a key part in interpreting physiological responses to changes in environmental conditions.

The aim of this review was to present exceptional properties of two different cyanobacteria, planktonic and benthic, their abilities to changing environmental condition, especially to irradiance. This information would be helpful in understanding the phenomenon of mass formation of cyanobacterial blooms worldwide, and would be very useful to interpret the domination of cyanobacteria in water ecosystem in summer months.

2. Light as a major factor controlling distribution, growth and functionality of photoautotrophic organisms

Light is one of the main trophic and morphogenetic factors in the life of photosynthetic organisms. Intensity, quality and the time of light impact affect photosynthesis, which is responsible for producing organic matter, cell division and the growth rate of organism. The positive effect of increasing light intensity states only to a point, after which stabilization or even a drop in cell division takes place (Ostroff et al., 1980; Latała & Misiewicz, 2000).

Effect of light on organisms can be investigated in two ways, firstly in ecological perspective i.e. paying special attention to the influence of this factor on distribution, and secondly in physiological point of view studying mechanisms of acclimation facilitating to survive in changing environmental conditions.   Light is one of the main factors controlling the distribution of photoautotrophs in water column demonstrating their light preferences. Similarly to vascular plants, algae and cyanobacteria may also indicate their shadow-tolerant features or heliophylous character (Falkowski & La Roche, 1991). This phenomenon is observed both in case of benthic forms attached to the bottom and planktonic form floated in water column. Shadow-tolerant autotrophs concentrate in deeper water, where intensity of light is significantly lower, whereas heliophylous ones, preferring higher light intensity to growth and functionality, live in shallows. However, at surface layer of water phototrophic organisms are exposed to harmful effect of very high intensity of light and also effect of ultraviolet radiation, what may cause photoinhibition. Moreover, at deep water body we can define the level, below which photoautotrophic life gradually disappeared. The depth, at which anabolic and catabolic processes are balanced, is named as compensation level. Below this level light condition are insufficient, so that photosynthetic organisms are not able to normally develop and reproduce (Falkowski & La Roche, 1991).

In natural conditions, phytoplankton which is vertically transferred within the euphotic zone experience not only changes of intensity but as well spectral quality (Rivkin, 1989). Spectral composition of downwelling light change progressively with increasing depth. Changes in spectral quality depend on the absorption spectra and scattering of the suspended particles within the water column and on absorption of water itself and the angle at which light impinges the water surface (Staal et al., 2002). In regions poor in nutrients the longer wavelengths are selectively absorbed within the upper ca 10 m and only blue-green light remains. However, in coastal regions rich in biogenic substances and yellow substances the shorter wavelengths are rapidly absorbed (Kirk, 1983; Rivkin, 1989).

When autotrophic organisms experience changes in light regime, both intensity and spectral quality, acclimation of photosynthetic apparatus to variable light condition is observed. Photoacclimation of these organisms is linked to alterations in total cellular concentration of light-harvesting and reaction centre pigments, and also in the ratio of different pigments. Generally, one observes an inverse correlation between light intensity and pigment content: the less light energy available, the more photosynthetic pigments are synthesized by the cells (Tandeau de Marsac & Houmard, 1988). According to the idea proposed by Engelmann in 1883 different groups of marine algae dominate the benthic vegetation at different depths because their pigment composition adapts them for absorption, and hence photosynthesis, in light quality that prevails at that depth. This phenomenon is known as chromatic adaptation. It has often been pointed out that this concept is based on extremely superficial generalizations about the vertical distribution of benthic algae, since the representatives of most of the major groups can be observed at most depths (Dring, 1992). Laboratory studies of marine algae grown at different light intensities and different light spectrum suggest that phenotypic variations in pigment composition with depth are controlled by the irradiance level and not by the quality of the light (Ramus, 1981; Dring, 1992). However, the effect of light wavelength on pigment content of the cells appears to be restricted to some cyanobacteria. Changes in cell pigmentation in response to spectral quality of light result from modifications of the relative amounts of phycoerythrin (PE) and phycocyanin (PC). These phycobiliproteins are the major light-harvesting pigments used to drive photosynthesis. Only cyanobacteria which are able to synthesize PE can undergo complementary chromatic adaptation (Tandeau de Marsac & Houmard, 1988).

Light quality may be important factor controlling and regulating metabolic processes in algae and cyanobacteria, however in this study we consider to discuss the effect of changes of light intensity on photoacclimation on photosynthesis level.

3. Photosynthesis and light response curves

The photosynthesis rate achieved by a photoautotroph cell depends on the rate of capture of quantum of light. Certainly, this is determined by the light absorption properties of the photosynthetic biomass, composition of photosynthetic pigments in the cell and by the intensity and spectral quality of the field (Kirk, 1996).

The relationship between photosynthesis rate and light intensity is well described by photosynthetic light response curves (P-E). P-Es are very useful tools to predict primary productivity, and their analysis provides a lot of valuable information about photoacclimation mechanisms of cells (Platt et al., 1980; Ramus, 1981; Richardson, 1983; Henley, 1993; Dring, 1998; MacIntyre et al., 2002). The changes in concentration of chlorophyll a and other photosynthetic pigments in cell influence the course of P-E curves, which illustrate maximum rate of photosynthesis (P_m), the initial slope of photosynthetic curves (α), the compensation (P_c) and saturation irradiances (E_k; the intercept between the initial slope of P-E and P_m) and dark respiration (R). The initial slope of photosynthetic curves at limiting light intensity (α) is a function of both light-harvesting efficiency and photosynthetic energy conversion efficiency (Henley, 1993). An increase of chlorophyll a concentration in the cell of photoautotrophic organism improves the effective absorption of light, what causes an increase of the P-E slope, in case the rate of photosynthesis is expressed in biomass unit. However, if photosynthesis is expressed in chlorophyll unit, it will not be observed the variability of α parameter. It is because of the photosynthetic rate at the limiting light intensity is proportional to the concentration of chlorophyll a (Dring, 1998). The chlorophyll a is arranged in the thylakoid membranes in photosynthetic units (PSU), each of which consists of 300-400 chlorophyll molecules associated with a single reaction centre. An increase in pigment concentration could be achieved either by building extra molecules into the existing PSUs without changing the number of reaction centers (called as a change of PSU size) or by building up complete new PSUs without changing their size (called as a change of PSU number) (Dring, 1998). These two mechanisms of photoacclimation have quite different effects on values of P_m. Since P_m is related to the number of reaction centers available, an increase in the size of PSUs have no effect on the maximum photosynthesis in biomass unit, that is why the photoautotrophs with different concentration of chlorophyll a have different P-E slope, but the same value of P_m. If, however, the photosynthesis rate is expressed in chlorophyll a, the value of P_m will drop with an increase of chlorophyll a concentration, but the P-E slope remains constant.

However, when the number of PSUs increases proportionally to the increase in chlorophyll *a* concentration, the maximum photosynthesis in biomass unit will also increase in proportion to the chlorophyll *a* concentration, but this means that P_m in chlorophyll unit will be constant (Dring, 1998). According to the Ramus (1981) model, if the number of PSUs increases, more quantum of light is necessary to saturate the photosynthesis (higher E_k and P_m), but if only the size of PSUs increases without any change of their number, not much light will be need to saturate the photosynthesis (lower E_k and P_m). In the second case, PSU will be working more effectively (Lobban and Harrison, 1997).

4. Photoacclimation strategies in two strains of cyanobacteria: Planktonic *Nodularia spumigena* and benthic *Geitlerinema amphibium*

Cyanobacteria are very interesting material to study relationships between biological activity of organism and various environmental factors, because they often exist in extreme circumstances and can acclimate efficiently to changing environmental conditions (Latała & Misiewicz, 2000).

The experiments were conducted on two different strains of Baltic cyanobacteria: planktonic *Nodularia spumigena* and benthic *Geitlerinema amphibium*. The investigated strains were isolated from southern Baltic Sea, exactly from the coastal zone of the Gulf of Gdańsk, and now they were maintained as an unialgal cultures in the Culture Collection of Baltic Algae (CCBA) (http://ccba.ug.edu.pl) at the Institute of Oceanography, University of Gdańsk, Poland (Latała et al., 2006). Buoyant cyanobacteria, like *N. spumigena*, previously mixed throughout the water column, float to the surface of water, where their cells are exposed to full sunlight, and this abrupt change in irradiance may induce photoinhibition (Ibelings, 1996). This species of cyanobacterium always occurs at water temperatures over 18°C and during weather conditions conductive to water column thermal stratification (Hobson et al., 1999). It is one of the filamentous cyanobacteria, which regularly occurs in the Baltic water in summer and often forms toxic blooms. *N. spumigena* grows especially well in the illuminated upper layer of the euphotic zone. In the Baltic Sea, it is most abundant to depths of 5 m (Hajdu et al., 2007), but it is also observed as deep as 18 m (Stal & Walsby 2000; Jodłowska & Latała, 2010). Similarly, mat-forming cyanobacterium, like *G. amphibium*, experience changes in light regime. These organisms may periodically be covered by a layer of sediment, which limits availability of light. On the other hands, sediment dispersion can lead to a rise in light intensity. *G. amphibium*, as a permanent element of summer microbial mats in the Gulf of Gdańsk (Southern Baltic) (Witkowski, 1986), may be an example of how organisms make use of their outstanding acclimation ability in the best possible way (Jodłowska & Latała, 2010).

According to literature data cyanobacteria are generally recognized to prefer low light intensity for growth and photosynthesis (Fogg & Thake, 1987; Ibelings, 1996). However, the investigated strain of *N. spumigena* was found to be well acclimated to relatively high light intensity (290 μmol photons·m^{-2}·s^{-1}), which was especially evident within the range of temperature 15-20°C (Fig. 1A). The factorial experiments with *N. spumigena* showed that irradiance had a promoting effect on cyanobacterial concentration, but the interaction between increasing irradiance of over about 100 μmol photons·m^{-2}·s^{-1} and increasing temperature over about 23°C inhibited the filament concentration. Prolonged exposure to high light intensity may cause photoinhibition, and induce harmful effects resulting from increased temperatures. Optimal growth conditions were noted at about 180-290 μmol photons·m^{-2}·s^{-1} and 15-17°C. The maximum number of filament units (about 5·10^5 filament units ·ml^{-1}) was about 10x greater than the minimum one. Similarly, the factorial experiments on *G. amphibium* showed that both irradiance and temperature have promoting effect on cyanobacterial culture concentration, but this positive effect was observed up to 120 μmol photons ·m^{-2}·s^{-1} at 35°C as well as up to 170 μmol photons ·m^{-2}·s^{-1} at 30°C (Fig. 1B). However, the interaction between temperature of 35°C and light intensity of 170 μmol photons ·m^{-2}·s^{-1} resulted in growth inhibition (Latała & Misiewicz, 2000). An excess of light energy absorbed by photosynthetic pigments together with high-temperature stress may accelerate photoinhibition by inhibiting the repair of photodamaged PSII (Allakhverdiev et al., 2008; Takahashi & Murata, 2008; Takahashi & Badger, 2011). Richardson & her co-workers (1983)

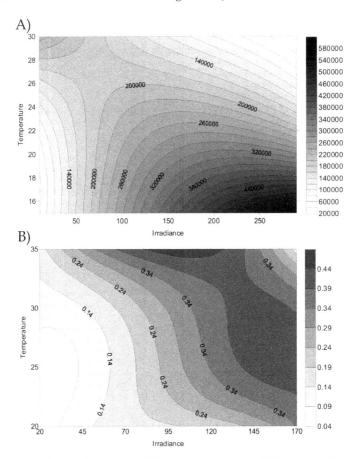

Fig. 1. Response-surface estimation at different temperatures (°C) and irradiances (μmol photons·m^{-2}·s^{-1}) of: A) *N. spumigena* filament concentration (1 filament unit = 100 μm), B) *G. amphibium* optical density at 750 nm. Experimental cultures were grown in three replicates and the measurements of culture were done on the last day of incubation in the exponential growth phase.

suggested that photoinhibition of the growth and photosynthesis occurred in the natural populations at above 200 µmol photons·m^{-2}·s^{-1}, although a lot of species can be subjected to this phenomenon at lower irradiance. *Aphanizomenon ovalisporum* indicated inhibition of the cell concentration even at 50 µmol photons·m^{-2}·s^{-1} and 20-35°C (Hadas et al., 2002).

The experiments on *N. spumigena* and *G. amphibium* showed that both irradiance and temperature were important factors contributing to the variation of chlorophyll *a* content at the investigated strains, but the influence of irradiance was higher (Latała & Misiewicz, 2000; Jodłowska & Latała, 2010) (Fig. 2). In two investigated strains, pigment content was negatively affected by irradiance and positively by temperature. The highest values of pigment content at low light treatment and high temperature treatment were almost 95% and 85% for *N. spumigena* and *G. amphibium*, respectively, higher than the lowest ones at high light treatment and low temperature treatment.

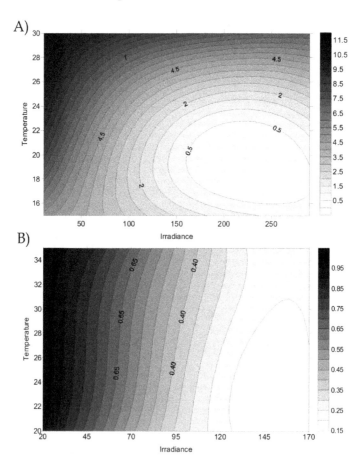

Fig. 2. Response-surface estimation of chlorophyll *a* content (pg filament unit^{-1}): A) *N. spumigena* and B) *G. amphibium* at different temperatures (°C) and irradiances (µmol photons·m^{-2}·s^{-1}).

The lower cellular content of chlorophyll *a* noted in the population acclimated to high light is associated with a decrease in the size and/or the number of photosynthetic units (PSUs) that can be reflected by P-E curves (Platt et al., 1980; Prézelin, 1981; Ramus, 1981; Richardson et al., 1983; Henley, 1993; Dring, 1998; Mouget et al., 1999; MacIntyre et al., 2002; Jodłowska & Latała, 2010). P-E curves obtained for *N. spumigena* and *G. amphibium* grown under different light and temperature conditions illustrate that these two strains conforms to more than one of the photoadaptive models used to categorize species (Fig. 3) (Jodłowska & Latała, 2010). Photosynthetic rates, normalized to biomass and chlorophyll unit, were plotted against irradiance. The P-E curves were fitted to the data with Statistica using the mathematical function by Platt & Jassby (1976), as follows: $P=P_m \cdot \tanh(\alpha \cdot E/P_m)+R_d$. To illustrate the course of P-E curves, the results recorded at 15°C were chosen (Fig. 3 and 4). In *N. spumigena* culturing at 15°C, the biomass-specific P_m was about 70% higher in the low light treatment than in the high light treatment (Fig. 3A), whereas in *G. amphibium* the same difference was about 60% (Fig. 3B). Since the biomass-specific P_m noted in the strains acclimated to low light was higher than that grown in high light, it indicates a change in the number of the PSUs. However, higher chl*a*-specific P_m in the strains acclimated to high light in comparison to that in the low light strain indicates there was a change in the size of the PSUs. In *N. spumigena* culturing at 15°C, the chlorophyll-specific P_m was about 45% higher in the high light treatment than in the low light treatment (Fig. 3B), whereas in *G. amphibium* the same difference was about 70% (Fig. 4B).

These two mechanisms of photoacclimation, concerning the changes in the size and the number of PSUs, explain significant changes in photosynthesis rate and its parameters upon the influence of different light intensities and temperatures (Fig. 5). It is noteworthy that *N. spumigena* and *G. amphibium* exhibit substantial changes in E_k and P_c within the range of irradiance tested. However, it is typical for E_k values to increase in phototrophic populations as irradiance increases (Richardson et al., 1983), whereas the link between the evolution of compensation light intensity (P_c) and increased irradiance is somewhat more characteristic of Chlorophyta (Falkowski & Owens, 1980). The results of variance analysis showed that both irradiance and temperature were important factors contributing to the variation of E_k and P_c parameters at the investigated strains, but the influence of irradiance was higher. In *N. spumigena*, the minimum values of E_k (about 145 µmol photons·m^{-2}·s^{-1}) were about 47% lower than the maximum ones (about 275 µmol photons·m^{-2}·s^{-1}) (Fig. 5A), but in *G. amphibium* the differences were higher, and were about 94% (Fig. 5B). In contrast, in cyanobacterium *N. spumigena*, the minimum values of P_c (about 10 µmol photons·m^{-2}·s^{-1}) were lower about 95% than the maximum ones (about 180 µmol photons·m^{-2}·s^{-1}) (Fig. 5C), whereas in *G. amphibium* the minimum P_c (about 5 µmol photons·m^{-2}·s^{-1}) were lower about 93% than the maximum ones (about 70 µmol photons·m^{-2}·s^{-1}) (Fig. 5D). The obtained minimum values of E_k and P_c are close to those reported for shadow-tolerant plants, while the maximum ones are close to those noted in heliophylous plants (Rabinowitch, 1951; Wallentinus, 1978). Achieved results for both parameters showed good acclimation capacity of the investigated species to irradiance changes. It is also noteworthy that P-E curves for investigated cyanobacteria did not indicate photosynthetic photoinhibition until approximately 700-1000 µmol photons m^{-2} s^{-1}, even if the cyanobacteria were acclimated to very low light intensity (5 -10 µmol photons m^{-2} s^{-1}). Moisander et al. (2002) did not also note this phenomenon in *N. spumigena* at irradiances exceeding 1000 µmol photons m^{-2} s^{-1}.

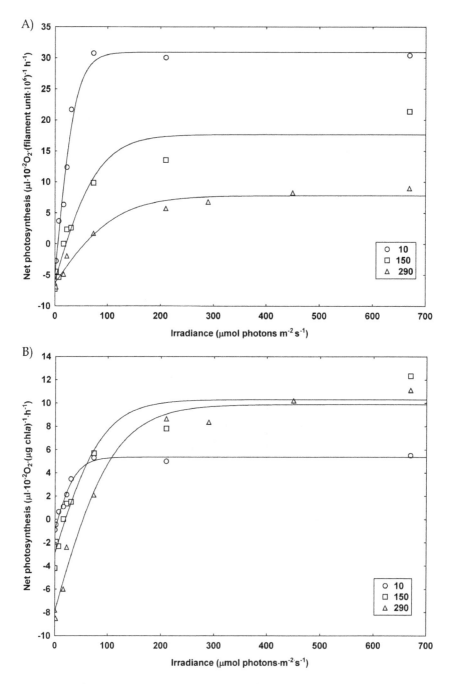

Fig. 3. P-E curves for *N. spumigena* growing at three light intensity (μmol photons·m⁻²·s⁻¹) and at 15°C: A) in filament unit, B) in chlorophyll unit (Jodłowska & Latała, 2010).

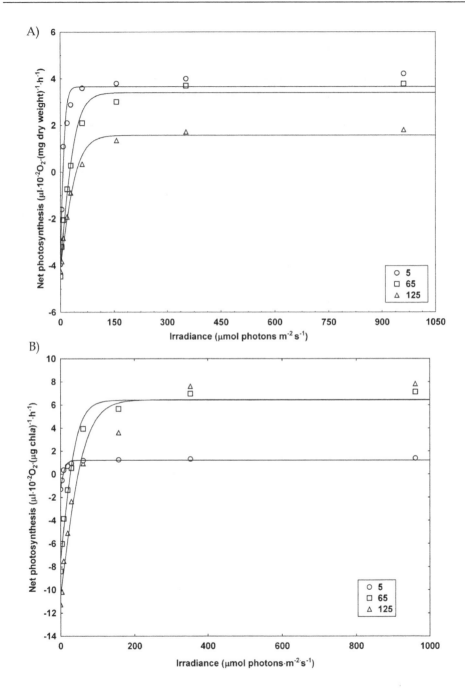

Fig. 4. P-E curves for *G. amphibium* growing at three light intensity (μmol photons·m⁻²·s⁻¹) and at 15°C: A) in biomass unit, B) in chlorophyll unit.

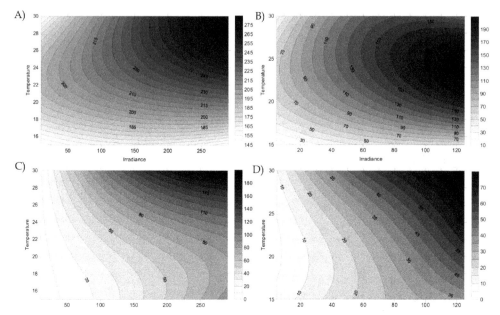

Fig. 5. Response-surface estimation at different temperatures (°C) and irradiances (μmol photons·m^{-2}·s^{-1}) of E$_k$ (μmol photons m^{-2} s^{-1}) for: A) *N. spumigena*, B) *G. amphibium* and P$_c$ (μmol photons m^{-2} s^{-1}) for: C) *N. spumigena*, D) *G. amphibium*.

5. Conclusions

Every phototrophic organisms, including cyanobacteria, differ in their light intensity range within which they grow and photosynthesize. This is determined by their optimal ecological requirements for light and controlled by metabolic properties of each species (Collier et al., 1978; Richardson et al., 1983). The P-E response curves, derived from short-term measurement, provides an information about condition of photosynthetic apparatus and yields insight into the regulation of energy and material balance of the cell (MacIntyre et al., 2002).

In this review, we give importance of interpreting P-Es in the context photoacclimation mechanisms in cyanobacteria. The current experiments on two different strains of cyanobacteria demonstrated their capacity to acclimate to irradiance, which is reflected in the wide range of changes in the growth, pigment composition as well as photosynthetic activity of the examined cyanobacteria. The studied parameters demonstrated their exceptional adaptability of changes, which explain why these strains grow successfully in well-illuminated habitats and also can grow well in position with limited access of light.

The identification of factors that regulate the growth and photosynthetic activity of cyanobacteria can be helpful for understanding the ecological triggers of cyanobacterial blooms.

6. Reference

Allakhverdiev, S.I.; Kreslavski, V.D.; Klimov, V.V.; Los, D.A.; Carpentier, R. & Mohanty, P. (2008). Heat stress: an overview of molecular responses in photosynthesis. *Photosynthesis Research* 98: 541-550.

Dring, M.J. (1998). *The biology of marine plants*. Cambridge University Press, Cambridge. 199 pp.

Falkowski, P.G. & Owens, T.G. (1980). Light-shade adaptation: two strategies in marine phytoplankton. *Plant Physiology* 66: 592–595.

Falkowski, P.G.; LaRoche, J. (1991). Acclimation to spectral irradiance in algae. *Journal of Phycology* 27: 8-14.

Fogg, G.E. & Thake, B. (1987). *Algal cultures and phytoplankton ecology*. University of Wisconsin Press, Madison and Milwaukee. 269 pp.

Hadas, O.; Pinkas, R.; Malinsky-Rushansky, N.; Shalev-Alon, G.; Delphine, E.; Berner, T.; Sukenik, A. & Kaplan, A. (2002). Physiological variables determined under laboratory conditions may explain the bloom of *Aphanizomenon ovalisporum* in Lake Kinneret. *European Journal of Phycology* 37: 259-267.

Hajdu, S.; Höglander, H. & Larsson, U. (2007). Phytoplankton vertical distribution and composition in Baltic Sea cyanobacterial blooms. *Harmful Algae* 6: 189-205.

Henley, W.J. (1993). Measurement and interpretation of photosynthetic light-response curves in algae in the context of photoinhibition and diel changes. *Journal of Phycology* 29: 729-739.

Hirschberg, J. & Chamovitz, D. (1994). Carotenoids in cyanobacteria. In: Bryant DA (ed) *The molecular biology of cyanobacteria*, Kluwer Academic Publishers, Netherlands, pp 559-579.

Hobson, P.; Burch, M. & Fallowfield, H.J. (1999). Effect of total dissolved solids and irradiance on growth and toxin production by *Nodularia spumigena*. *Journal of Applied Phycology* 11: 551-558.

Ibelings, B.W. (1996). Changes in photosynthesis in response to combined irradiance and temperature stress in cyanobacterial surface waterblooms. *Journal of Phycology* 32: 549-557.

Jodłowska, S. & Latała, A. (2010). Photoacclimation strategies in the toxic cyanobacterium *Nodularia spumigena* (Nostocales, Cyanobacteria). *Phycologia* 49: 203-211.

Krik, J.T.O. (1994). *Light and photosynthesis in aquatic ecosystems*. Cambridge University Press, Cambridge. 509 pp.

Lakatos, M.; Bilger, W.; Büdel, B. (2001). Carotenoid composition of terrestrial Cyanobacteria: response to natural light conditions in open rock habitats in Venezuela. *European Journal of Phycology* 36: 367-375.

Latała, A.; Misiewicz, S. (2000). Effects of light, temperature and salinity on the growth and chlorophyll-a content of Baltic cyanobacterium *Phormidium* sp. *Archiv für Hydrobiologie* 136; *Algological Studies* 100: 157-180.

Latała, A.; Jodłowska, S. & Pniewski, F. (2006). Culture Collection of Baltic Algae (CCBA) and characteristic of some strains by factorial experiment approach. *Archiv für Hydrobiologie* 165, *Algological Studies* 122: 137-154.

Lobban, C. S. & Harrison, P. J. (1997). *Seaweed Ecology and Physiology*. Cambridge University Press, Cambridge, UK, 366 pp.

MacIntyre, H.L.; Kana, T.M.; Anning, T. & Geider, R.J. (2002). Photoacclimation of photosynthesis irradiance response curves and photosynthetic pigments in microalgae and cyanobacteria. *Journal of Phycology* 38: 17-38.

Moisander, P.H.; McClinton III, E. & Paerl, H.W. (2002). Salinity effects on growth, photosynthetic parameters, and nitrogenase activity in estuarine planktonic cyanobacteria. *Microbial Ecology* 43: 432-442.

Mouget, J.-L.; Tremblin, G.; Morant-Manceau, A.; Morancais, M. & Robert, J.-M. (1999). Long-term photoacclimation of *Haslea ostrearia* (Bacillariophyta): effect of irradiance on growth rates, pigment content and photosynthesis. *European Journal of Phycology* 34: 109-115.

Ostroff, C.R.; Karlander, E.P. & Van Valkenburg, S.D. (1980). Growth rates of *Pseudopedinella pyriforme* (Chrysophyceae) in response to 75 combination of light, temperature and salinity. *Journal of Phycology* 16: 421-423.

Platt, T. & Jassby, A.D. (1976). The relationship between photosynthesis and light for natural assemblages of coastal marine phytoplankton. *Journal of Phycology* 12: 421-430.

Platt, T.; Gallegos, C.L. & Harrison, W.G. (1980). Photoinhibition of photosynthesis in natural assemblages of marine phytoplankton. *Journal of Marine Research* 38: 687-701.

Prézelin, B.B. (1981). Light reactions in photosynthesis. In: Platt T (ed) Physiological bases of phytoplankton ecology, *Canadian Bulletin of Fisheries and Aquatic Sciences*, no. 210, Ottawa, 1-43 pp.

Rabinowitch, E.I. (1951). *Photosynthesis and Related Processes*. Vol. II, part 1. Interscience Publishers, New York. 1208 pp.

Ramus, J. (1981). The capture and transduction of light energy. In: *The biology of seaweeds* (Ed. by C.S. Lobban & M.J. Wynne), Blackwell Scientific, Oxford. 458-492 pp.

Richardson, K.; Beardall, J. & Raven, J.A. (1983). Adaptation of unicellular algae to irradiance: an analysis of strategies. *New Phytologist* 93: 157-191.

Rivkin, R.B. (1989). Influence of irradiance and spectral quality on the carbon metabolism of phytoplankton. I. Photosynthesis, chemical composition and growth. *Marine Ecology Progress Series* Vol. 55 (1989), pp. 291–304.

Staal, M.; te Lintel Hekkert, S.; Herman, P. & Stal, L.J. (2002). Comparison of model describing light dependence of N_2 fixation in heterocystous cyanobacteria. *Applied and Environmental Microbiology* 68(9): 4679-4683.

Stal, L.J. & Walsby, A.E. (2000). Photosynthesis and nitrogen fixation in a cyanobacterial bloom in the Baltic Sea. *European Journal of Phycology* 35: 97-108.

Stal, L.J.; Albertano, P.; Bergman, B.; Bröckel, K.; Gallon, J.R.; Hayes, P.K.; Sivonen, K. & Walsby, A.E. (2003). BASIC: Baltic Sea cyanobacteria. An investigation of the structure and dynamic of water blooms of cyanobacteria in the Baltic Sea – response to a changing environment. *Continental Shelf Research* 23: 1695-1714.

Steiger, S.; Schäfer, L. & Sandmann, G. (1999). High-light-dependent upregulation of carotenoids and their antioxidative properties in the cyanobacterium *Synechocystis* PCC 6803. *Journal of Photochemistry and Photobiology* 52: 14-18.

Takahashi, S. & Badger, M.R. (2011). Photoprotection in plants: a new light on photosystem II damage. *Trends in Plant Science* 16(1): 53-60. *Plant Science* 13(4): 178-182. Tandeau de Marsac, N. & Houmard, J. (1988). Complementary chromatic adaptation: physiological conditions and action spectra. In: *Methods in Enzymology* (Ed. by L. Packer & A.N. Glazer), Academic Press, New York. 318-328 pp.

Wallentinus, I. (1978). Productively studies on Baltic macroalgae. *Botanica Marina* 21: 365-380.

Witkowski, A. (1986). Microbial mat with an incomplete vertical structure, from brackish-water environment, the Puck Bay, Poland, a possible analog of an "advanced anaerobic ecosystem"?. *Origins of Life and Evolution of Biospheres* 16(3-4): 337-338.

6

Fast Kinetic Methods with Photodiode Array Detection in the Study of the Interaction and Electron Transfer Between Flavodoxin and Ferredoxin NADP⁺-Reductase

Ana Serrano and Milagros Medina

Department of Biochemistry and Molecular and Cellular Biology and Institute of Biocomputation and Physics of Complex Systems, University of Zaragoza
Spain

1. Introduction

The primary function of Photosystem I (PSI)[1] during photosynthesis is to provide reducing equivalents, in the form of NADPH, that will then be used in CO_2 assimilation (Golbeck, 2006). In plants, this occurs via reduction of the soluble [2Fe-2S] Ferredoxin (Fd) by PSI. Subsequent reduction of NADP⁺ by Fd_{rd} is catalysed by the FAD-dependent Ferredoxin-NADP⁺ reductase (FNR) ($E_{ox/hq}$ = -374 mV at pH 8.0 and 25°C) through the formation of a ternary complex (Arakaki *et al.*, 1997; Nogués *et al.*, 2004; Sancho *et al.*, 1990). In most cyanobacteria and some algae under low iron conditions the FMN-dependent Flavodoxin (Fld), particularly its Fld_{sq}/Fld_{hq} couple ($E_{ox/sq}$ = -256 mV, $E_{sq/hq}$ = -445 mV at pH 8.0 and 25°C), substitutes for the Fd_{ox}/Fd_{rd} pair in this reaction (Bottin & Lagoutte, 1992; Goñi *et al.*, 2009; Medina & Gómez-Moreno, 2004). Thus, two Fld_{hq} molecules transfer two electrons to one FNR_{ox} that gets fully reduced after formation of the FNR_{sq} intermediate. FNR_{hq} transfers then both electrons simultaneously to NADP⁺ (Medina, 2009) (Figure 1).

Additionally to their role in photosynthesis, Fld and FNR are ubiquitous flavoenzymes that deliver low midpoint potential electrons to redox-based metabolisms in plastids, mitochondria and bacteria in all kingdoms (Müller, 1991). They are also basic prototypes for a large family of di-flavin electron transferases that display common functional and structural properties, where the electron transfer (ET) flow supported by the Fld/FNR modules occurs in reverse direction to the photosynthesis (Brenner *et al.*, 2008; Wolthers & Scrutton, 2004). The better understanding of the ET flavin-flavin mechanism in this system should witness a greater comprehension of the many physiological roles that Fld and FNR, either free or as modules in multidomain proteins, play. In these chains there are still many

---

[1] PSI, Photosystem I; FNR, ferredoxin-NADP⁺ reductase; FNR_{ox}, FNR in the fully oxidised state; FNR_{sq}, FNR in the semiquinone state; FNR_{hq}, FNR in the hydroquinone (fully reduced) state; Fld, Fladoxin; Fld_{ox}, Fld in the fully oxidised state; Fld_{sq}, Fld in the semiquinone state; Fld_{hq}, Fld in the hydroquinone (fully reduced) state; Fd, ferredoxin; Fd_{ox}, Fd in the oxidised state; Fd_{rd}, Fd in the reduced state; $E_{ox/sq}$, midpoint reduction potential for the ox/sq couple; $E_{sq/hq}$, midpoint reduction potential for the sq/hq couple; ET, electron transfer; WT, wild-type; $k_{A \to B}$, $k_{B \to C}$, $k_{C \to D}$, apparent rate constants obtained by global analysis of spectral kinetic data; I, ionic strength; UV/Vis, ultraviolet/visible; PDA, photodiode array detector; SVD, single value decomposition.

open questions in understanding not only the ET mechanisms, but also the role that the flavin itself might play.

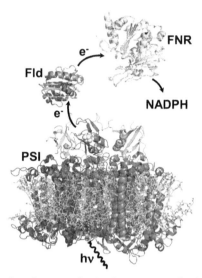

Fig. 1. Proteins involved in the photosynthetic electron transfer from PSI to NADP$^+$ via Flavodoxin (Fld, orange) and Ferredoxin-NADP$^+$ reductase (FNR, blue).

Recent studies on the FNR:Fld interaction and ET indicate that the orientation driven by the alignment of the Fld molecular dipole moment with that of FNR contributes to the formation of a bunch of alternative binding modes competent for the efficient ET reaction (Frago et al., 2010; Goñi et al., 2009; Medina et al., 2008). FNR, Fld and NADP$^+$ are able to form a ternary complex, indicating that NADP$^+$ is able to occupy a site on FNR without displacing Fld (Martínez-Júlvez et al., 2009; Velázquez-Campoy et al., 2006). The two binding sites are not completely independent, and the overall reaction is expected to work in an ordered two-substrate process, with the pyridine nucleotide binding first, as reported for the Fd system (Batie & Kamin, 1984a, 1984b). Complex formation of Fd_{rd} with FNR:NADP$^+$ was found to increase the ET rate by facilitating the rate-limiting step of the process, the dissociation of the product (Fd_{ox}) (Carrillo & Ceccarelli, 2003). Binding equilibrium and steady-state studies in WT and mutant proteins envisaged similar mechanisms for Fld, but the less specific FNR:Fld interaction might alter this behaviour (Medina, 2009).

Fast kinetic stopped-flow methods have been used for the analysis of the mechanisms involving binding and ET between FNR and Fld. So far only single wavelength stopped-flow spectrophotometry studies (mainly at 600 nm) have been reported (Casaus et al., 2002; Goñi et al., 2008; Goñi et al., 2009; Nogués et al., 2003; Nogués et al., 2005), not involving the effects of NADP$^+$, the ionic strength of the media, or the evaluation of the intermediate and final compounds of the equilibrium mixture. In this work, we use stopped-flow with photodiode array detection (PDA) to better evaluate the intermediate and final species in the equilibrium mixture during the ET process in the binary Fld:FNR and ternary Fld:FNR:NADP$^+$ systems.

2. Experimental methods improving the kinetic analysis of pre-steady-state electron transfer processes

ET has been analysed at different ionic strengths (I) and in both directions; the physiological photosynthetic ET, from Fld_{hq} to FNR_{ox}, and the reverse ET, from FNR_{hq} to Fld_{ox}, used to provide reducing equivalents to different metabolic pathways. Different *Anabaena* FNR and Fld variants have been used in the kinetic characterization of these ET processes using stopped-flow with photodiode detection.

2.1 Biological material and steady-state spectroscopic measurements

The different FNR and Fld variants were purified from *Luria-Bertani* IPTG-induced *E. coli* cultures containing, respectively, the pTrc99a-Fld and pET28a-FNR plasmids encoding *Anabaena* WT or E301A FNRs and WT, E16K/E61K or E16K/E61K/D126K/D150K Flds as previously described (Casaus *et al.*, 2002; Goñi *et al.*, 2009; Martínez-Júlvez *et al.*, 2001; Medina *et al.*, 1998; Tejero *et al.*, 2003). UV/Vis spectra were recorded in a Cary 100 spectrophotometer at 25 °C. Unless otherwise stated, all measurements were recorded in 50 mM Tris/HCl, pH 8.0. Binding ability between WT FNR_{ox} and WT Fld_{ox} was determined using difference absorption spectroscopy as previously described (Goñi *et al.*, 2009; Martínez-Júlvez *et al.*, 2001; Medina *et al.*, 1998).

2.2 Stopped-flow pre-steady-state kinetic measurements

Fast ET processes between Fld and FNR were followed by stopped-flow under anaerobic conditions in 50 mM Tris/HCl, pH 8.0 at 12°C. Reactions were analysed by collecting multiple wavelength absorption data (360-700 nm) using an Applied Photophysics SX17.MV stopped-flow and a photodiode array detector (PDA) with the X-Scan software (*App. Photo. Ltd.*). Typically, 400 spectra *per* second were collected for each reaction. Anaerobic conditions were obtained by several cycles of evacuation and bubbling with O_2-free argon. FNR_{hq} and Fld_{hq} samples were obtained by photoreduction in the presence of 1 mM EDTA and 5 μM 5-deazariboflavin (Medina *et al.*, 1998). The use of PDA allowed detecting that under our experimental conditions the produced Fld_{hq} samples usually contained a small amount of Fld_{sq} (detected at ~580 nm), taken into account in subsequent analysis. Unless otherwise stated, the mixing molar ratio was 1:1 with a final concentration of 10 μM for each protein. Molar ratios of 1:1, 1:2, 1:4 and 1:8 were also assayed for the reaction of FNR_{hq} and Fld_{ox}, with a final FNR_{hq} concentration of 10 μM. Reactions were also analysed at different ionic strengths, obtained using variable NaCl concentrations ranging from 0 to 300 mM in 50 mM Tris/HCl, pH 8.0. The reaction between FNR_{ox} and Fld_{hq} was also studied in presence of ~300 μM NADP⁺ at the indicated ionic strength conditions.

2.3 Kinetic analysis of multiple-wavelength absorption data

Spectral intermediates formed during reactions were resolved by singular value decomposition (SVD) using Pro-Kineticist (*App. Photo. Ltd.*). Data collected were fit either to a single step, A→B, to a two steps A→B→C, or to a three steps, A→B→C, model allowing estimation of the conversion rate constants ($k_{A→B}$, $k_{B→C}$, $k_{C→D}$) (Tejero *et al.*, 2007). Estimated errors in the determined rate constant values were ±15%. It should be stressed that SVD analysis of PDA spectra over a selected time domain resolves the spectra in the minima number of spectral intermediates species that are formed during the reaction, reflecting a distribution of protein intermediates (reactants, complexes, products...) at a certain point along the reaction time course, and not discrete enzyme intermediates. Moreover, none of them represent individual species and, their spectra cannot be included as fixed values in the global-fitting. Consequently, a spectral intermediate, in particular one that is formed in the middle of a reaction sequence, is an equilibrium distribution of protein species that are formed in a resolvable kinetic phase. Model validity was assessed by lack of systematic deviation from the residual plot at different wavelengths, inspection of calculated spectra and consistence among the number of significant singular values with the fit model. Simulations using Pro-Kineticist were also performed in order to validate the determined ET kinetics constants for the direct and reverse processes.

2.4 Determination of the absorbance spectra of FNR_{sq} and Fld_{sq}, and of protein discrete species contained in the spectral intermediates

Due to the high maximum level for Fld_{sq} stabilisation, the spectrum from this species was easily determined from photoreduction experiments (Fig. 2A) (Frago et al., 2007). The spectrum of FNR_{sq} was determined by following the one-electron oxidation of FNR_{hq} with ferricyanide in the spectral range between 360 and 700 nm (Batie & Kamin, 1984a). Molar ratios FNR_{hq}:ferricyanide of 1:10, 1:15 and 1:20 were used, with a final FNR_{hq} concentration of 10 µM (Fig. 2B). Analysis of the spectroscopic data along the time was performed by Multivariate Curve Resolution-Alternating Least Squares (MCR-ALS).

Intermediate A, B, C and D species were deconvoluted initially considering they are produced by the lineal combination of FNR_{ox}, FNR_{sq}, FNR_{hq}, Fld_{ox}, Fld_{sq} and Fld_{hq} spectra (Fig. 2), and having into account the mass balance for total FNR and Fld. Deviations from the linear combination of the different redox states of FNR and Fld were observed in all cases. Visible spectra of FNR-Fld reaction mixtures slightly differ from the lineal combination of the individual components due to modulation of the flavin spectroscopic properties when changing its environment polarity upon complex formation. This indicates contribution of FNR:Fld complexes to the spectra of intermediate and final species (Casaus et al., 2002; Martínez-Júlvez et al., 2001; Nogués et al., 2003). Nevertheless, simulations proved that the method was adequate to estimate the percentage of the different redox states of each protein.

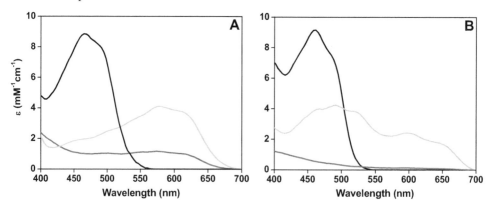

Fig. 2. Extinction coefficient of (A) Fld and (B) FNR in their different oxidation states. Oxidised, semiquinone and, reduced species are shown in black, green and red, respectively.

3. Spectral evolution of the electron transfer from Fld$_{hq}$ to FNR$_{ox}$

Mixing Fld$_{hq}$ with FNR$_{ox}$ at I ≤ 125 mM produced a slight increase in the global amount of neutral semiquinone species within the instrumental dead-time. Then, absorbance decreased in the region of the flavin band-I (458 and 464 nm), while additionally increased in the 507-650 nm range (Fig. 3A and 3B). The overall reaction best fits a single ET step (described by an apparent $k_{B \rightarrow C}$ rate constant), without detection of a protein-protein interaction step ($k_{A \rightarrow B}$). Resolution of the spectroscopic properties of B was consistent with FNR$_{ox}$ and Fld$_{hq}$ as the main components, while those of the final species C indicated Fld accumulated mainly as Fld$_{sq}$, while FNR consisted mainly of FNR$_{ox}$ (80%) in equilibrium with smaller amounts of FNR$_{sq}$ and FNR$_{hq}$ (~10% each). Deviation of the lineal combination of the

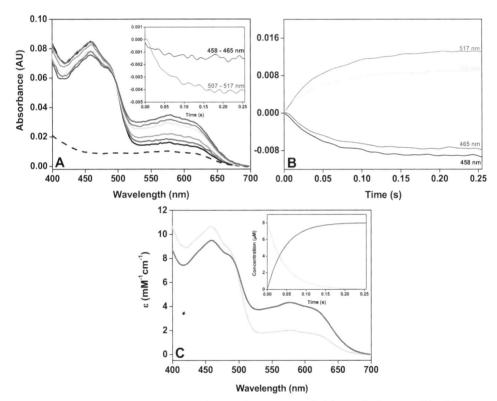

Fig. 3. Evolution of spectral changes during the reaction of Fld$_{hq}$ with FNR$_{ox}$ in 50 mM Tris/HCl, pH 8.0, 100 mM NaCl at 12°C. (A). Time course for the reaction. Spectra recorded at 0.00128, 0.0064, 0.01664, 0.03712, 0.05504, 0.0832 and, 0.2547 s after mixing. The spectrum of Fld$_{hq}$ before mixing is shown as a dashed line, and the first spectrum after mixing is in black. The inset shows differences between kinetic traces at the indicated wavelengths. (B). Kinetics of the absorbance changes at 458, 464, 507, 517 and 577 nm. (C). Absorbance spectra for the pre-steady-state kinetically distinguishable species obtained by global analysis. Intermediate B and C species are shown in green and red lines, respectively. The inset shows the evolution of these species along the time.

different FNR and Fld redox spectra was observed for B and C (Fig. 2), indicating mutual modulation of the spectroscopic properties of each one of the flavins, and, therefore, of its environment, by the presence of the second flavoprotein. This indicates a number of molecules must be forming FNR:Fld interactions, which might even be in different oxido-reduction states.

Noticeably, increasing ionic strength up to 125 mM had a drastic deleterious impact in the ET apparent $k_{B \to C}$ rate constant (Fig. 4), consistent with the lack of interaction observed by difference spectroscopy when WT FNR_{ox} was titrated with WT Fld_{ox} at I = 125 mM (not shown). These results indicated a considerable decrease in the FNR:Fld affinity upon increasing the ionic strength of the media.

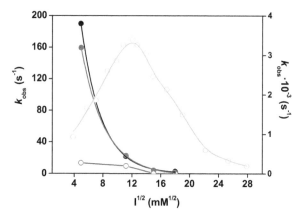

Fig. 4. Ionic strength dependence of the apparent rate constants for the reduction of FNR_{ox} by Fld_{hq} in absence (black) and presence of $NADP^+$ (red). $k_{B \to C}$ rate constants in absence (black closed circles) and in presence (red closed circles) of $NADP^+$. $k_{C \to D}$ rate constant in presence of $NADP^+$ (red open circles). The ionic strength profile reported for the reduction of FNR_{ox} by Fd_{rd} is plotted on the right axis of the figure (open green circles) (Medina et al., 1998).

Further increases in the ionic strength (I > 225 mM) produced the appearance of an additional final slow absorbance increase in the band-I, and, therefore, two ET steps described the process (Fig. 5). Thus, a slight absorbance bleaching of the band-I was observed from species B into C (as before described at lower I), but C further evolved with absorbance increases of this band and minor changes in the semiquinone one (Fig. 5A and 5C). Resolution of the spectroscopic properties of B was also consistent with FNR_{ox} and Fld_{hq} as the main components, those of C indicated Fld_{sq} and FNR_{ox} in equilibrium with some amounts of FNR_{sq} (~10%), and FNR_{hq} (~20%), while D was mainly composed by Fld_{sq} and FNR_{ox} with ~15% of Fld_{ox} and FNR_{hq}. The ionic strength additionally contributed to considerably diminish the apparent $k_{B \to C}$ ET rates (Fig. 4).

ET from Fld_{hq} to the FNR_{ox}:$NADP^+$ preformed complex was consistent with two ET steps at all the assayed ionic strengths, and the presence of the pyridine nucleotide apparently modulated the electronic properties of the flavins (compare Fig. 6 with 3A and 3C). These observations correlated with the changes produced in the FNR spectrum upon $NADP^+$ complexation (Tejero et al., 2005). Despite the presence of $NADP^+$ barely affected apparent $k_{B \to C}$ between Fld and FNR (Table 1, Fig. 4), C showed increments in both the semiquinone

Fast Kinetic Methods with Photodiode Array Detection in the Study of the Interaction
and Electron Transfer Between Flavodoxin and Ferredoxin NADP⁺-Reductase

115

and the band-I of the flavin with regard to B. The different spectroscopic properties of the intermediate species in the absence and presence of the nucleotide suggests that the nicotinamide portion of NADP⁺ must contribute to the catalytically competent active site, probably by modulating the orientation and/or distance between the reacting flavins. This is in agreement with the negative cooperative effect observed for simultaneous binding of Fld and NADP⁺ to FNR (Velázquez-Campoy *et al.*, 2006). Additionally, C slowly evolved with a slight absorbance increase at the band-I of the flavin with minor changes in the semiquinone. These observations still indicate initial production of Fld_{sq} and reduction of FNR_{ox}, but suggest an additional step consistent with further FNR reoxidation. Analysis of the kinetic traces at 340 nm, where absorbance increase upon reduction of NADP⁺ to

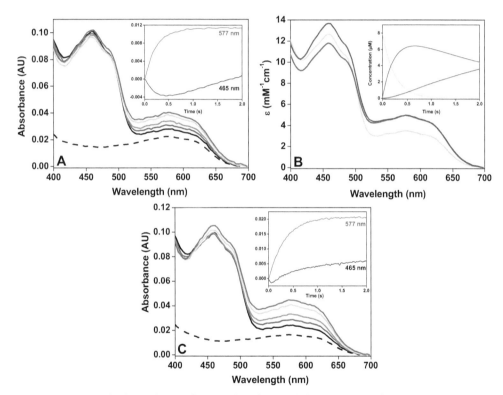

Fig. 5. Ionic strenght dependence of spectral evolution. (A). Time course for the reaction Fld_{hq} with FNR_{ox} in 50 mM Tris/HCl, pH 8.0, 200 mM NaCl at 12°C. Spectra recorded at 0.00384, 0.007552, 0.1626, 0.265, 0.5514 and, 2.047 s after mixing. The inset shows kinetic traces at the indicated wavelengths. (B). Absorbance spectra for the pre-steady-state kinetically distinguishable species obtained by global analysis of the reaction in A. The inset shows the species evolution along the time. B, C, and D species are shown in green, red and purple lines, respectively. (C). Time course for the reaction at 300 mM NaCl. Spectra recorded at 0.00384, 0.08576, 0.2035, 0.3315, 0.5158 and, 2.047 s after mixing. The inset shows kinetic traces at the indicated wavelengths. In A and C, the spectra corresponding to the Fld_{hq} before mixing are shown as a dashed line, and the first spectra after mixing are in black.

NADPH is expected, shows that amplitudes are considerably larger for samples containing the coenzyme (not shown). This indicates the last detected step must be related with the hydride transfer from FNR_{hq} to $NADP^+$ to produce its reduced form. Additionally, while the ionic strength of the media decreases the amplitudes at 340 nm for the reactions in the absence of $NADP^+$, those in its presence are nearly unaffected (not shown).

This process was also analysed for some FNR or Fld mutants previously produced and showing altered interaction or ET properties. This includes the FNR variant produced when Ala substituted for E301, a residue involved in the FNR catalytic mechanism as proton donor as well as for the stabilisation of reaction intermediates (Dumit *et al.*, 2010; Medina *et al.*, 1998). Analysis of the reduction of this variant by WT Fld_{hq} using the PDA detector showed that most of the ET took place within the instrumental dead time (not shown), confirming this is a very fast process as previously suggested (Medina *et al.*, 1998). Moreover, in contrast to the WT process, no changes in rate constants or species were observed upon increasing the ionic strength of the media (Table 1). Thus, only information of the final spectra was obtained, being this similar to those for the WT reaction with main accumulation of Fld_{sq} and FNR_{ox}. In the presence of $NADP^+$ production of NADPH was observed at 340 nm with similar rates to those reported for the WT reaction.

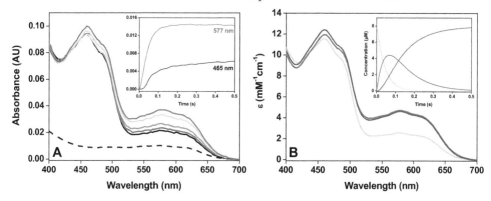

Fig. 6. Evolution of spectral changes during the reaction of Fld_{hq} with FNR_{ox} in 50 mM Tris/HCl, pH 8.0, 100 mM NaCl at 12°C and $NADP^+$ ~300 μM. (A). Time course for the reaction. Spectra recorded at 0.00128, 0.0064, 0.01664, 0.03712, 0.05504 and, 0.5107 s after mixing. The inset shows kinetic traces at the indicated wavelengths. (B). Absorbance spectra for the pre-steady-state kinetically distinguishable species obtained by global analysis. The inset shows the evolution of these species along the time. Species B, C and D are shown in green, red and purple lines, respectively.

The process was similarly analysed for two Fld mutants for which the interaction step with FNR was reported to be modified by the introduction of multiple mutations (Goñi *et al.*, 2009). Spectral evolution for the ET process between E16K/E61K Fld_{hq} and FNR_{ox} at low ionic strength (I = 25 mM) presented similar features to those above described for WT Fld (not shown). Despite a similar starting behaviour was observed for the reaction of E16K/E61K/D126K/D150K Fld_{hq}, final absorbance increase and decrease in the flavin I and semiquinone bands, respectively, were observed (not shown). In both cases, but particularly with E16K/E61K/D126K/D150K Fld_{hq}, evolution of the system was considerably hindered.

Fast Kinetic Methods with Photodiode Array Detection in the Study of the Interaction
and Electron Transfer Between Flavodoxin and Ferredoxin NADP⁺-Reductase

117

Despite process for E16K/E61K Fld$_{hq}$ fits a single ET step, with B showing similar characteristics to those in the WT reaction, a 17-fold decrease in $k_{B\to C}$ was observed (Table 1, Fig. 7A). Moreover, the final species C showed the appearance of Fld$_{ox}$ (25%) in equilibrium with Fld$_{sq}$ (75%), while the amount of FNR$_{ox}$ (40%) decreased with regard to the WT reaction with the consequent increase in the proportions of FNR$_{sq}$ (20%) and FNR$_{hq}$ (40%). The process for E16K/E61K/D126K/D150K Fld$_{hq}$ fits a two ET model, with $k_{B\to C}$ being hindered up to 190-fold with regard to WT (Table 1, Fig. 7B). Surprisingly, C shows important amounts of the reduced species of both proteins and its additional transformation is consistent with Fld$_{sq}$ and FNR$_{hq}$ as the main products of the reaction. For both Fld mutants the increase in the ionic strength again had deleterious kinetic effects in the overall ET (Table 1).

FNR form	Fld form	I (mM)	FNR$_{ox}$ + Fld$_{hq}$ $k_{B\to C}$ (s^{-1})	FNR$_{ox}$ + Fld$_{hq}$ $k_{C\to D}$ (s^{-1})	FNR$_{hq}$ + Fld$_{ox}$ $k_{B\to C}$ (s^{-1})	FNR$_{hq}$ + Fld$_{ox}$ $k_{C\to D}$ (s^{-1})
WT	WT	25	190[a]		2.0	0.5
WT	WT	125	22		0.7	0.3
E301A	WT	25	>300[a]		0.4	0.1
E301A	WT	125	>300[a]		0.4	0.03
WT:NADP⁺	WT	25	160	14		
WT:NADP⁺	WT	125	23	10		
WT	E16K/E61K	25	11.3		0.9	0.3
WT	E16K/E61K	125	<0.005		1.4	0.3
WT	E16K/E61K/D126K/D150K	25	1.0	0.4	1.2	0.1
WT	E16K/E61K/D126K/D150K	125	<0.005		<0.001	

Table 1. Kinetic parameters for the electron transfer between *Anabaena* Fld and FNR variants determined by stopped-flow and PDA detection. [a]Most of the reaction took place in the instrumental dead time.

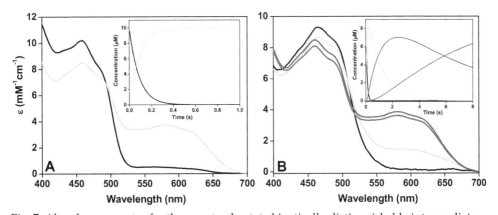

Fig. 7. Absorbance spectra for the pre-steady-state kinetically distinguishable intermediate species in the reaction of FNR$_{ox}$ with (A) E16K/E61K Fld$_{hq}$ and (B) E16K/E61K/D126K/D150K Fld$_{hq}$. The insets show the time evolution of these species. Species A, B, C and D species are shown in black, green, red and purple lines, respectively. Processes studied in 50 mM Tris/HCl, pH 8.0 at 12°C.

4. Spectral evolution of the electron transfer from FNR$_{hq}$ to Fld$_{ox}$

When FNR$_{hq}$ reacted with Fld$_{ox}$ an initial bleaching and displacement of the maximum from 464 nm (typical of Fld$_{ox}$) to 458 nm (maximum for FNR$_{ox}$) was observed, together with the appearance of a neutral semiquinone band (Fig. 8A). Then, absorbance increased in both the band-I and the semiquinone one (Fig. 8A and 8B). During the overall process only minor absorbance changes were detected in the Fld$_{ox/sq}$ isosbestic point (517 nm) at all the ionic strengths assayed, but an absorbance bleaching was initially observed at the FNR$_{ox/sq}$ one (507 nm) (Fig. 8B).

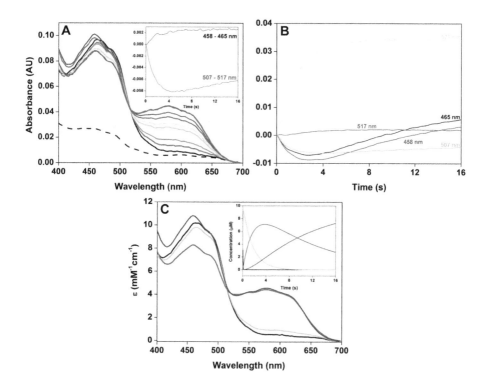

Fig. 8. Evolution of spectral changes observed during the reaction of Fld$_{ox}$ with FNR$_{hq}$. (A). Time course with spectra recorded at 0.03968, 0.2035, 0.4083, 0.695, 1.187, 1.555, 2.456, 3.521, 16.38 and 12.61 s after mixing. The spectrum of FNR$_{hq}$ before mixing is shown as a dashed line, and the first spectrum after mixing is in black. The inset shows kinetic traces at the indicated wavelengths. (B). Kinetics of absorbance changes observed at 458, 464, 507, 516 and 577 nm. (C). Absorbance spectra for the pre-steady-state kinetically distinguishable species obtained by global analysis. The inset shows the evolution of these species along the time. A, B, C and, D species are shown in black, green, red and purple lines, respectively. Processes studied in 50 mM Tris/HCl, pH 8.0, 100 mM NaCl at 12°C.

When globally analysed this reaction best fits to a three steps process (A → B → C → D) (Fig. 8C). Conversion of species A into B was relatively fast ($k_{A \to B}$ >50 s⁻¹) with very little absorbance changes. Both species had FNR_{hq} and Fld_{ox} as the main components, but deviation of mathematical combination of individual redox spectra (Fig. 2) indicated their spectroscopic properties were modulated by the presence of each other. This suggests a number of FNR_{hq} and Fld_{ox} molecules must be forming a FNR_{hq}:Fld_{ox} complex. This step appears, therefore, rather indicative of a protein-protein interaction or reorganisation event than of an ET one. B evolved to C in a relatively slow process (Table 1), in which the intensity of the band-I of the flavin gets decrease by ~20% and displaced to shorter wavelengths, whereas accumulation of semiquinone is produced (Fig. 8B). These observations are compatible with Fld_{ox} being consumed and FNR_{sq}, Fld_{sq} and FNR_{ox} as the main components of species C. On conversion of C into D, there is an increment in the band-I absorption intensity (with the maximum at ~458 nm) and minor changes in the semiquinone band amplitude. This is consistent with FNR_{ox} and Fld_{sq} as the main components of D. The faster ET rates are observed at the lower ionic strength, but the ionic strength effect is much more moderated than for the reverse reaction (Fig. 9A). Reaction of FNR_{hq} with Fld_{ox} was also analysed at different protein-protein ratios showing similar behaviour and a linear dependence of $k_{B \to C}$ and $k_{C \to D}$ rates (Fig. 9B).

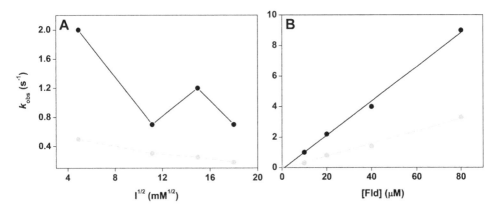

Fig. 9. (A). Ionic strength dependence of the apparent rate constants for the reduction of Fld_{ox} by FNR_{hq}. The reaction was fit to a two-step process, $k_{B \to C}$ (black) and $k_{C \to D}$ (green). (B). Fld_{ox} concentration dependence of the apparent rate constants for the reduction of Fld_{ox} by FNR_{hq} at I = 25 mM.

Reactions of FNR_{hq} with E16K/E61K or E16K/E61K/D126K/D150K Fld_{ox} variants showed similar features to the WT Fld (Fig. 10). Transformation of A into B appeared slightly slower for E16K/E61K ($k_{A \to B}$ ~11 s⁻¹), while those species could not be resolved for E16K/E61K/D126K/D150K. These observations are in agreement with the 23-fold decrease and the absence of interaction reported for the complexes of E16K/E61K or E16K/E61K/D126K/D150K Fld_{ox}, respectively, with FNR_{ox} (Goñi et al., 2009). B evolved to C in an ET process only 2-fold slower than for WT Fld_{ox} (Table 1) and similarly consistent with FNR_{sq} and Fld_{sq} as the main components of C. On conversion of C into D, absorbance increments are observed in the band-I and the semiquinone band of the flavin, particularly in the E16K/E61K/D126K/D150K Fld variant. This might be consistent with FNR_{sq} deproportionation into FNR_{ox} and FNR_{hq}, with the FNR_{hq} produced being able to reduce another Fld_{ox} molecule to the semiquinone state. Again processes of E301A FNR_{ox} with Fld_{hq} resembled those for the native proteins, with rate constants just slightly decreased and in agreement with previous reported data (Medina et al., 1998).

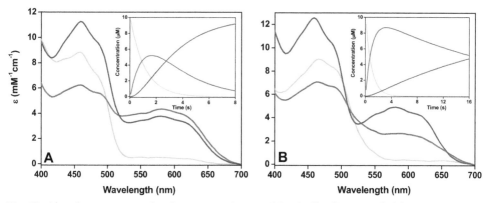

Fig. 10. Absorbance spectra for the pre-steady-state kinetically distinguishable species obtained by global analysis of the reaction of FNR_{hq} with (A) E16K/E61K Fld_{ox} and (B) E16K/E61K/D126K/D150K Fld_{ox}. Insets show the evolution of these species along the time. Intermediate B, C and D species are shown in green, red and, purple lines, respectively. Processes were studied in 50 mM Tris/HCl, pH 8.0, at 12°C.

5. Insights into the electron transfer processes in the Fld:FNR system

The physiological ET from WT Fld_{hq} to WT FNR_{ox} was reported as a very fast process difficult to follow by single-wavelength stopped-flow methods. Nevertheless, this methodology resulted useful to study the processes for some FNR or Fld mutants with altered interaction or ET properties, but interpretation of the data was usually confuse due to the spectral similarities between the different oxido-reduction states of both proteins (Fig. 2) (Casaus et al., 2002; Goñi et al., 2008; Goñi et al., 2009; Medina et al., 1998; Nogués et al., 2003; Nogués et al., 2005). Analysis of mutants suggested that the reaction might take place in two steps interpreted as formation of the semiquinones of both proteins followed by further reduction of FNR_{sq} to the fully reduced state, with further accumulation of Fld_{sq} at the end of the reaction. Therefore, a similar behaviour was assumed for the WT system. In this work, we have taken advantage of stopped-flow with PDA detection to better evaluate the intermediate and final species in the equilibrium mixture during these ET processes for the reactions using the WT proteins and some of their mutants.

Analysis of the process for the reduction of WT FNR_{ox} by WT Fld_{hq}, under similar conditions to those so far reported at single-wavelengths, indicates that even using PDA it is difficult to extract conclusions for this ET process. The data suggest a very fast interaction (or collisional) step unable to be observed ($k_{A\rightarrow B}$), followed by an ET step than cannot be resolved from the subsequent equilibration to finally produce Fld_{sq} and FNR_{ox} (process (1) in Scheme 1). However, despite FNR_{sq} is hardly detected along the reaction, both semiquinones, Fld_{sq} and FNR_{sq}, must be initially produced. Moreover, FNR_{sq} does not appear to be further reduced to FNR$_{hq}$, and apparently quickly relaxes to FNR$_{ox}$. Surprisingly, such behaviour was observed even when the overall ET reaction results slowed down by increasing the ionic strength. Thus, fast relaxation of the equilibrium distribution after the initial ET is produced with the consequent accumulation of FNR$_{ox}$ and Fld$_{sq}$. Previous fast kinetic studies using laser flash photolysis indicated that ET from WT FNR$_{sq}$ to WT Fld$_{ox}$ is an extremely fast process ($k_{obs} \sim 7000$ s^{-1}), suggesting the produced FNR$_{sq}$ will quickly react with any traces of Fld$_{ox}$ producing Fld$_{sq}$ and FNR$_{ox}$ (Medina *et al.*, 1992). Moreover, the proper nature of the PDA experiment might also contribute to this effect, since the high intensity of the lamp simultaneously exciting a wavelength range might produce side energy transfer reactions.

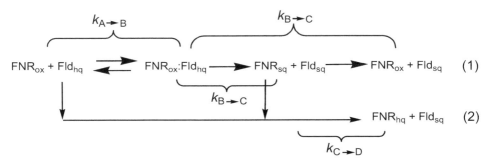

Scheme 1. Reaction pathways describing the processes observed for the reaction of FNR$_{ox}$ with Fld$_{hq}$ in the WT system (1) and with some of the Fld$_{hq}$ mutants (2).

When reduction of E301A FNR$_{ox}$ by Fld$_{hq}$ was analysed a very similar behaviour to the WT one described the process, again suggesting quick FNR$_{sq}$ deproportionation. When using the same methodology to analyse the process for two Fld mutants, with interaction and ET parameters considerably hindered (Goñi *et al.*, 2009), intermediates and products of the reaction were in agreement with the mechanism previously proposed using single-wavelength detection and with the final production of FNR$_{hq}$ and Fld$_{sq}$ under the assayed conditions (Scheme 1 reaction (2)). The PDA analysis additionally allowed improving the determination of the ET rates (Table 1). In these cases the initially produced FNR$_{sq}$ appears unable to quickly react with traces of the Fld$_{ox}$ mutants, preventing the quick relaxation after the initial ET. This effect can be explained since the $E_{ox/sq}$ for these Fld variants is more negative than in WT Fld, getting closer to the FNR $E_{ox/sq}$ and making ET from FNR$_{sq}$ to Fld$_{ox}$ less favourable from the thermodynamic point of view than in the WT reaction (Table 2). Therefore, our observations suggest that the stopped-flow methodology, independently of the detector, does not allow to identify the initial acceptance by WT or E301A FNR$_{ox}$ of a single electron from WT Fld$_{hq}$, since under the experimental conditions (even upon increasing the ionic strength of the media) the subsequent relaxation of the putatively initially produced FNR$_{sq}$ is faster than the initial ET process. Moreover, the products of this relaxation consist of a mixture of species that might not have physiological relevance within the *Anabaena* cell, where FNR$_{sq}$ must be able to accept electrons from a second Fld$_{hq}$ molecule.

The reverse ET reaction, FNR$_{hq}$ with Fld$_{ox}$ was reported as a slow process (when compared with the photosynthetic one) taking place in two sequential ET steps; production of both flavoprotein semiquinones (reaction (3) in Scheme 2), followed by the reduction of a second Fld_{ox} molecule by the FNR_{sq} produced in the first step (reaction (4) in Scheme 2) (Casaus *et al.*, 2002; Medina *et al.*, 1998; Nogués *et al.*, 2005). The spectral evolutions acquired using the PDA detector confirm such mechanism, and allowed to improve the assignment of intermediates and the major contribution of apparent rate constants to particular steps of the process.

Fld form	$E_{ox/sq}$ (mV)	$E_{sq/hq}$ (mV)	$\frac{\Delta E_{ox/sq}}{\Delta E_{ox/sq}{}^{WT}}$ (mV)	$-\frac{\Delta E_{sq/hq}}{\Delta E_{sq/hq}{}^{WT}}$ (mV)	$-K_d{}^{FNRox:Fldox}$ (μM)
WT[a]	-256	-445	---	---	2.6
E16K/E61K[a]	-301	-390	-45	55	46
E16K/E61K/D126K/D150K[a]	-297	-391	-41	54	---
FNR form					
WT[b]	-325	-338	---	---	3[c]
E301A	-284[b]	-358[b]	41	-20	4[c]

Table 2. Midpoint reduction potentials for the different *Anabaena* Fld and FNR forms. Data obtained in 50 mM Tris/HCl at pH 8.0 and 25 °C for Fld[a] and at 10°C for FNR[b]. [a]Data from (Goñi *et al.*, 2009). [b]Data from (Faro *et al.*, 2002b). [c]Data from (Medina *et al.*, 1998).

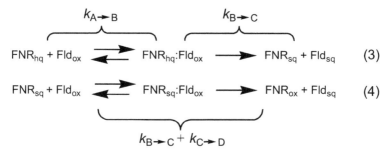

Scheme 2. Reaction pathways describing the processes observed for the reaction of the FNR_{hq} variants with the Fld_{ox} variants.

Thus, for the process with the E16K/E61K Fld variant, the step corresponding to complex formation-reorganisation was erroneously related with an ET step in a previous study. At the lowest ionic strength assayed our data indicate that E16K/E61K Fld_{ox} and, particularly, E16K/E61K/D126/D150K Fld_{ox} are still able to accept electrons from FNR_{ox} with apparent rates that only decreased by 2-fold and with final production of FNR_{ox} occurring in higher degree that in the WT system (Table 1, Fig. 8 and 10). Their slightly more negative $E_{ox/sq}$ values (Table 2) makes them poorer electron acceptors from FNR_{hq} than WT Fld and might explain the small differences in rates (Goñi *et al.*, 2009). Additionally, E301A FNR_{hq} is also able to pass electrons to Fld_{ox} with a rate 5-fold slower than WT (Table 1). This behaviour might be related with the very low stability of the semiquinone form in this FNR mutant (Table 2), that makes the formation of this intermediate state non-favourable (Medina *et al.*, 1998).

A biphasic dependence of the observed rate constants on the protein concentration has been reported for the ET reaction from Fd to FNR (Fig. 4), and associated with the appearance of an optimum ionic strength value and with the electrostatic stabilisation at low ionic strengths of non-optimal orientations within the intermediate ET complex. Thus, specific electrostatic and hydrophobic interactions play an important role in these association and dissociation processes, as well as in the rearrangement of the complex (Faro *et al.*, 2002a; Hurley *et al.*, 2006; Martínez-Júlvez *et al.*, 1998; Martínez-Júlvez *et al.*, 1999; Martínez-Júlvez *et al.*, 2001; Medina & Gómez-Moreno, 2004; Morales *et al.*, 2000; Nogués *et al.*, 2003). Despite some residues on the FNR surface are critical for the interaction with Fld and it is accepted that FNR interacts using the same region with Fld and Fd (Hurley *et al.*, 2002; Martínez-Júlvez *et al.*, 1999), the bell-shaped profile for the ionic strength dependence is not reproduced for ET reactions between FNR and Fld (Fig. 4). A strong deleterious influence of the ionic strength is observed on the overall ET process between Fld and FNR, particularly in the photosynthetic direction (Fig. 4 and 9A). This suggests re-arrangement of the initial FNR:Fld interaction either does not take place or does not increase the efficiency of the system, while at lower ionic strength the electrostatic interactions contribute to produce more efficient orientations between the flavin cofactors.

Biochemical and docking studies suggested that the FNR:Fld interaction does not rely on a precise complementary surface of the reacting molecules. Thus, WT Fld might adopt different orientations on the FNR surface without significantly altering the relative disposition and contact between the FMN and FAD groups of Fld and FNR and, therefore, the distance between their methyl groups (Fig. 11A) (Goñi *et al.*, 2009; Medina *et al.*, 2008; Medina, 2009). Those studies suggested the molecular dipole moment alignment as one of the major determinants for the efficiency of this system (Fig. 11B). However, kinetic

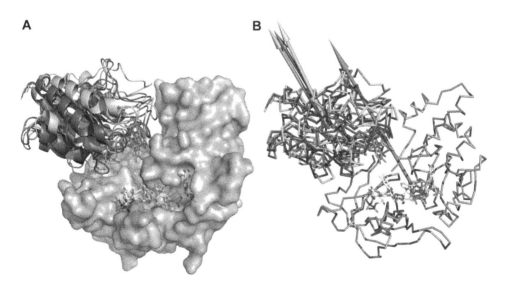

Fig. 11. (A). Model for the interaction of Fld and FNR. The figure shows several positions determined by docking of Fld onto the FNR surface. (B). Magnitude and orientation of the dipole moment of Fld and FNR in the model.

parameters reported to the date for these processes were only obtained at low ionic strength (~0.03 M), conditions far away from those found in the thylakoid (0.15-0.3 M (Durán et al., 2006)). Our data suggest that at physiological ionic strengths the ET efficiency, particularly in the photosynthetic direction, will be considerably hindered with regard to the data reported in vitro at low ionic strengths. The ionic strength will shield the dipole moment alignment contribution, making it just one additional contribution to proteins encounter. Among those contributions we might include electrostatic and hydrophobic interactions imposed by the thylakoid membrane, physical diffusional parameters and molecular crowding inside the cell. It is also worth to note that increasing the ionic strength makes reduction of FNR by Fld_{hq} only 4-8 times faster than the reverse process (Compare Fig. 4A and 9A). Thus, shielding the effect of the dipole moment appears to have a larger impact in producing the competent ET orientation between the redox centres in the $FNR_{ox}:Fld_{hq}$ complex than in the $FNR_{hq}:Fld_{ox}$ one. In other words, it reduces the probability of obtaining the best $FNR_{ox}:Fld_{hq}$ orientations for ET.

The Fld mutants here studied, particularly E16K/E61K/D126/D150K, have lost the ability to efficiently reduce FNR (Table 1). More positive $E_{sq/hq}$ values (Table 2) might somehow contribute to this behaviour, but previous studies suggested the introduced mutations induced changes in the Fld electrostatic potential surfaces, as well as in the orientation and magnitude of the Fld molecular dipole moment (Goñi et al., 2009). Despite the thermodynamic parameters favour the process, the observed reaction might only correlate with a collisional-type reaction. Therefore, it could exit the possibility that steering of the dipole moment contribution might produce a positive effect on the overall ET process. However, our analysis also shows that the increasing of the ionic strength again had a deleterious effect in the ET processes from these Fld_{hq} mutants. Thus, electrostatic and hydrophobic interactions and the dipole moment still must contribute to the formation of productive interactions between both proteins at physiological ionic strengths. In vivo the presence of other proteins competing for Fld_{hq} might also result in changes to electron channelling into distinct pathways. When going to physiological conditions Fld interaction with FNR is confirmed to be less specific than that of Fd. Subtle changes at the isoalloxazine environment influence the Fld binding abilities and modulate the ET processes by producing different orientations and distances between the redox centres. Therefore, ET reactions involving Fld might not have as much specific interaction requirements as other reactions involving protein-protein interactions. Thus, the bound state could be formed by dynamic ensembles instead of single conformations as has already been proposed for this system (Fig. 11) (Goñi et al., 2009; Medina et al., 2008) and also observed in other ET systems (Crowley & Carrondo, 2004; Worrall et al., 2003). This further confirms previous studies suggesting that Fld interacts with different structural partners through non-specific interactions, which in turn decreased the potential efficiency in ET that could achieve if unique and more favourable orientations were produced with a reduced number of partners. Heterogeneity of ET kinetics is an intrinsic property of Fld oxido-reduction processes, and can be most probably ascribed to different conformations of FNR:Fld complexes (Medina et al., 2008; Sétif, 2001). During Fld-dependent photosynthetic ET the Fld molecule must pivot between its docking sites in PSI and in FNR. Formation of transient complexes of Fld with FNR in vivo is useful during this process, though not critical, for promoting efficient reduction of Fld and FNR and for avoiding reduction of oxygen by the donor redox centres (Goñi et al., 2008; Goñi et al., 2009; Hurley et al., 2006; Sétif, 2001, 2006).

Fast Kinetic Methods with Photodiode Array Detection in the Study of the Interaction
and Electron Transfer Between Flavodoxin and Ferredoxin NADP+-Reductase
125

6. Conclusion

Single-wavelength fast kinetic stopped-flow methods have been widely used for the analysis of the mechanisms involving transient binding and ET between Fld and FNR. However, this methodology did not allow to un-ambiguously identifying the intermediate and final compounds of the reactions. PDA detection combined with fast kinetic stopped-flow methods results useful to better understand the mechanisms involving transient binding and ET between Fld and FNR. Despite the high similarity among the spectra for the same redox states within both proteins, the methodology here used allowed identifying the composition of the intermediate species and final species of the reactions, as long as the kinetics fits in the measurable instrumental time. The mechanism of these inter-flavin ET reactions is revisited, evaluating the evolution of the reaction along the time within a wavelength spectral range by using a PDA detector. Additionally, our analysis of the dependence of the inter-flavin ET mechanism on the ionic strength suggest that, under physiological conditions, the electrostatic alignment contributes to the overall orientation but it is not anymore the major determinant of the orientation of Fld on the protein partner surface. Additionally, the presence of the coenzyme reveals a complex modulation of the process.

7. Acknowledgment

This work has been supported by Ministerio de Ciencia e Innovación, Spain (Grant BIO2010-14983 to M.M.). We thank to Dr. G. Goñi for the production of the Fld mutants and Dr. R. Tauler for his help in initial spectral deconvolution of intermediate species.

8. References

Arakaki, A. K., Ceccarelli, E. A. & Carrillo, N. (1997). Plant-type ferredoxin-NADP+ reductases: a basal structural framework and a multiplicity of functions. *Faseb J.* Vol.11, No.2, pp. 133-140, ISSN 0892-6638.

Batie, C. J. & Kamin, H. (1984a). Electron transfer by ferredoxin:NADP+ reductase. Rapid-reaction evidence for participation of a ternary complex. *J Biol Chem.* Vol.259, No.19, pp. 11976-11985, ISSN 0021-9258.

Batie, C. J. & Kamin, H. (1984b). Ferredoxin:NADP+ oxidoreductase. Equilibria in binary and ternary complexes with NADP+ and ferredoxin. *J Biol Chem.* Vol.259, No.14, pp. 8832-8839, ISSN 0021-9258.

Bottin, H. & Lagoutte, B. (1992). Ferredoxin and flavodoxin from the cyanobacterium *Synechocystis* sp PCC 6803. *BBA-Bioenergetics.* Vol.1101, No.1, pp. 48-56, ISSN 0005-2728.

Brenner, S., Hay, S., Munro, A. W. & Scrutton, N. S. (2008). Inter-flavin electron transfer in cytochrome P450 reductase - effects of solvent and pH identify hidden complexity in mechanism. *Febs J.* Vol.275, No.18, pp. 4540-4557, ISSN 1742-464X.

Carrillo, N. & Ceccarelli, E. A. (2003). Open questions in ferredoxin-NADP+ reductase catalytic mechanism. *Eur J Biochem.* Vol.270, No.9, pp. 1900-1915, ISSN 0014-2956.

Casaus, J. L., Navarro, J. A., Hervás, M., Lostao, A., De la Rosa, M. A., Gómez-Moreno, C., Sancho, J. & Medina, M. (2002). *Anabaena* sp. PCC 7119 flavodoxin as electron carrier from photosystem I to ferredoxin-NADP+ reductase. Role of Trp(57) and Tyr(94). *J Biol Chem.* Vol.277, No.25, pp. 22338-22344, ISSN 0021-9258.

Crowley, P. B. & Carrondo, M. A. (2004). The architecture of the binding site in redox protein complexes: implications for fast dissociation. *Proteins.* Vol.55, No.3, pp. 603-612, ISSN 0887-3585.

Dumit, V. I., Essigke, T., Cortez, N. & Ullmann, G. M. (2010). Mechanistic insights into ferredoxin-NADP(H) reductase catalysis involving the conserved glutamate in the active site. *J Mol Biol.* Vol.397, No.3, pp. 814-825, ISSN 0022-2836.

Durán, R. V., Hervás, M., de la Cerda, B., de la Rosa, M. A. & Navarro, J. A. (2006). A laser flash-induced kinetic analysis of in vivo photosystem I reduction by site-directed mutants of plastocyanin and cytochrome c_6 in *Synechocystis sp.* PCC 6803. *Biochemistry.* Vol.45, No.3, pp. 1054-1060, ISSN 0006-2960.

Faro, M., Frago, S., Mayoral, T., Hermoso, J. A., Sanz-Aparicio, J., Gómez-Moreno, C. & Medina, M. (2002a). Probing the role of glutamic acid 139 of *Anabaena* ferredoxin-NADP+ reductase in the interaction with substrates. *Eur J Biochem.* Vol.269, No.20, pp. 4938-4947, ISSN 0014-2956.

Faro, M., Gómez-Moreno, C., Stankovich, M. & Medina, M. (2002b). Role of critical charged residues in reduction potential modulation of ferredoxin-NADP+ reductase. *Eur J Biochem.* Vol.269, No.11, pp. 2656-2661, ISSN 0014-2956.

Frago, S., Goñi, G., Herguedas, B., Peregrina, J. R., Serrano, A., Pérez-Dorado, I., Molina, R., Gómez-Moreno, C., Hermoso, J. A., Martínez-Júlvez, M., Mayhew, S. G. & Medina, M. (2007). Tuning of the FMN binding and oxido-reduction properties by neighboring side chains in *Anabaena* flavodoxin. *Arch Biochem Biophys.* Vol.467, No.2, pp. 206-217, ISSN 0003-9861.

Frago, S., Lans, I., Navarro, J. A., Hervás, M., Edmondson, D. E., de la Rosa, M. A., Gómez-Moreno, C., Mayhew, S. G. & Medina, M. (2010). Dual role of FMN in flavodoxin function: Electron transfer cofactor and modulation of the protein-protein interaction surface. *BBA-Bioenergetics.* Vol.1797, No.2, pp. 262-271, ISSN 0005-2728.

Golbeck, J. H., Ed. (2006). *Photosystem I. The light-driven platocyanin:ferredoxin oxidoreductase,* Springer, ISBN 1-4020-4255-8, Dordrecht, The Netherlands.

Goñi, G., Serrano, A., Frago, S., Hervás, M., Peregrina, J. R., de la Rosa, M. A., Gómez-Moreno, C., Navarro, J. A. & Medina, M. (2008). Flavodoxin-mediated electron transfer from photosystem I to ferredoxin-NADP+ reductase in *Anabaena*: role of flavodoxin hydrophobic residues in protein-protein interactions. *Biochemistry.* Vol.47, No.4, pp. 1207-1217, ISSN 0006-2960.

Goñi, G., Herguedas, B., Hervás, M., Peregrina, J. R., de la Rosa, M. A., Gómez-Moreno, C., Navarro, J. A., Hermoso, J. A., Martínez-Júlvez, M. & Medina, M. (2009). Flavodoxin: A compromise between efficiency and versatility in the electron transfer from Photosystem I to Ferredoxin-NADP+ reductase. *BBA-Bioenergetics.* Vol.1787, No.3, pp. 144-154, ISSN 0005-2728.

Hurley, J. K., Morales, R., Martínez-Júlvez, M., Brodie, T. B., Medina, M., Gómez-Moreno, C. & Tollin, G. (2002). Structure-function relationships in *Anabaena* ferredoxin/ferredoxin-NADP+ reductase electron transfer: insights from site-directed mutagenesis, transient absorption spectroscopy and X-ray crystallography. *BBA-Bioenergetics.* Vol.1554, No.1-2, pp. 5-21, ISSN 0005-2728.

Hurley, J. K., Tollin, G., Medina, M. & Gómez-Moreno, C. (2006). Electron transfer from ferredoxin and flavodoxin to ferredoxin-NADP⁺ reductase, In: *Photosystem I. The light-driven platocyanin:ferredoxin oxidoreductase*, J. H. Golbeck (Ed.), pp. 455-476, Springer, ISBN 1-4020-4255-8, Dordrecht, The Netherlands.

Martínez-Júlvez, M., Medina, M., Hurley, J. K., Hafezi, R., Brodie, T. B., Tollin, G. & Gómez-Moreno, C. (1998). Lys75 of *Anabaena* ferredoxin-NADP⁺ reductase is a critical residue for binding ferredoxin and flavodoxin during electron transfer. *Biochemistry*. Vol.37, No.39, pp. 13604-13613, ISSN 0006-2960.

Martínez-Júlvez, M., Medina, M. & Gómez-Moreno, C. (1999). Ferredoxin-NADP⁺ reductase uses the same site for the interaction with ferredoxin and flavodoxin. *J Biol Inorg Chem*. Vol.4, No.5, pp. 568-578, ISSN 0949-8257.

Martínez-Júlvez, M., Nogués, I., Faro, M., Hurley, J. K., Brodie, T. B., Mayoral, T., Sanz-Aparicio, J., Hermoso, J. A., Stankovich, M. T., Medina, M., Tollin, G. & Gómez-Moreno, C. (2001). Role of a cluster of hydrophobic residues near the FAD cofactor in *Anabaena* PCC 7119 ferredoxin-NADP⁺ reductase for optimal complex formation and electron transfer to ferredoxin. *J Biol Chem*. Vol.276, No.29, pp. 27498-27510, ISSN 0021-9258.

Martínez-Júlvez, M., Medina, M. & Velázquez-Campoy, A. (2009). Binding thermodynamics of ferredoxin:NADP⁺ reductase: two different protein substrates and one energetics. *Biophys J*. Vol.96, No.12, pp. 4966-4975, ISSN 0006-3495.

Medina, M., Gómez-Moreno, C. & Tollin, G. (1992). Effects of chemical modification of *Anabaena* flavodoxin and ferredoxin-NADP⁺ reductase on the kinetics of interprotein electron transfer reactions. *Eur J Biochem*. Vol.210, No.2, pp. 577-583, ISSN 0014-2956.

Medina, M., Martínez-Júlvez, M., Hurley, J. K., Tollin, G. & Gómez-Moreno, C. (1998). Involvement of glutamic acid 301 in the catalytic mechanism of ferredoxin-NADP⁺ reductase from *Anabaena* PCC 7119. *Biochemistry*. Vol.37, No.9, pp. 2715-2728, ISSN 0006-2960.

Medina, M. & Gómez-Moreno, C. (2004). Interaction of ferredoxin-NADP⁺ reductase with its substrates: Optimal interaction for efficient electron transfer. *Photosynth Res*. Vol.79, No.2, pp. 113-131, ISSN 0166-8595.

Medina, M., Abagyan, R., Gómez-Moreno, C. & Fernández-Recio, J. (2008). Docking analysis of transient complexes: interaction of ferredoxin-NADP⁺ reductase with ferredoxin and flavodoxin. *Proteins*. Vol.72, No.3, pp. 848-862, ISSN 0887-3585.

Medina, M. (2009). Structural and mechanistic aspects of flavoproteins: Photosynthetic electron transfer from photosystem I to NADP⁺. *Febs J*. Vol.276, No.15, pp. 3942-3958, ISSN 1742-464X.

Morales, R., Charon, M. H., Kachalova, G., Serre, L., Medina, M., Gómez-Moreno, C. & Frey, M. (2000). A redox-dependent interaction between two electron-transfer partners involved in photosynthesis. *EMBO Rep*. Vol.1, No.3, pp. 271-276, ISSN 1469-221X.

Müller, F., Ed. (1991). *Chemistry and Biochemistry of Flavoenzymes*, CRC Press, ISBN 0-8493-4393-3 Boca Raton, Florida.

Nogués, I., Martínez-Júlvez, M., Navarro, J. A., Hervás, M., Armenteros, L., de la Rosa, M. A., Brodie, T. B., Hurley, J. K., Tollin, G., Gómez-Moreno, C. & Medina, M. (2003). Role of hydrophobic interactions in the flavodoxin mediated electron transfer from photosystem I to ferredoxin-NADP⁺ reductase in *Anabaena* PCC 7119. *Biochemistry*. Vol.42, No.7, pp. 2036-2045, ISSN 0006-2960.

Nogués, I., Tejero, J., Hurley, J. K., Paladini, D., Frago, S., Tollin, G., Mayhew, S. G., Gómez-Moreno, C., Ceccarelli, E. A., Carrillo, N. & Medina, M. (2004). Role of the C-terminal tyrosine of ferredoxin-nicotinamide adenine dinucleotide phosphate reductase in the electron transfer processes with its protein partners ferredoxin and flavodoxin. *Biochemistry*. Vol.43, No.20, pp. 6127-6137, ISSN 0006-2960.

Nogués, I., Hervás, M., Peregrina, J. R., Navarro, J. A., de la Rosa, M. A., Gómez-Moreno, C. & Medina, M. (2005). *Anabaena* flavodoxin as an electron carrier from photosystem I to ferredoxin-NADP⁺ reductase. Role of flavodoxin residues in protein-protein interaction and electron transfer. *Biochemistry*. Vol.44, No.1, pp. 97-104, ISSN 0006-2960.

Sancho, J., Medina, M. & Gómez-Moreno, C. (1990). Arginyl groups involved in the binding of *Anabaena* ferredoxin-NADP⁺ reductase to NADP⁺ and to ferredoxin. *Eur J Biochem*. Vol.187, No.1, pp. 39-48, ISSN 0014-2956.

Sétif, P. (2001). Ferredoxin and flavodoxin reduction by photosystem I. *BBA-Bioenergetics*. Vol.1507, No.1-3, pp. 161-179, ISSN 0005-2728.

Sétif, P. (2006). Electron transfer from the bound iron-sulfur clusters to ferredoxin/flavodoxin: kinetic and structural properties of ferredoxin/flavodoxin reduction by photosystem I, In: *Photosystem I. The light-driven platocyanin:ferredoxin oxidoreductase*, J. H. Golbeck (Ed.), pp. 439-454, Springer, ISBN 1-4020-4255-8, Dordrecht, The Netherlands.

Tejero, J., Martínez-Júlvez, M., Mayoral, T., Luquita, A., Sanz-Aparicio, J., Hermoso, J. A., Hurley, J. K., Tollin, G., Gómez-Moreno, C. & Medina, M. (2003). Involvement of the pyrophosphate and the 2'-phosphate binding regions of ferredoxin-NADP⁺ reductase in coenzyme specificity. *J Biol Chem*. Vol.278, No.49, pp. 49203-49214, ISSN 0021-9258.

Tejero, J., Pérez-Dorado, I., Maya, C., Martínez-Júlvez, M., Sanz-Aparicio, J., Gómez-Moreno, C., Hermoso, J. A. & Medina, M. (2005). C-terminal tyrosine of ferredoxin-NADP⁺ reductase in hydride transfer processes with NAD(P)⁺/H. *Biochemistry*. Vol.44, No.41, pp. 13477-13490, ISSN 0006-2960.

Tejero, J., Peregrina, J. R., Martínez-Júlvez, M., Gutierrez, A., Gómez-Moreno, C., Scrutton, N. S. & Medina, M. (2007). Catalytic mechanism of hydride transfer between NADP⁺/H and ferredoxin-NADP⁺ reductase from *Anabaena* PCC 7119. *Arch Biochem Biophys*. Vol.459, No.1, pp. 79-90, ISSN 0003-9861.

Velázquez-Campoy, A., Goñi, G., Peregrina, J. R. & Medina, M. (2006). Exact analysis of heterotropic interactions in proteins: Characterization of cooperative ligand binding by isothermal titration calorimetry. *Biophys J*. Vol.91, No.5, pp. 1887-1904, ISSN 0006-3495.

Wolthers, K. R. & Scrutton, N. S. (2004). Electron transfer in human methionine synthase reductase studied by stopped-flow spectrophotometry. *Biochemistry*. Vol.43, No.2, pp. 490-500, ISSN 0006-2960.

Worrall, J. A., Reinle, W., Bernhardt, R. & Ubbink, M. (2003). Transient protein interactions studied by NMR spectroscopy: the case of cytochrome *c* and adrenodoxin. *Biochemistry*. Vol.42, No.23, pp. 7068-7076, ISSN 0006-2960.

7

Photosynthesis in Microalgae as Measured with Delayed Fluorescence Technique

Maja Berden-Zrimec[1], Marina Monti[2] and Alexis Zrimec[1]
[1]*Institute of Physical Biology,*
[2]*Istituto Nazionale di Oceanografia e Geofisica Sperimentale,*
[1]*Slovenia*
[2]*Italy*

1. Introduction

Two to three percent of the absorbed sun energy is re-emitted from the pigment systems as fluorescence. Delayed fluorescence (DF) represents only 0.03 % of that emission (Jursinic, 1986). But although DF reflects an insignificant loss of the total energy stored by photosynthesis, it is a sensitive indicator of the many steps in photosynthesis processes (Jursinic, 1986). This sensitivity makes DF an extremely complex phenomenon, however with awareness and control of the variables, DF becomes an important intrinsic probe (Jursinic, 1986).

Delayed fluorescence, also termed delayed luminescence or delayed light emission, is a long-lived light emission by plants, algae and cyanobacteria after being illuminated with light and placed in darkness (Strehler & Arnold, 1951). It can last from milliseconds to several minutes, which is quite a long time in a nanosecond world of classical fluorescence.

The main source of DF is the photosystem II (PSII) (Jursinic, 1986), whereas the photosystem I (PSI) contributes much less to the emission. Delayed fluorescence emission spectrum resembles the fluorescence emission spectrum of chlorophyll a (Arnold & Davidson, 1954; Jursinic, 1986; Van Wijk et al., 1999). The main difference from prompt fluorescence is in the origin of the excited single state of the emitting pigment molecule (Jursinic, 1986). Delayed fluorescence originates from the repopulation of excited states of chlorophyll from the stored energy after charge separation (Jursinic, 1986), whereas prompt fluorescence reflects the radiative de-excitation of excited chlorophyll molecules before charge separation. This is why delayed and prompt fluorescence contain information about different fundamental processes of the photosynthetic apparatus (Goltsev et al., 2003).

An important feature of DF is that it is emitted only by a functionally active chlorophyll – in other words, when photosynthesis is active (Bertsch, 1962). The emission depends on the number of PSII centers and the rate of back reactions in the photosynthetic chain, which are influenced by the membrane potential and pH gradient (Avron & Schreiber, 1979; Joliot & Joliot, 1980).

Delayed fluorescence is affected by many chemical and physical variables, such as ATP (Avron & Schreiber, 1979), proton gradient in the thylakoids (Wraight & Crofts, 1971), chill stress (Melcarek & Brown, 1977), different xenobiotics (Berden-Zrimec et al., 2007; Drinovec et al., 2004a; Katsumata et al., 2006), excitation light spectral and intensity characteristics (Wang et al., 2004; Zrimec et al., 2005), cell culture growth stage (Berden-Zrimec et al., 2008b; Monti et al., 2005), and nutrient status (Berden-Zrimec et al., 2008a). The changes in chemical and physical parameters affect the reduction state of the plastoquinone pool or its coupling with PSII by modulating the reversed electron flow (Avron & Schreiber, 1979; Mellvig & Tillberg, 1986).

In the field studies, the intensity of delayed fluorescence is used as a measure of photosynthetic activity and living algal biomass (Berden-Zrimec et al., 2009; Krause & Gerhardt, 1984; Kurzbaum et al., 2007; Schneckenburger & Schmidt, 1996). Additionally, DF excitation spectra can be utilized for the analysis of taxonomical changes in the algal communities (Greisberger & Teubner, 2007; Hakanson et al., 2003; Istvanovics et al., 2005; Yacobi et al., 1998) (Figure 1).

2. Basic characteristics of delayed fluorescence decay kinetics

Delayed fluorescence shows monotonic decay kinetics in the first seconds, sometimes followed by a more or less pronounced transient peak (Bertsch & Azzi, 1965). The emission is composed of several components, characterized by different decay rates (Bjorn, 1971; Desai et al., 1983). The faster decaying components (first few seconds) provide information about the fate of energy absorbed by PSII (Desai et al., 1983). The slow components (from few seconds to minutes or hours) originate in back reactions in the photosynthetic chain as well as between the S2 and S3 states of the oxygen evolving complex (OEC) and quinones Q_A and Q_B (Joliot et al., 1971). OEC reacts with quinone molecules and their reduction state is in the equilibrium with PQ. The increase of PQH_2 concentration induces reverse electron flow, producing Q_A^- and Q_B^- states (Joliot & Joliot, 1980), which contribute to DF. The

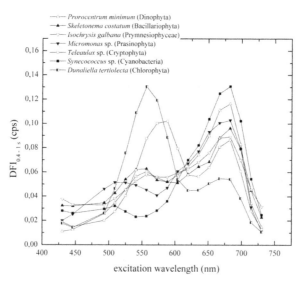

Fig. 1. Delayed fluorescence excitation spectra of marine phytoplankton species. cps – counts per second, DFI $_{0.4-1s}$ – delayed fluorescence intensity in the interval 0.4 – 1 s after sample illumination (Berden-Zrimec et al., 2010).

reduction state of PQ pool is influenced by many reactions including electron transport in PSI and dark reactions of photosynthesis.

The reaction rate is additionally influenced by changes in pH gradient and electric field in thylakoid membrane of chloroplasts. In the case of monotonous DF decay kinetics (without the transition peak), both electric field and pH gradient decay slowly after the initial increase caused by single light pulse excitation.

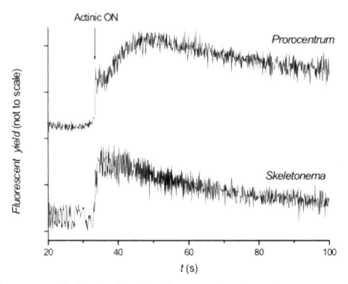

Fig. 2. Slow fluorescence induction kinetics of two marine algae (*Prorocentrum minimum* and *Skeletonema costatum*) after switching on actinic light (Drinovec et al., 2004b).

The slow components probably provide information on temporary energy storage during the photosynthetic electron transport (Desai et al., 1983), because they depend on electron distribution in the plastoquinone pool (PQ) and photosystem I (PSI) (Katsumata et al., 2006). By a measurement of the fluorescent yield it is possible to monitor the reduction state of quinone Q_A in PSII. We measured slow florescence induction kinetics by switching on an actinic source of the same intensity as was used for illumination in the DF measurement (Drinovec et al., 2004b). The fluorescence induction curves of marine dinoflagellate *Prorocentrum minimum* had a distinct transient peak and in marine diatom *Skeletonema costatum* a monotonous decay kinetics (Figure 2). In *S. costatum*, a slow decay of fluorescent yield after the initial sharp increase was observed. This is a consequence of progressive oxidation of quinones as Calvin cycle is initiated. *P. minimum* produces a maximum fluorescent yield about 15 seconds after the start of actinic illumination. At this point, the reduction state of the quinones is at the maximum and starts decreasing slowly. This experiment showed the occurrence of transient peaks represents an important physiological parameter for investigation of photosynthetic processes.

The occurrence of the transient peak in DF decay kinetics probably depends on the rates of back reactions and possibly the organization of the thylakoid membrane (Desai et al., 1983). The exact physiological interpretation of transient peaks is quite difficult due to complex electron pathways and their interactions, however they appear to be formed when the metabolic conditions affect the redox status of Q_A and Q_B directly or indirectly. It has been reported that ATP and NADPH can reduce quinones (Joliot & Joliot, 1980). Degradation of starch to PGA begins immediately after switching off the light. ATP and NADPH thus formed may enhance the reduction of quinones and induce the formation of charge pairs with higher S states as long as they exist (Mellvig & Tillberg, 1986). ATP concentration oscillates after switching cells from light to darkness which is a consequence of feedback mechanisms in the reactions of photosynthesis. A strong coupling of biochemical reactions in thylakoid membrane is also an essential prerequisite for hyperbolic decay.

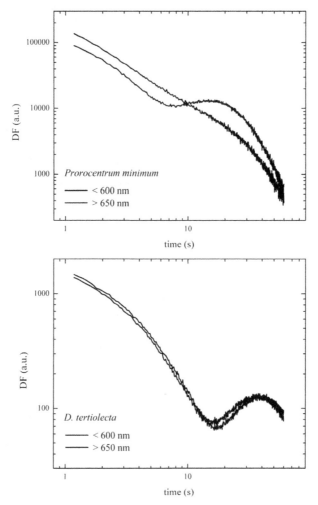

Fig. 3. Delayed fluorescence decay kinetics after illumination with two different wavelengths. a) *Prorocentrum minimum*, b) *Dunaliella tertiolecta*. a.u. – arbitrary units.

The transient peak is preferentially stimulated by far-red excitation (Desai et al., 1983; Hideg et al., 1991), but in some species it can also be induced by shorter wavelengths

(Berden-Zrimec et al., 2008a; Zrimec et al., 2005) (Figure 3). In our experiments, *Dunaliella tertiolecta* Butcher (Chlorophyta) exhibited a peak at the illuminations below 600 nm and above 650 nm (Zrimec et al., 2005) (Figure 3b), whereas *Desmodesmus (Scenedesmus) subspicatus* Chodat 1926 (Chlorophyta) did not exhibit the peak at all (Berden-Zrimec et al., 2007). *Prorocentrum minimum* (Pavillard) Schiller (Dinophyta) exhibited a peak only when illuminated with wavelengths above 650 nm (Figure 3a). Bertsch (1962) observed the peak in *Chlorella sp.* (Chlorophyta) at the illumination wavelength of 700 nm, but not at 650 nm.

The presence of a peak in DF decay curves after a pulse of light of longer wavelengths indicates PSI involvement in DF generation (Bertsch, 1962; Desai et al., 1983; Hideg et al., 1991; Mellvig & Tillberg, 1986), because far-red light is predominantly absorbed by PSI. If cyclic electron flow produces excess ATP over NADPH, back electron flow from PSI can be generated, resulting in the transient peak from a few to tens of seconds after their being illuminated (Mellvig & Tillberg, 1986).

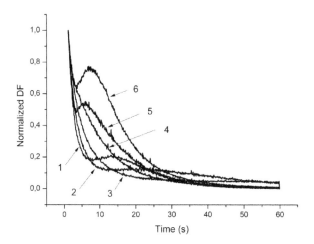

Fig. 4. Normalized delayed fluorescence decay curves of different algal species after a 3 s white-light illumination pulse. (1) *Prorocentrum minimum* (Dinophyta), (2) *Scrippsiella trochoidea* (Dinophyta), (3) *Gyrodinium* sp. (Dinophyta), (4) *Skeletonema costatum* (Bacillariophyceae), (5) *Pyrenomonas* sp. (Chryptophyta), (6) *Isochrysis galbana* (Chrysophyta). (Berden-Zrimec et al., 2010).

The peak position varies greatly between species (Figure 4). In algae, the peak is usually positioned in first minute after the illumination (Figure 4) (Mellvig & Tillberg, 1986), whereas in higher plants it lies in the range of minutes after excitation (Desai et al., 1983). DF decay kinetics can differ even among strains (Berden-Zrimec et al., 2008b; Monti et al., 2005), due to the kinetic rate constants of the electron back reactions depending on the physiological and organizational state of the entire photosynthetic apparatus. Several peaks appear when algae are put in low CO_2 conditions or as a consequence of phosphorous starvation (Mellvig & Tillberg, 1986). In both cases, dark reactions of photosynthesis are affected.

3. Modelling of delayed fluorescence kinetics

There are several proposed phenomenological models of long-term DF decay kinetics: the "multiexponential models" (Schmidt & Schneckenburger, 1992), the "hyperbolic models" (Lavorel & Dennery, 1982; Scordino et al., 1993), the "coherence models" (Yan et al., 2005). Recently, Qiang Li and co-workers presented a very interesting mathematical–physical analysis where they modeled the electron reflux for photosynthetic electron transport chain (Li et al., 2007). Unfortunately, all the presently published models fail to include all the experimental data available in the literature. Especially problematic is the modeling of DF decay kinetics with the transient peaks.

The monotonous decay kinetics (Figure 5) is simply described as a hyperbolic decay using a function $I=I_0/(t+t_0)^m$ (Scordino et al., 1996) (Figure 6). In usual decay processes the relaxation kinetics is exponential. A hyperbolic decay kinetics was reported to be a sign of a coupled system (Lavorel & Dennery, 1982).

We modeled the non-monotonous relaxation kinetics of DF – the decay curves with the transient peak – as a multiexponential relaxation among three pools of electrons in metastable states. The first pool are the excited electrons, stabilized on quinones, $q(t)$, that relaxes as delayed fluorescence emission or by transfer to the plastoquinone pool, $p(t)$, which in turn preferably relaxes further to the slower reactions, $d(t)$, or back to the quinones (Eqs. 1, 2 and 3):

$$q'(t) = -k_1 q(t) + k_2 p(t),$$ (1)

$$p'(t) = -k_2 p(t) + k_3 d(t),$$ (2)

$$d'(t) = -k_3 d(t),$$ (3)

where t is time in seconds, and k_i are the kinetic constants, and we assume only the pool of electrons that eventually relaxes back to the ground state by emitting delayed fluorescence.

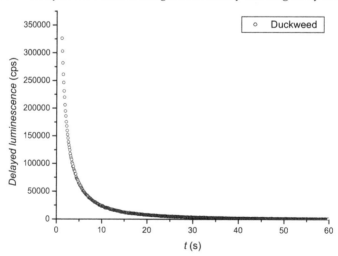

Fig. 5. A monotonous delayed fluorescence of duckweed (*Lemna minor*). cps – counts per second.

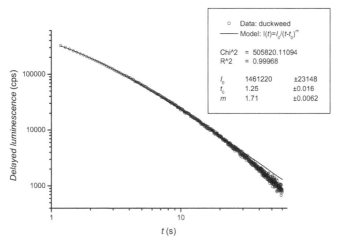

Fig. 6. Delayed fluorescence kinetics of duckweed in log- log scale. A hyperbolic curve was fitted to the measured data. cps – counts per second (Drinovec et al., 2004b).

To model the different positions and amplitudes of the transient peak, we introduced a variable parameter, a, that continuously variates the different equation parameters. The variation in the peak amplitude (Figure 7) is best ascribed to the different pumping of the system, therefore a variates the initial distribution of excited electrons among the three pools – where $a=1$ when the quinone pool is maximized. The location of the peak (Figure 8) depends on the reactions rate, namely the kinetic constants – where $a=1$ when the k_i are maximized. The combined effect of both, the pumping rate and the relaxation rate, best models the variations in temperature dependence of DF (Figure 9). In this case, the initial distribution of electrons among the three pools as well as the kinetic constants variate in parallel. Higher temperatures are modeled with higher values of the parameter a.

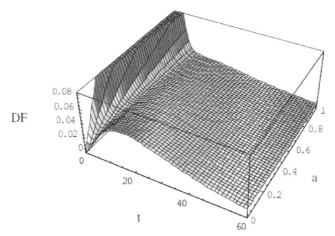

Fig. 7. Dependence of the transient peak amplitude on the initial pumping of the electron pools. DF is in relative units, t is in seconds, and a is the variable parameter, where $a=1$ when the quinone pool is maximized.

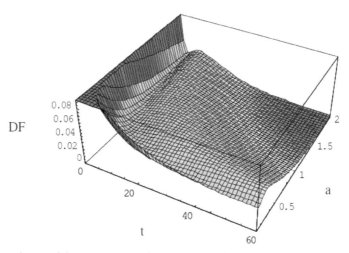

Fig. 8. Dependence of the transient peak position on the reaction rates. DF is in relative units, t is in seconds, and a is the variable parameter, where $a=1$ when the kinetic constants are maximized.

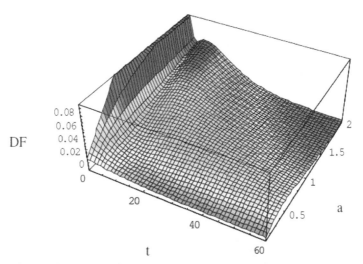

Fig. 9. Dependence of the transient peak on temperature. Both, peak position and amplitude change with temperature, with higher a standing for higher temperatures. DF is in relative units, t is in seconds.

4. Physiology

In delayed fluorescence, specific changes of physiological state are reflected in its intensity and kinetics. Delayed fluorescence intensity (DFI) is represented by an integral under the DF decay curve. In many cases, DFI can be utilized as a measure of living cell concentration (Berden-Zrimec et al., 2009). It also reflects the number of PSII centers, the fluorescence yield, and the rate of back reactions, which are influenced by the membrane potential and pH gradient (Avron & Schreiber, 1979; Joliot et al., 1971; Joliot & Joliot, 1980; Wraight & Crofts, 1971). In DF kinetics, the changes are most obvious when observing the position and intensity of the transient peak. The peak is culture-state dependent – the peak position and intensity change during culture growth (Berden-Zrimec et al., 2008a; Monti et al., 2005).

The results presented here were acquired by a 3 seconds long illumination and a sensor with a red-light-sensitive photomultiplier tube (Hamamatsu R1104) with a Hamamatsu C3866 Photon Counting Unit for signal conditioning and amplification (Monti et al., 2005; Zrimec et al., 2005).

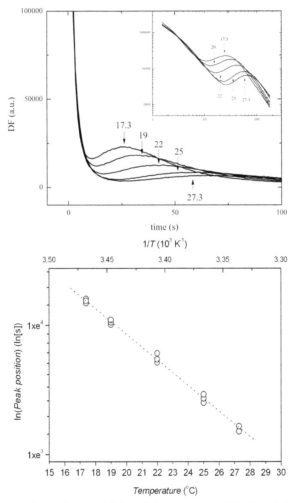

Fig. 10. Temperature dependence of delayed fluorescence. a) Delayed fluorescence decay kinetics at different temperatures in *Prorocentrum minimum* (Dinophyta); inset: log-log scale of DF decay kinetics, a.u. – arbitrary units; b) temperature dependence of the peak position: the line represents a linear fit.

4.1 Temperature and illumination intensity dependence

The temperature and illumination intensity strongly influence the transient peak position and the intensity. With increasing temperature, peak position is moving towards the beginning of the relaxation curve (Figure 10a) and average delayed fluorescence intensity increases until a maximum around 28 - 30°C (Wang et al., 2004; Yan et al., 2005; Zrimec et al., 2005). Due to changes in the kinetics, the temperature dependence of DFI is more complicated because it also depends on the time interval on which it is averaged.

The peak position has a typical temperature dependency for metabolic biochemical reactions (Zrimec et al., 2005). In the Arrhenius plot the natural logarithm of the peak position is linearly dependent on temperature (Figure 10b). Q10 of 2.6 and the activation energy of 71.5 kJ/mol calculated from the plot are in the expected range of plastoquinone-PSII reactions (Zrimec et al., 2005).

The illumination intensity profoundly influences only the peak intensity and less the peak position (Figure 11). Delayed emission can already be observed at relatively low illumination intensities. In the experiment with *Dunaliella tertiolecta* Butcher (Chlorophyta), DF was saturated already by a 3 s excitation pulse of 15 μmol m^{-2} s^{-1} PAR (Zrimec et al., 2005), which is even lower than obtained by Wang et al. (2004) for isolated spinach chloroplasts. At the excitation light intensity of 3.75 μmol m^{-2} s^{-1} PAR, DF showed slight differences in decay kinetics in the region of the transient peak, probably due to the changed oxidation state of the plastoquinone pool (Zrimec et al., 2005). DFI at 3.75 μmol m^{-2} s^{-1} PAR was only approximately 3% lower than at 15 and 60 μmol m^{-2} s^{-1} PAR (Zrimec et al., 2005). The maximal peak intensity was observed at the light intensity which was used for growing of batch cultures.

Dependence on light intensity can be explained by its influence on the ratio of the light captured by PSI and PSII: at low light intensity more light is absorbed by PSI, because its absorption spectrum has a maximum at longer wavelengths compared to the PSII. Thus cyclic electron transport producing only ATP is stronger than linear transport and an excess of ATP over NADPH is produced. The electron flow through PSI also causes oxidation of the plastoquinone pool. At higher light intensities the PSI get saturated and the ratio of the light absorbed by PSI compared to PSII is decreased.

4.2 Salinity

Changes in salinity influence photosynthesis in several ways. Increased salinity studies in the red alga *Porphyra perforata* showed there are at least three sites in the photosynthetic apparatus that are affected (Satoh et al., 1983). The first site, photoactivation and dark-inactivation of electron flow on the reducing side of PSI, was completely inhibited at high salinity. The second site, electron flow on the oxidizing side (water side) of PSII, was inhibited as was the re-oxidation of Q in the presence of 3-(3,4-dichlorophenyl)-1,1-dimethylurea. The third site affected by high salinity was the transfer of light energy probably from pigment system II to I. High salinity also reduced the amount of light energy that reached the reaction centers of PSII.

Photosynthetic activity was reduced by lowered salinity in two brown algae *Ascophyllum nodosum* and *Fucus versiculosus* (Connan & Stengel, 2011). Chlorophyll and phycobiliprotein concentrations were lower in changed salinity conditions in red alga *Gelidium coulteri* (Macler, 1988).

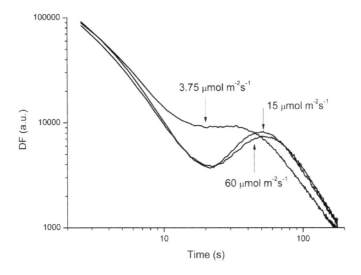

Fig. 11. Delayed fluorescence decay kinetics of *Dunaliella tertiolecta* after 3 s illumination with white light of different intensities.

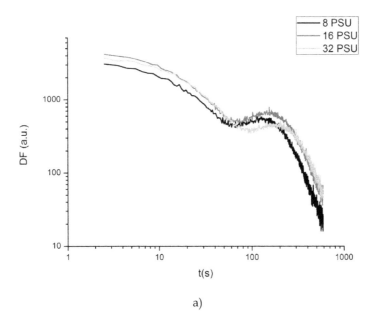

a)

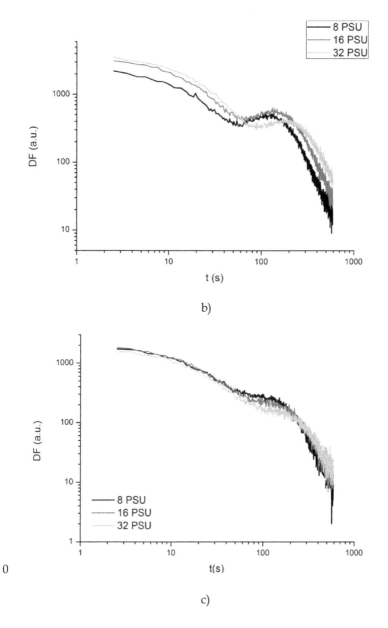

0

Fig. 12. Delayed fluorescence decay kinetics at different salinities in three strains of *Prorocentrum minimum*. a) strain from the Adriatic Sea, original salinity 32 PSU; b) strain from the Chesapeake Bay, USA, original salinity 16 PSU; c) strain from the Baltic Sea, original salinity 8 PSU. a.u. – arbitrary units.

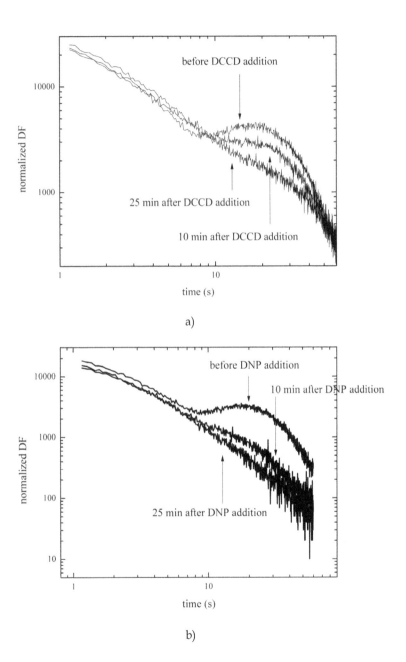

a)

b)

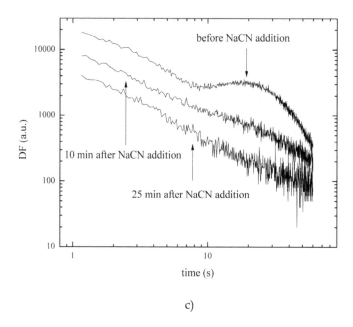

c)

Fig. 13. Influence of photosynthetic inhibitors on *P. minimum* DF decay kinetics: a) 15 µM dicyclohexylcarbodiimide (DCCD) – inhibition of ATP synthesis inhibition; b) 2 µM dinitrophenol (DNP) - destruction of proton gradient and inhibition of electron flow in thylacoid membrane; c) 30 mM NaCN - inhibition of respiration and photosynthesis.

In DF kinetics, higher salinity delayed the transient peak in three strains of *Prorocentrum minimum* (Figure 12). The strains were originally growing at different salinities, namely the Baltic strain (BAL) at 8 PSU, the Chesapeake Bay strain (D5) at 16 PSU, and the Adriatic strain (PmK) at 32 PSU (Monti et al., 2005). In our study, all strains were incubated in all three salinities for their whole growth period. BAL had in general earliest peak onset and PmK the latest in all salinities. The transient peak ceased at the end of the growth period, disappearing sooner at 32 PSU than in the lower salinities.

4.3 Influence of toxins

Toxic effects of photosynthesis inhibitors can be measured by DF already after a few minutes of incubation (Figure 13) (Berden-Zrimec et al., 2007). The transient peak can disappear soon after the addition of photosynthetic inhibitors like ATP synthesis inhibitior dicyclohexylcarbodiimide (DCCD) – (Figure 13a), dinitrophenol (DNP), which destroys the proton gradient and inhibits electron flow in thylacoid membranes (Figure 13b), or NaCN, an inhibitior of respiration and photosynthesis (Figure 13c) (Berden-Zrimec et al., 2010). These toxins influence photosynthesis in different ways, but at the end they affect the reduction state of the plastoquinone pool or its coupling with PSII by inhibiting the reversed electron flow and thus DF (Wang et al., 2004).

Delayed fluorescence response to toxins is dose-dependent (Figure 14), like in the case of herbicide diuron (DCMU, 3-(3,4-dichlorophenyl)-1,1-dimethylurea), which competes with plastoquinone and plastoquinol for the Q_B binding site, preventing the electron flow between PSII and the plastoquinone pool, or 3,5-dichlorophenol (3,5-DCP), which is used as an unspecific reference toxicant in toxicity tests (Berden-Zrimec et al., 2007). DNP locks the ATP-ase in open state, thus allowing H^+ ions to pass freely. The backreactions in PSII are enhanced by pH gradient in thylakoid membrane and there is a report that a permanent pH gradient in thylakoid membrane is present even in the dark (Joliot and Joliot 1980). The reason for the reduction of DF intensity in the region before peak formation is most likely the collapse of pH gradient caused by DNP. The disappearance of transient peak after 10 minutes of DNP action confirms the idea that pH gradient is directly or indirectly responsible for peak formation.

DF is a very good parameter in rapid toxicity tests (Berden-Zrimec et al., 2007; Katsumata et al., 2006). DFI was equally or more sensitive to the tested toxicants compared with the cell concentration and absorbance, which are standard parameters in algal growth inhibition tests (Berden-Zrimec et al., 2007). The advantage of the delayed fluorescence is that only living cells are measured, the sensitivity in toxicity tests thus being increased. Additionally, minimal disturbance to the cells, small sample volumes enabling homogenous illumination of all samples, and short test duration minimize a negative influence of changing physico-chemical properties of the medium on the results, being the most important features for the quality toxicity tests.

4.4 Nutrients

The nutrient status of algal cells markedly influences delayed fluorescence decay kinetics (Berden-Zrimec et al., 2008a; Burger & Schmidt, 1988; Mellvig & Tillberg, 1986). Phosphorus starvation can induce one or several transient peaks (Mellvig & Tillberg, 1986) or change the peak position (Berden-Zrimec et al., 2008a) (Figure 15). Nitrogen limitation also influences the peak position as well as causes the peak cessation (Figure 15) (Burger & Schmidt, 1988). DFI per cell significantly increases due to both, phosphorus and nitrogen limitation (Figure 16).

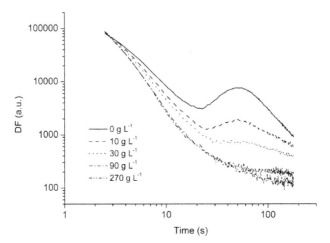

Fig. 14. The response of delayed fluorescence decay kinetics to geometrical series of diuron (DCMU) concentrations.

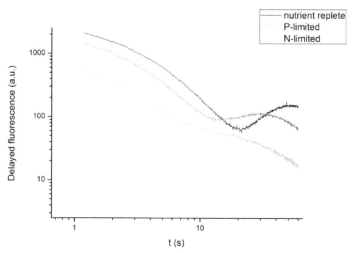

Fig. 15. Peak position dependence on nutrient status in *Dunaliella tertiolecta* cells. P – phosphorus, N – nitrogen, a.u. – arbitrary units.

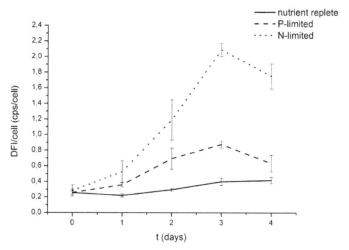

Fig. 16. DFI per cell dependence on nutrient status in *Dunaliella tertiolecta* cells. P – phosphorus, N – nitrogen, a.u. – arbitrary units.

Nitrogen and phosphorus starvation influence DF decay kinetics by changing the back reactions rates in the electron backflow. Nitrogen starvation causes reduction in the number of active PSII reactive centers and linear electron flow, but does not influence active PSI leading to relatively higher rates of cyclic photophosphorylation (Berges et al., 1996). Nitrogen starvation also influences thylakoid organization, facilitating trapped energy transfer from PSII to PSI (Berges et al., 1996). In marine phytoplankton, nitrogen limitation affects photosynthesis by reducing the efficiency of energy collection due to loss of chlorophyll a and increases non-photochemically active carotenoid pigments (Berges et al., 1996; Geider et al., 1998). Phosphate availability is connected to regulation of Calvin cycle activity or by interdependence of light and dark reactions via ATP/ADP, with consequent reductions in the efficiency of photosynthetic electron transfer (Geider et al., 1998).

DF decay kinetics is influenced differently by nitrogen and phosphorus starvation, making it a potential method of discriminating various nutrient conditions. Such a discriminating technique is still missing in the monitoring of oceanic phytoplankton population changes.

5. Conclusions

Delayed fluorescence has been used for researching photosynthesis since 1951. Nevertheless, not many publications can be found in the literature, mostly due to lack of commercially available measuring devices. Delayed fluorescence provides different information about photosynthesis as prompt fluorescence. It is emitted only from living cells therefore the problems with fluorescent background in field samples are omitted. The measurements can be successfully utilized in toxicity tests, biomass monitoring, primary productivity measurements and following changes in phytoplankton composition. Some more research, however, must be done on better understanding of the complex processes influencing the delayed fluorescence kinetics.

6. Acknowledgements

Our colleague Luka Drinovec has assembled the delayed fluorescence measuring apparatus and prepared the software. Alfred Beran has provided us with the algal cultures. Lidija Berden revised the language. The work was financed by the Slovenian Research Agency (grants #P1-0237, L4-6222, and V4-0106).

7. References

Arnold W., Davidson J.B. (1954). The identity of the fluorescent and delayed light emission spectra in *Chlorella*. *Journal of General Physiology*. Vol.37, pp. 677-684

Avron M., Schreiber U. (1979). Properties of ATP induced chlorophyll luminescence in chloroplasts. *Biochimica et Biophysica Acta*. Vol.546, pp. 448-454

Berden-Zrimec M., Drinovec L., Molinari I., Zrimec A., Fonda S., & Monti M. (2008a). Delayed fluorescence as a measure of nutrient limitation in *Dunaliella tertiolecta*. *Journal of Photochemistry and Photobiology B-Biology*. Vol.92, pp. 13-18

Berden-Zrimec M., Drinovec L., & Zrimec A. (2010). Delayed fluorescence, In: *Chlorophyll a fluorescence in aquatic sciences: methods and applications*, D.J. Sugget, O. Prasil& M. Borowitzka, (Eds.), 293-309, Springer,ISBN 978-90-481-9267-0, Dordrecht

Berden-Zrimec M., Drinovec L., Zrimec A., & Tisler T. (2007). Delayed fluorescence in algal growth inhibition tests. *Central European Journal of Biology*. Vol.2, pp. 169-181

Berden-Zrimec M., Flander V., Drinovec L., Zrimec A., & Monti M. (2008b). Growth, delayed fluorescence and pigment composition of four *Prorocentrum minimum* strains growing at two salinities. *Biological Research*. Vol.41, pp. 11-23

Berden-Zrimec M., Kozar-Logar J., Zrimec A., Drinovec L., Franko M., & Malej A. (2009). New approach in studies of microalgal cell lysis. *Central European Journal of Biology*. Vol.4, pp. 313-320

Berges J.A., Charlebois D.O., Mauzerall D.C., & Falkowski P.G. (1996). Differential effects of nitrogen limitation on photosynthetic efficiency of photosystems I and II in microalgae. *Plant Physiology.* Vol.110, pp. 689-696

Bertsch W.F. (1962). Two photoreactions in photosynthesis: evidence from delayed light emission of *Chlorella. Proceedings of the National Academy of Sciences of the United States of America.* Vol.48, pp. 2000-2004

Bertsch W.F., Azzi J.R. (1965). A relative maximum in the decay of long-term delayed light emission from the photosynthetic apparatus. *Biochimica et Biophysica Acta.* Vol.94, pp. 15-26

Bjorn L.O. (1971). Far-red induced, long-lived afterglow from photosynthetic cells - Size of afterglow unit and paths of energy accumulation and dissipation. *Photochemistry and Photobiology.* Vol.13, pp. 5-20

Burger J., Schmidt W. (1988). Long-term delayed luminescence - A possible fast and convenient assay for nutrition deficiencies and environmental pollution damages in plants. *Plant and Soil.* Vol.109, pp. 79-83

Connan S., Stengel D.B. (2011). Impacts of ambient salinity and copper on brown algae: 1. Interactive effects on photosynthesis, growth, and copper accumulation. *Aquatic Toxicology.* Vol.104, pp. 94-107

Desai T.S., Rane S.S., Tatake V.G., & Sane P.V. (1983). Identification of far-red-induced relative increase in the decay of delayed light emission from photosynthetic membranes with thermoluminescence peak V appearing at 321 K. *Biochimica et Biophysica Acta.* Vol.724, pp. 485-489

Drinovec L., Drobne D., Jerman I., & Zrimec A. (2004a). Delayed fluorescence of *Lemna minor*: a biomarker of the effects of copper, cadmium, and zinc. *Bulletin of Environmental Contamination and Toxicology.* Vol.72, pp. 896-902

Drinovec, L., Zrimec, A., & Berden-Zrimec, M.(2004b). Analysis and interpretation of delayed luminescence kinetics in the seconds range., 1961-238-333-2, Grosuplje, Slovenija,

Geider R.J., MacIntyre H.L., Graziano L.M., & Mckay R.M.L. (1998). Responses of the photosynthetic apparatus of *Dunaliella tertiolecta* (Chlorophyceae) to nitrogen and phosphorus limitation. *European Journal of Phycology.* Vol.33, pp. 315-332

Goltsev V., Zaharieva I., Lambrev P., Yordanov I., & Strasser R. (2003). Simultaneous analysis of prompt and delayed chlorophyll a fluorescence in leaves during the induction period of dark to light adaptation. *Journal of Theoretical Biology.* Vol.225, pp. 171-183

Greisberger S., Teubner K. (2007). Does pigment composition reflect phytoplankton community structure in differing temperature and light conditions in a deep alpine lake? An approach using HPLC and delayed fluorescence techniques. *Journal of Phycology.* Vol.43, pp. 1108-1119

Hakanson L., Malmaeus J.M., Bodemer U., & Gerhardt V. (2003). Coefficients of variation for chlorophyll, green algae, diatoms, cryptophytes and blue-greens in rivers as a basis for predictive modelling and aquatic management. *Ecological Modelling.* Vol.169, pp. 179-196

Hideg E., Kobayashi M., & Inaba H. (1991). The far red induced slow component of delayed light from chloroplasts is emitted from photosystem II - Evidence from emission spectroscopy. *Photosynthesis Research.* Vol.29, pp. 107-112

Istvanovics V., Honti M., Osztoics A., Shafik H.M., Padisak J., Yacobi Y., & Eckert W. (2005). Continuous monitoring of phytoplankton dynamics in Lake Balaton (Hungary) using on-line delayed fluorescence excitation spectroscopy. *Freshwater Biology.* Vol.50, pp. 1950-1970

Joliot P., Joliot A. (1980). Dependence of delayed luminescence upon adenosine triphosphatase activity in *Chlorella. Plant Physiology.* Vol.65, pp. 691-696

Joliot P., Joliot A., Bouges B., & Barbieri G. (1971). Studies of system-II photocenters by comparative measurements of luminescence, fluorescence, and oxygen emission. *Photochemistry and Photobiology.* Vol.14, pp. 287-305

Jursinic P.A. (1986). Delayed fluorescence: Current concepts and status, In: *Light emission by plant and bacteria,* Govindjee& D.C. Fork, (Eds.), 291-328, Academic Press, New York

Katsumata M., Koike T., Nishikawa M., Kazumura K., & Tsuchiya H. (2006). Rapid ecotoxicological bioassay using delayed fluorescence in the green alga *Pseudokirchneriella subcapitata. Water Research.* Vol.40, pp. 3393-3400

Krause H., Gerhardt V. (1984). Application of delayed fluorescence of phytoplankton in limnology and oceanography. *Journal of Luminescence.* Vol.31-2, pp. 888-891

Kurzbaum E., Eckert W., & Yacobi Y. (2007). Delayed fluorescence as a direct indicator of diurnal variation in quantum and radiant energy utilization efficiencies of phytoplankton. *Photosynthetica.* Vol.45, pp. 562-567

Lavorel J., Dennery J.M. (1982). The slow component of photosystem II luminescence - A process with distributed rate-constant. *Biochimica et Biophysica Acta.* Vol.680, pp. 281-289

Li Q., Xing D., Ha L., & Wang J.S. (2007). Mechanism study on the origin of delayed fluorescence by an analytic modeling of the electronic reflux for photosynthetic electron transport chain. *Journal of Photochemistry and Photobiology B-Biology.* Vol.87, pp. 183-190

Macler B.A. (1988). Salinity effects on photosynthesis, carbon allocation, and nitrogen assimilation in the red alga, *Gelidium coulteri. Plant Physiology.* Vol.88, pp. 690-694

Melcarek P.K., Brown G.N. (1977). Effects of chill stress on prompt and delayed chlorophyll fluorescence from leaves. *Plant Physiology.* Vol.60, pp. 822-825

Mellvig S., Tillberg J.E. (1986). Transient peaks in the delayed luminescence from *Scenedesmus obtusiusculus* induced by phosphorus starvation and carbon dioxide deficiency. *Physiologia Plantarum.* Vol.68, pp. 180-188

Monti M., Zrimec A., Beran A., Berden Zrimec M., Drinovec L., Kosi G., & Tamberlich F. (2005). Delayed luminescence of *Prorocentrum minimum* under controlled conditions. *Harmful Algae.* Vol.4, pp. 643-650

Satoh K., Smith C.M., & Fork D.C. (1983). Effects of salinity on primary processes of photosynthesis in the red alga *Porphyra perforata. Plant Physiology.* Vol.73, pp. 643-647

Schmidt W., Schneckenburger H. (1992). Time-resolving luminescence techniques for possible detection of forest decline .1. Long-term delayed luminescence. *Radiation and Environmental Biophysics.* Vol.31, pp. 63-72

Schneckenburger H., Schmidt W. (1996). Time-resolved chlorophyll fluorescence of spruce needles after different light exposure. *Journal of Plant Physiology.* Vol.148, pp. 593-598

Scordino A., Grasso F., Musumeci F., & Triglia A. (1993). Physical aspects of delayed luminescence in *Acetabularia acetabulum*. *Experientia*. Vol.49, pp. 702-705

Scordino A., Triglia A., Musumeci F., Grasso F., & Rajfur Z. (1996). Influence of the presence of atrazine in water on the in-vivo delayed luminescence of *Acetabularia acetabulum*. *Journal of Photochemistry and Photobiology B: Biology*. Vol.32, pp. 11-17

Strehler B.L., Arnold W. (1951). Light production by green plants. *Journal of General Physiology*. Vol.34, pp. 809-820

Van Wijk R., Scordino A., Triglia A., & Musumeci F. (1999). 'Simultaneous' measurements of delayed luminescence and chloroplast organization in *Acetabularia acetabulum*. *Journal of Photochemistry and Photobiology B-Biology*. Vol.49, pp. 142-149

Wang C., Xing D., & Chen Q. (2004). A novel method for measuring photosynthesis using delayed fluorescence of chloroplast. *Biosens Bioelectron*. Vol.20, pp. 454-459

Wraight C.A., Crofts A.R. (1971). Delayed fluorescence and the high-energy state of chloroplasts. *Eur J Biochem*. Vol.19, pp. 386-397

Yacobi Y.Z., Gerhardt V., Gonen-Zurgil Y., & Sukenik A. (1998). Delayed fluorescence excitation spectroscopy: A rapid method for qualitative and quantitative assessment of natural population of phytoplankton. *Water Research*. Vol.32, pp. 2577-2582

Yan Y., Popp F.A., Sigrist S., Schlesinger D., Dolf A., Yan Z.C., Cohen S., & Chotia A. (2005). Further analysis of delayed luminescence of plants. *Journal of Photochemistry and Photobiology B-Biology*. Vol.78, pp. 235-244

Zrimec A., Drinovec L., & Berden-Zimec M. (2005). Influence of chemical and physical factors on long-term delayed fluorescence in *Dunaliella tertiolecta*. *Electromagnetic Biology and Medicine*. Vol.24, pp. 309-318

8

Photosynthesis in Lichen:
Light Reactions and Protective Mechanisms

Francisco Gasulla[2], Joaquín Herrero[1], Alberto Esteban-Carrasco[1],
Alfonso Ros-Barceló[3], Eva Barreno[2],
José Miguel Zapata[1] and Alfredo Guéra[1]

[1]University of Alcalá
[2]University of Valencia
[3]University of Murcia
Spain

1. Introduction

Lichens are symbiotic associations (holobionts) established between fungi (mycobionts) and certain groups of cyanobacteria or unicellular green algae (photobionts). This symbiotic association has been essential in establishing the colonization of terrestrial and consequently dry habitats. About 44 genera of algae and cyanobacteria have been reported as lichen photobionts. Due to the uncertain taxonomy of many of these photobionts, these numbers were considered as approximations only. Ahmadjian (1993) estimates that only 25 genera were typical lichen photobionts. The most common cyanobionts are *Nostoc*, *Scytonema*, *Stigonema*, *Gloeocapsa*, and *Calothrix*, in order of frequency (Büdel, 1992). Green algal photobionts include *Asterochloris*, *Trebouxia*, *Trentepohlia*, *Coccomyxa*, and *Dictyochloropsis* (Gärtner, 1992). These authors assessed that more than 50% of all lichen species are associated with *Trebouxia* and *Asterochloris* species. However, this is just estimation since in only 2% of the described lichen species the photobiont genus is reported (Tschermak-Woess, 1989), mostly by the difficulties to isolate and then characterize the algae from the lichen thalli. Lichens are well known for their slow growth and longevity. Their radial growth is measured in millimetres per year (Hale, 1973), while individual lichens live for hundreds or even thousands of years. It is assumed that in lichens the photobiont population is under mycobiont control. Lichenologists have proposed some control mechanisms such as, cell division inhibitors (Honegger, 1987), phytohormones (Backor & Hudak, 1999) or nutrients competition (Crittenden et al., 1994; Schofield et al., 2003).

Similar to plants, all lichens photosynthesise. They need light to provide energy to make their own matter. More specifically, the algae in the lichen produce carbohydrates and the fungi take those carbohydrates to grow and reproduce. The amount of light intensity needed for optimal lichen growth varies widely among species. The optimum light intensity range of most algal photobionts in axenic cultures is very low, between 16-27 $\mu mol\ m^{-2}\ s^{-1}$. If the response of cultured photobionts to light is similar to that of the natural forms (lichen), then there must be additional mechanisms protecting the algae in the lichen that are not developed under culture conditions. Pigments and crystal of secondary metabolites in the upper cortex are supposed to decrease the intensity of light reaching the photobionts especially under desiccated conditions by absorbing certain wavelengths and by reflecting light (Heber et al,. 2007; Scott, 1969; Veerman et al., 2007). Apparently, the balance between energy conservation and energy dissipation is tilted towards dissipation in many poikilohydric autotrophs, whereas, in higher plants, energy conservation assumes dominance over energy dissipation. It thus appears that sensitivity to excess light is higher in the mosses and the lichens than in higher plants (Heber, 2008).

Lichens are found among poikilohydric organisms, those that cannot actively regulate their water content, but are capable of surviving long periods in a desiccated state (Kappen & Valladares, 2007). In the dry state many lichens exhibit an enhanced resistence to other stress. For instance, heat resistance up to 70-75 °C in species from sheltered microhabitats and up to 90-100 °C in species from exposed microhabitats (Lange, 1953). Desiccation tolerance was described in nematodes and in rotifers observed by van Leeuwenhoek in 1702, and has since been discovered in four other phyla of animals, algae, fungi, bacteria, in ca. 350 species of flowering plants and ferns and in most bryophytes and seeds of flowering plants (Alpert, 2006; Proctor & Tuba, 2002). Among them, algae, lichen and bryophytes can be considered fully desiccation-tolerant plants because can survive very rapid drying events (less than 1 h) and recover respiration and photosynthesis within a few minutes (Oliver & Wood, 1997; Proctor & Smirnoff, 2000). Most lichen-forming fungi and their photobionts are fully adapted to daily wetting and drying cycles, but die off under continuously moist conditions (Dietz & Hartung, 1999; Farrar, 1976a, 1976b). It is well known that photosynthesis in homoiohydric plants is very sensitive to water stress conditions (Heber et al., 2001), especially under high irradiance. Under these conditions, reactive oxygen species (ROS) generation associated to photosynthetic electron transport is enhanced. The question arises of how lichen algae can maintain the function of their photosynthetic machinery under continuous desiccation-rehydration processes. We will review in this chapter the possible mechanisms which should allow maintaining of photosynthesis performance under the life style of poikilohydric organisms.

2. Methods for isolating lichen photobionts

One of the main problems to study the mechanisms of photosynthesis in lichens under well-controlled conditions is to develop an appropriate method for isolating the lichen photobionts. Many chlorolichens contains more than one photobiont. For instance, *Ramalina farinacea* includes two different *Trebouxia* photobionts (TR1 and TR9) and isolation of these algae allowed to characterise physiological differences between both of them (Casano et al., 2010; del Hoyo et al., 2011). There are different methods in function of the objective of investigation. We can distinguish between those which allow and not allow obtaining axenic cultures.

Axenic cultures are useful to study the taxonomy, biochemical, molecular or physiological behaviour of microscopic algae outside the symbiosis. There are lots of methods, but the most popular isolation method was developed by Ahmadjian (1967a, 1967b) and consists of cutting the lichen photobiont layer into thin slices, then grinding it between two glass slides and finally spreading the homogenate on a solid agar medium. There are several variations to this method, but the main common problem to all of them is the long time required after isolation to obtain clones.

On the other hand, non-axenic cultures can be used for studying algal metabolites or enzymatic activities of the lichenized photobionts. These methods consist in homogenization of the lichen thalli, followed by separation of the photobiont from the mycobiont and fragments of thalli using differential centrifugation (Richardson, 1971), gradient centrifugation on $CsCl_2$/KI (Ascaso, 1980) or on Percoll® gradients (Calatayud et al., 2001), and/or filtering (Weissman et al., 2005).

Here we resume the fast and simple methods developed in our laboratories (Gasulla et al., 2010): a low-scale isolation method (micromethod) and a large scale one (macromethod).

The micromethod for isolation of lichen photobions starts from 15–25 mg dry weight (DW) of lichen material that is washed first in tap water, and then slowly stirred in sterile distilled water in a bucket for 5 min. The fragments of thalli are homogenised in an sterile eppendorf tube with a pellet pestle and resuspended in sterile 1 ml of isotonic buffer (0.3 M sorbitol in 50 mM HEPES pH 7.5). After filtration through sterile muslin, the filtrate is centrifuged at 490×g in a bench-top microcentrifuge (Micro 20, Hettich, Germany) for 5 min. The pellet is resuspended in 200 µl of sterile isotonic buffer and then loaded on 1.5 ml, of sterile 80 % Percoll ® in isotonic buffer. After centrifugation at 10000×g for 10 min a clear green layer must be present near the top of the eppendorf tube and some grey particles and pellet at the bottom of the tube (Fig. 1). The green layer is recovered (ca. 400 µl), avoiding to take any drop of the upper interphase. Then, the green layer is diluted 2-fold with sterile distilled water and centrifuged at 1000×g for 10 min. The supernatant is discarded; the pellet is resuspended in 2 ml of sterile distilled water and a drop of Tween 20 is added. The resulting suspension is sonicated at 40 KHz (Elma Transsonic Digital 470 T, 140% ultrasound power) for 1 min and again centrifuged at 490×g for 5 min. This treatment is repeated five times. The final pellet containing the isolated algal cells is resuspended in 1 ml of sterile distilled water. This micromethod can be scaled up to a macromethod, which allows preparation of large amounts of photobiont cells.

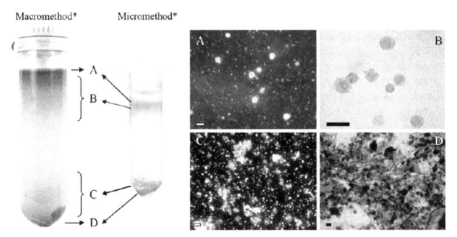

Fig. 1. Separation of *Ramalina farinacea* fractions after centrifugation of the extract of thalli at 10,000×g for 10 min on 40 ml (macromethod), or 1.5 ml (micromethod) of 80 % Percoll ® in isotonic buffer. A–D, each optical micrograph refers to the corresponding Percoll layer; A and C phase contrast microscopy. Scale = 15 µm. * Not real size. Photograph from Gasulla et al., (2010).

In the macromethod, one to two g DW of lichen thallus are homogenised with a mortar and pestle in 20 ml of sterile isotonic buffer. The steps following are similar to the micromethod, but the volume for resuspension of the first pellet is 1 ml. The second centrifugation step is carried out on 40 ml of sterile 80% Percoll in isotonic buffer using a fixed-angle rotor (221.22 V01/V02, Hermle, Germany). After this centrifugation step in the macromethod, four layers are visible: a) a 2–3 ml dark green supernatant at the top of the tube on the Percoll layer; b) a large and diffuse light green layer in the upper part of the Percoll gradient; c) a thick layer at the bottom of the tube and d) a grey pellet (Fig. 1). Five millilitres of the "b" layer are recovered and the subsequent isolation steps for this layer are identical to those described for the micromethod.

The algal suspensions isolated with any one of both methods, are diluted 100 folds with sterile distilled water and 50 µl of this suspension are spread on sterile 1.5% agar 3xN (meaning three times more nitrogen content in the form of $NaNO_3$) Bold's Basal Media (3NBBM) (Bischoff & Bold, 1963) in each of five Petri dishes using the streak method and sterile technique. The isolated algae are cultured under 15 µmol $m^{-2}s^{-1}$ (PPFD) with a 12 h photoperiod at 17°C. The number of algal colonies growing on each plate is counted after 45 days. Several colonies must be selected under the stereo-microscope and subcultured onto Petri dishes containing 1.5% agar 3NBBM medium supplemented with glucose (20 gl^{-1}) and casein (10 gl^{-1}) (Ahmadjian, 1967a) using a sterile toothpick.

3. Effects of water content on the carbon budget of lichens

The thallus water content is mainly determined by the water availability of the environment. When desiccated, their water status is frequently in the range of 10-20 % in respect to their fresh weight (Rundel, 1988). This state would be lethal for most of the vascular plants and organisms, however the vast majority of lichens are desiccation-tolerant and can survive in a suspended animation until water becomes available again, then they revive and resume normal metabolism (Kappen & Valladares, 2007). Upon rehydration they recover normal photosynthetic rates within a short time span, 15-60 min or less (Fos et al., 1999; Jensen et al., 1999; Tuba et al., 1996). Therefore, lichens may be the predominant life-form in extreme environments like cold and hot deserts. In lichens, photosynthetic activity of the photobiont partner is restricted to a short time when thalli are at least partly hydrated and solar radiation is available at temperatures within the range suitable for photosynthesis. Frequent drying and wetting cycles and the correlated in- and re-activation of photosynthesis is a pattern observed in most terrestrial habitats and produced by the nocturnal dewfall or fog (del Prado & Sancho, 2007; Kershaw, 1985; Lange, 1970; Lange et al., 2006). Typically, dewfall occurred in the night when temperatures had declined substantially from their daytime maximum value. Lichens readily absorb water from dewfall, and this water activates dark respiration (CO_2 exchange below the zero line) through the remaining night time hours. Sunrise activates net photosynthesis (CO_2 exchange above the zero line) but the peak was not reached after 1-2 h when the water content started to decrease. The net photosynthesis rate of lichens depends in large part on the water content of their thalli (Green et al., 1985; Lange & Matthes, 1981). In many lichens, when the thallus is fully saturated with water, diffusion of CO_2 to the phycobiont is hindered and maximum rates of CO_2 assimilation do not occur (Lange & Tenhunen, 1981). Furthermore, at maximum water saturation in continuous light, the photobiont eventually dies because all of its products are translocated to the fungus (Harris & Kershaw, 1971). It is only when the thallus dries to a 65-

90 % of the maximum water content that peak photosynthesis occurs. Thereafter, with increasing temperatures and light intensities, both water content and net photosynthesis decline. Desiccation occurs reasonably slowly, over hours rather than minutes (Kappen, 1974). Lichen photobionts are able to maintain maximum photosynthetic activity until the water content reach the 20 % (Gasulla et al., 2009), thus, during this period lichens photosynthetize at rates that are sufficient to allow a net positive carbon gain over the year. Thalline growth rates depend on the frequency and length of this period per day and year (Lange & Matthes, 1981). On the other hand, although carbon fixation is inhibited during desiccation, electron flow through the photosystems continues, and excitation energy can be transferred from photo-excited chlorophyll pigments to $^3O_2$, forming singlet oxygen ($^1O_2$), while superoxide and hydrogen superoxide can be produced at photosystem II and photosystem I by the Mehler reaction (Halliwell, 2006; Kranner & Lutzoni, 1999; Peltier et al., 2010). Likewise, rehydration of lichens produces a burst of ROS during the first minutes and then decrease (Minibayeva & Beckett, 2001; Weissman et al., 2005). Thus, although lichens have adapted their carbon assimilation necessities to their living conditions, they will need specific mechanisms to avoid the development of oxidative damage during the desiccation and rehydration processes. We can follow two levels of protection mechanisms at the photobiont cellular level. First, processes directed to the dissipation of excess light energy as heat, which can be considered as oxidative stress avoidance mechanisms. Second, enzymatic or non-enzymatic antioxidant systems that can constitute oxidative stress tolerance mechanisms.

4. Production of reactive oxygen species (ROS) and reactive nitrogen species (RNS) during desiccation/rehydration

Aerobic organisms generate ROS as a side-product product of metabolism. In healthy cells occurs at a controlled rate, but many abiotic and biotic stress conditions lead to cellular redox imbalance and accumulation of ROS (Foyer & Noctor, 2003; Halliwell & Gutteridge, 1999; Mittler, 2002; Sharma & Dietz, 2009; Smirnoff, 1993) that causes molecular and cellular damage. Free radicals are atoms or molecules with an unpaired electron, which is easily donated, thus, most free radicals are very reactive (Elstner & Osswald, 1994; Halliwell & Gutteridge, 1999). Oxygen is a highly oxidizing molecule that forms free radicals and participates in other oxidative chemical reactions (Abele, 2002; Finkel & Holbrook, 2000). Oxygen radicals include singlet oxygen ($^1O_2$), superoxide ($O_2^{•-}$), the hydroxyl radical ($OH^•$) (Elstner & Osswald, 1994; Finkel & Holbrook, 2000; Halliwell & Gutteridge, 1999). Together with hydrogen peroxide (H_2O_2) that is not a free radical but is also highly reactive. ROS accumulation is the most likely source of damage to nucleic acids, proteins and lipids that can, as a final result, conduct to cell death (Zapata et al., 2005).

Every free radical formed in a living organism can initiate a series of chain reactions that will continue until they are eliminated (Halliwell, 2006). Free radicals disappear from the organism only by reactions with other free radicals or, more important, due to the actions of the antioxidant system that will be treated in section 6 of this review. Any imbalance in the redox state, which altered equilibrium in the direction of pro-oxidant molecules production, may result in univalent reduction of molecular oxygen to the potentially dangerous radical anion superoxide (Foyer & Noctor, 2003). Its formation is an unavoidable consequence of aerobic respiration (Møller, 2001) and photosynthesis (Halliwell, 2006), but imbalances may also occur under changing environmental conditions such as desiccation-rehydration cycles experienced by lichens, causing oxidative damage in cells of the photobiont and mycobiont.

Several reactions in chloroplast and mithochondria generate the free superoxide radical ($O_2{}^{\bullet-}$) which can in turn react with hydrogen peroxide (H_2O_2) to produce singlet oxygen ($^1O_2$) and the hydroxyl radical ($OH^\bullet$) (Casano et al., 1999; Elstner, 1982). Moreover, in the presence of Fe or Cu (II), $OH^\bullet$ radicals are formed as quickly that can attack and damage almost every molecule found in living cells as lipids, amino acids and even nucleic acids by direct attack or activation of endonucleases (Kranner & Birtic, 2005; Yruela et al., 1996). They can, for example, hydroxylate purine and pyrimidine bases in DNA (Aruoma et al., 1989), thus enhancing mutation rates. Oxidative damage to proteins changes their configuration, mostly by oxidizing the free thiol residues of cysteine to produce thiyl radicals. These can form disulphide bonds with other thiyl radicals, causing intra- or inter-molecular cross-links. After oxidative modification, proteins become sensitive to proteolysis and/or may be inactivated, or may show reduced activity (Kranner & Birtic, 2005). Singlet oxygen and $OH^\bullet$ can also initiate peroxidation chain reactions in lipids. Lipid peroxides decompose to give volatile hydrocarbons and aldehydes (Esterbauer et al., 1991; Valenzuela, 1991). Accumulation of malondialdehyde, an indicator of lipid peroxidation, has been observed in lichens during dehydration (Kranner & Lutzoni, 1999). The latter can act as secondary toxic messengers that disseminate initial free radical events (Esterbauer et al., 1991). On the other hand, hydrogen peroxide is a ubiquitous constituent of plant cells under a fine homeostatic control which prevents its accumulation (Foyer & Noctor, 2003; Ros Barceló, 1998).

In most organisms, desiccation is associated with production of ROS and associated deleterious effects (Weissman et al., 2005). ROS may modify the properties of the thylakoids, thereby changing the yield of Chl a fluorescence, leading to photoinhibition (Ort, 2001). Under normal growth conditions, the production of ROS in cells is low (240 μM s^{-1} $O_2{}^-$ and a steady-state level of 0.5 μM H_2O_2 in chloroplasts, Polle, 2001). Many stresses such as drought stress and desiccation disrupt the cellular homeostasis of cells and enhance the production of ROS (240–720 μM s^{-1} $O_2{}^-$ and a steady-state level of 5–15 μM H_2O_2, Polle, 2001). The production of ROS during desiccation results from pathways such as photorespiration, from the photosynthetic apparatus and from mitochondrial respiration (Mittler et al., 2002). Whether lichens have photorespiration process is not clear and likely it depends on the species (Ahmadjian, 1993). We hypothesize (from genomic data), that probably, in algae, the glycolate oxidase reaction in peroxisomes is prevented by the presence of the alternative enzyme glycolate dehydrogenase, which does not produce hydrogen peroxide.

As we have said, lichens can tolerate dehydration, but what happen in the chloroplast when water is unavailable as reductant? When desiccated in the light, chlorophyll molecules continue to be excited, but the energy not used in carbon fixation will cause formation of singlet oxygen (Kranner et al., 2005). In these cases, oxygen can be the electron acceptor forming superoxide and here can begin the generation of ROS (Heber et al., 2001). Desiccation stress in lichens shows features similar to reversible photoinhibition (Chakir & Jensen, 1999; Jensen & Feige, 1991) because ROS production is enhanced during dessication process (Weissman et al., 2005). In the same manner, excess of illumination causes either excessive reduction on the acceptor side or oxidation on the donor side (Anderson & Barber, 1996) producing ROS in the thylakoids. Therefore, under both high illumination and dehydration, reactive oxygen species are a major cause of damage in photosynthetic organisms (Demmig-Adams & Adams, 2000; Halliwell, 1984).

On the other hand, less studied in algae and lichen but potentially important is Nitric Oxide (NO). NO is a relatively stable paramagnetic free radical molecule involved in many physiological processes in a very broad range of organisms. These functions include signal transduction, cell death, transport, basic metabolism, ROS production and degradation (Almagro et al., 2009; Curtois et al., 2008; Ferrer & Ros Barcelo, 1999; Palmieri et al., 2008). It is now clear that $NO^{\bullet}$ and, in general, most of the Reactive Nitrogen Species (RNS) ($NO^{\bullet}$, NO^+, NO^-, $NO^{\bullet}_2$, and $ONOO^-$), are major signalling molecules in plants (Durner & Klessig, 1999) which can be synthesized during stress responses at the same time as H_2O_2 (Almagro et al., 2009). Feelisch & Martin (1995) suggested a role for NO in both the early evolution of aerobic cells and in symbiotic relationships involving NO efficacy in neutralizing ROS. In addition, NO is involved in the abiotic stress response of green algae such as *Chlorella pyrenoidosa* Pringsheim, by reducing the damage produced by photo-oxidative stress (Chen et al., 2003).

The first work that focused on NO production in lichens was published, by Weissman et al. (2005), who carried out a microscopy study of *Ramalina lacera* (With.) J.R. Laundon. These authors described the occurrence of intracellular oxidative stress during rehydration together with the release of NO by the mycobiont, but not by the photobiont. We have recently reported evidence that NO is involved in oxidative stress in lichens exposed to the oxidative agent cumene hydroperoxide (Catalá et al., 2010) Our group has studied the role of NO during rehydration of the lichen *Ramalina farinacea* (L.) Ach., its isolated photobiont partner *Trebouxia sp.* and *Asterochloris erici* (formerly *Trebouxia erici*). The results showed that lichen NO plays an important role in the regulation of lipid peroxidation and photobiont photo-oxidative stress during rehydration. Its role is similar in plants and animals where NO is known to modulate the toxic potential of ROS and to limit lipid peroxidation, acting as a chain-breaking antioxidant to scavenge peroxyl radicals (Darley-Usmar et al., 2000; Kroncke et al., 1997; Miranda et al., 2000). Our data showed that rehydration is accompanied by ROS and NO generation and thus confirmed the results of Weissman et al. (2005). Moreover, the inhibition of NO action altered the photosynthetic activity of the photobionts suggesting that NO is involved in PSII stabilization and could be related with the limited role of classical antioxidant systems during desiccation-rehydration cycles in *Asterochloris* photobionts. These results point to the importance of NO in the early stages of lichen rehydration. However, there are needed further studies on NO function, or on the occurrence of NO in other lichens.

5. Mechanisms of light energy dissipation in lichen algae

Plants have developed mechanisms to prevent the formation of ROS by dissipating the excess of energy as heat. This phenomenon comprises several processes, which applying a nomenclature derived from Chl *a* fluorescence theory are globally known as non-photochemical quenching (denoted as NPQ or q_N). The light energy when reach the photosystems is transformed in heat, fluorescence or photochemistry. These processes compete among them and then make possible estimate the proportion of light energy employed in photosynthetic electron transport or dissipated as heat from direct measurement of the variation of fluorescence emission. The electron transfer from the reaction center chlorophyll of PSII (P680) to the primary quinone acceptor of PSII (Q_A) produces losses in fluorescence emission in a process known as photochemical quenching. Photochemical quenching (q_P) reaches maximal values when all the quinone pool is oxidized and electron transport is not impaired (open reaction centers) and minimal values if all the quinone pool is reduced (closed reaction centers). The value of q_p usually maintains a non-linear relationship with the actual proportion of reduced quinone (Kramer et al., 2004). The rate of heat loss is correlated with the non-photochemical quenching of fluorescence; this is, with the loss of fluorescence emission that is independent of photochemical events and results in heat dissipation of light energy (Baker, 2008). In consequence, the maximum fluorescence emission (termed F_m) will be registered when all reaction centers are closed (what is obtained by a saturating flash of light) and mechanisms of thermal energy dissipation are inactivated (what is assumed to happen within a dark period of about 20 min for most vascular plants). After new exposure to light, the mechanisms of heat dissipation become active and after a new saturating flash the value of the fluorescence peak (F_m') will be lower. Hence, the expression NPQ= (F_m/F_m')-1 (Bilger & Björkman, 1990) describes the extent of energy dissipation. When a leaf is kept in the dark, Q_A becomes maximally oxidized and the heat dissipation mechanisms are relaxed. Exposure of a dark-adapted leaf to a weak modulated measuring beam (photosynthetically active photon flux density -PPFD- of $ca.0.1$ μmol m^{-2} s^{-1}) results in the minimal level of fluorescence, called F_o (Baker, 2008). If taken into account the maximal variable fluorescence after dark adaptation (F_m-F_o) and after a period in the light $(F_m'-F_o')$, an alternative expression (Schreiber et al., 1986) for non-photochemical quenching can be calculated like q_N= 1-$(F_m'-F_o')/(F_m-F_o)$.

According to Krause & Weis (1991), non-photochemical quenching can be related to three different events: high trans-thylakoidal pH gradient (q_E), to state I-state II transitions (q_T) and to photosystem II photoinhibition (q_I). The "energy-dependent" quenching (q_E), originated by the formation of a proton gradient across the thylakoidal membranes, is the main mechanism implied in the development of NPQ (Krause & Weis, 1991; Müller et al., 2001). This q_E is characterized by its rapid relaxation kinetics, occurring within 3 min of darkness (Li et al., 2002; Munekage et al., 2002). Under excessive light, an elevated proton concentration in the thylakoid luminal space activates violaxanthin de-epoxidase, generating antheraxanthin and then zeaxanthin. The increase of violaxanthin deepoxidation (DPS) leads to an increase of thermal energy dissipation that is correlated with the NPQ parameter of chlorophyll fluorescence (Demmig-Adams et al., 1996). The generation of NPQ requires also the participation of a protein associated to PSII, the protein Psbs of the light harvesting complex antenna, which can change its conformation at lower pH and increase its affinity for zeaxanthin. De-epoxidated xanthophylls then binds to the subunit S where could accept the excitation energy transferred from chlorophyll and thereby act as a direct quencher in NPQ (Holt et al., 2004; Krause & Jahns 2004; Niyogi et al., 2004; Ruban et al., 2002; Spinall-O'Dea et al., 2002). An alternative explanation is that PsbS alone can cause the quenching and zeaxanthin acts as an allosteric activator, but not as the primary cause of the process (Crouchman et al., 2006; Horton et al. 2005).

The transthylakoidal proton gradient is generated by water splitting in the thylakoid lumen, but also by the Q cycle of quinones around cytochrome b_6/f complex. The latter allows that the proton gradient and NPQ generation (like ATP production) can be maintained by cyclic electron flow and probably by a thylakoidal NADH dehydrogenase complex (Guéra et al., 2005). In some mosses and chlorolichens has been described the possibility of the activation of energy dissipation by CO_2 dependent protonation. Carbon dioxide, which, at a pK of 6.31, acts as a very weak protonating agent, is capable of promoting NPQ in a light independent way, provided some zeaxanthin is present (Bukhov et al., 2001; Heber et al., 2000; Heber, 2008; Kopecky et al.,2005).

In vascular plants NPQ changes in response to diurnal variations in the light environment. Large changes in the composition of the xanthophyll cycle are observed over the course of the day. This consisted of increases and subsequent decreases in the zeaxanthin and antheraxanthin content of the leaves that paralleled the changes in incident PFD rather closely (Adams & Demmig-Adams, 1992). There are also acclimations to the plant light regime. Sun plants or sun acclimated leaves possess a higher capacity for the use of light in photosynthesis and also for rapid increases in xanthophyll cycle-dependent energy dissipation. Sun-grown leaves typically exhibit a larger total pool size of the xanthophyll cycle components as well as a greater ability to convert this pool to antheraxanthin and zeaxanthin rapidly under high light (Demmig-Adams & Adams, 1993, 1996). Finally, seasonal changes in NPQ have been described for several evergreen species (Adams et al., 2001; Zarter et al., 2006a, 2006b).

Most green algal photobiont cultures require low light intensities of 10-30 μmol m^{-2} s^{-1} (Ahmadjian, 1967a; Friedl & Büdel, 2008). One reason which could justify the necessity of low intensities for lichen algae culture is that compared to surficial light measurements, the light reaching the photobiont is reduced by 54-79% when dry, and 24-54% when fully hydrated (Büdel & Lange, 1994; Dietz et al., 2000; Ertl, 1951; Green et al., 2008). During desiccation a decrease in fluorescence emission is observed in lichens (Veerman et al., 2007). This decrease is likely associated with their phototolerance and could be caused by structural changes in the thallus that induce changes in light-scattering and shading properties or by changes in shape and aggregation of algae (de los Rios et al., 2007; Scheidegger et al., 1995; Veerman et al., 2007). The thallus also offers some protection against photodamage through the use of light-absorbing pigments (Gauslaa & Solhaug, 1999; Holder et al., 2000). All this kind of features decreases the exposure of the photosynthetic apparatus of the photobiont to light and can be described as sunshade mechanisms (Veerman et al., 2007). However, shading or the production of sun-protectant pigments reduce but cannot prevent photooxidation. Photosynthetic pigments absorb light, whether the organisms are hydrated or desiccated, but energy conservation by carbon assimilation is possible only in the presence of water. Photosynthetic reaction centers can remain intact during desiccation in lichens and then photoreactions threaten to cause severe photooxidative damage (Heber & Lüttge, 2011).

Piccotto & Tetriach (2010) found some similarities of lichens with the acclimation of vascular plants to the light regime, as the photobiont activity of lichen should be significantly modulated by the growth light regime of the thallus. According to these authors, lichen chlorobionts show the same general variation patterns described in "sun" and "shade" leaves, with optimisation of light absorption and harvesting capacity in chlorobionts of thalli grown in shaded habitat, and increased photosynthetic quantum conversion and lower Chl a fluorescence emission in chlorobionts of thalli grown under direct solar irradiation. However, this comparison should be complicated by the influence in lichens of extrinsic factors, as nitrogen availability, or intrinsic factors, as light transmittance through the peripheral mycobiont layer. These "sun" and "shade" patterns are maintained by the chlorobionts independently of the symbiosis. Our group isolated the two species of phycobionts that are always coexisting in the lichen Ramalina farinacea, and we observed that one species, so called TR9, was better adapted to high irradiances than the other one, so called TR1 (Casano et al., 2010). The proportion of the two phycobionts in the lichen changes depending on the local conditions, allowing the lichen behaves either as a "sun" or as "shade" species. The ecophysiological plasticity of this symbiosis allows the lichen proliferates in a wide variety of habitats.

In lichens, the exposure to high light in the hydrated state produces photoinhibition in chlorolichens and cyanolichens, an effect much more pronounced in cyanolichens, which do not reverse photosynthetic depression after a recovery period as chlorolichens do. Photoinhibition can be largely diminished if both kinds of lichens are in the desiccated state or become desiccated during the period of high light exposure (Demmig-Adams et al., 1990a). Demmig-Adams et al. (1990b) described significant increases of zeaxanthin content after two hours of exposure to high light of six chlorolichens previously adapted to a low light regime. This effect was not observed in cyanolichens. Following treatment of the thalli with an inhibitor of the violaxanthin de-epoxidase, dithiothreitol, the response of green algal lichens to light became very similar to that of the blue-green algal lichens. Thus, Demmig-Adams et al. (1990b) proposed that the higher light stress in blue-green algae lichens is related to the absence of an effective accumulation of zeaxanthin (lack of the xanthophyll cycle) in cyanobacteria. On the other hand, Heber & Shuvalov (2005) and Heber (2008) report differences in the recovery under different light tretments between mosses collected in autumn-winter or in spring-summer, indicating seasonal acclimation of poikilohydric organisms to the light regime, but they found that in this case it was independent of zeaxanthin accumulation. In lichens the xanthophyll cycle activation cannot either be assumed as a general response to light dissipation under stress conditions because some lichen species, like *Pseudevernia furfuracea* or the isolated lichen photobiont *Trebouxia excentrica*, do not increase the xanthophyll DPS upon dehydration or rehydration (Kranner et al., 2003, 2005).

Relationship between NPQ and desiccation tolerance in lichens has been studied during the last years. For instance, Calatayud et al. (1997) found that in the thallus of *Parmelia quercina*, NPQ increases during dehydration, which seems to be related with the conversion from violaxantin to zeaxanthin observed during the desiccation. These effects were also observed in *Ramalina maciformis* (Zorn et al., 2001). Fernández Marín et al. (2010) described in the lichen *Lobaria pulmonaria* that violaxantin deepoxidase needs a time to be activated during desiccation and then de-epoxidation of violaxanthin to zeaxanthin occurs only when the tissue has lost most of its water and dehydration is slow. Kranner et al. (2005) described that activation of the xanthophyll cycle in *Cladonia vulcani* is dependent of the symbiosis because an effective accumulation of zeaxanthin can be observed in the lichen during desiccation and rehydration processes, but not in its isolated photobiont. In accordance, during a study carried out in the isolated photobiont *Asterochloris erici* (Gasulla et al., 2009) we did not found significant differences in the deepoxidation state of xanthophylls after rapid or slow desiccation treatments when compared with controls. On the other hand, a significant increase in the deepoxidation state was registered after exposure of the isolated *A. erici* to high light intensities. Kosugi et al. (2009) found that the responses to air drying and hypertonic treatments of *Ramalina yasudae* and its isolated *Trebouxia sp.* photobiont are different in three ways: 1) PSII activity is completely inhibited in the desiccated lichen but not in the isolated photobiont; 2) the dehydration induced quenching of PSII fluorescence was lower in the lichen than in the photobiont; and 3) the isolated *Trebouxia* was more sensitive to photoinhibition than the *R. yasudae* thalli. The authors proposed that a lichen substance or mechanism lost during the photobiont isolation could be implied in light dissipation. In any case, all these results indicate that the activity of the algae is modulated by the symbiotic association with funghi, but also that other factors, as the speed of desiccation can influence on the photosynthetic activity after rehydration.

Alternative mechanisms of energy dissipation have been described to take place in mosses and lichens, where new quenching centres are functional during desiccation (Heber et al., 2006a, 2006b, 2007; Heber, 2008). In a work based on changes on the emission spectra of chlorophylls, Bilger et al. (1989) proposed that with green algal symbionts desiccation induces a functional interruption of energy transfer between the light harvesting Chl a/b pigment complex and PSII and that this can be largely restored by rehydration with humidified air. Heber et al. (2006a, 2006b) described that the recovery of fluorescence levels after drying was better in dark dried mosses than in sun-dried mosses, an effect that the authors can not ascribe to zeaxanthin and classical NPQ protection. Heber & Shuvalov (2005) found the existence in briophytes and lichens of an alternative quencher of chlorophyll fluorescence characterised by a long wavelenght (720 nm) emission. Helped by results obtained in dessicated spinach leaves, Heber (2008) proposed that electrons can be redirected in the dried state from pheophitin to a secondary chlorophyll placed very close to P_{680}, implying the formation of the radical pair $P_{680}^+Chl^-$. A further and slower recombination between Chl^- and a carotene molecule positively charged should complete a photoprotective cycle in desiccated leaves. The authors argued that the similarity of fluorescence emission spectra of dessicated leaves with those obtained in a desiccated fern and a desiccated moss should justify the extrapolation of this model to poikilohydric plants. A stronger support for the hypothesis of alternative sinks for energy or electrons during desiccation was obtained by Veerman et al. (2007) with steady-state, low temperature, and time-resolved chlorophyll fluorescence spectroscopy. These authors presented a model in which a pigment molecule with a peak emission at 420 nm should act as a sink for energy accumulated on P680 in the lichen *Parmelia sulcata* under desiccation. Heber et al. (2007) afford more evidences for the existence of these alternative mechanisms, as they found in dehydrated mosses and lichens a quenching mechanism independent of light activation. This mechanism is probably dependent of conformational changes in a protein-pigment complex as is inhibited by glutaraldehyde or heat treatments. This mechanism is also dependent of the speed of desiccation, as fast drying is less effective in decreasing chlorophyll fluorescence than slow drying. Gasulla et al. (2009) found that the basal fluorescence (F_o) values in desiccated *Asterochloris erici* were significantly higher after rapid dehydration, than after slow dehydration, suggesting higher levels of light energy dissipation in slow-dried algae. Higher values of PSII electron transport were recovered after rehydration of slow-dried *A. erici* compared to rapid-dried algae. The authors suggest that there is probably a minimal period required to develop strategies which will facilitate transition to the desiccated state in chlorobionts. In this process, the xanthophyll cycle and classical antioxidant mechanisms play a very limited role. More recently, Komura et al. (2010) have found proof that the quencher of chlorophyll fluorescence under desiccation conditions is not a chlorophyll molecule and suggest a new kind of quenching in PSII antenna or aggregation in PSII. The mechanisms implicated in protection under desiccation should be dependent of the desiccation rate, independent of light and probably associated to conformational changes in a chlorophyll-protein complex (Heber et al., 2007; Heber, 2008; Heber et al., 2010; Heber & Lüttge, 2011). Finally, Heber et al. (2011) propose that photoprotection is achieved by the drainage of light energy out of the reaction centers.

6. Antioxidant systems in lichen algae

Photobionts, as aerobic organisms, have to prevent and control oxidative stress damages through a complex antioxidant system capable to maintain controlled the levels of ROS, what is known as redox homeostasis. This defence system has co-evolved with aerobic metabolism to counteract oxidative damage from ROS (Gülçin et al., 2002). It has been studied in detail in plants, and includes enzyme activities of the so-called ascorbate-glutathione cycle (Asada, 1994; Foyer & Halliwell, 1976; Mittler, 2002), superoxide dismutase (SOD; Bowler et al., 1992), peroxidases (POX; Esteban-Carrasco et al., 2000, 2001; Ros Barceló et al., 2007; Zapata et al., 1998), catalase (CAT; Mittler, 2002) along with redox metabolites, like ascorbic acid, and glutathione (GSH; Noctor & Foyer, 1998). In the ascorbate-glutathione cycle, the two major antioxidant molecules, ascorbate and glutathione, play an important role as reductants, and they are involved in the scavenging of H_2O_2 produced by SOD (Kranner & Birtic, 2005; Lascano et al., 1999). Indeed, Superoxide dismutases catalyze the dismutation of $O_2^{\cdot-}$ to H_2O_2 and prevent the further and dangerous conversion into $OH^{\cdot}$ (Casano et al., 1997). Peroxidases catalyse the oxidation of a wide range of substrates at the expense of H_2O_2 (Ros Barceló et al., 2007; Zapata et al., 2005). Finally, catalases break down H_2O_2 very rapidly producing H_2O and O_2, but are much less effective than peroxidases at removing H_2O_2 because of their lower affinity (high Km) to H_2O_2 (Kranner & Birtic, 2005).

Concerning to antioxidant metabolites, glutathione and ascorbate are the main antioxidants present in organisms. Glutathione are involved in scavenging of the highly reactive $OH^{\cdot}$ by a cycle that includes the glutathione reductase enzyme (GR) to recover the biologically active glutathione molecule (Noctor & Foyer, 1998). On the other hand, ascorbate reacts rapidly with $OH^{\cdot}$, $O_2^{\cdot-}$ and $^1O_2$ (Halliwell & Gutteridge, 1999), forming monodehydroascorbate (MDA) and then dehydroascorbate (DHA). Regeneration of Asc may occur via a Mehler peroxidase reaction sequence or through the Asc-GSH cycle (Foyer & Halliwell, 1976).

In desiccation tolerant organisms, as lichens, the desiccated state is characterised by little intracellular water and almost no metabolic activity. Many deleterious effects are associated to this state, such as, irreversible damage to lipids, proteins and nucleic acids through Maillard reactions and ROS (Kranner et al., 2002). During their lifetime, lichens undergo continuous cycles of dehydration-rehydration and therefore they have to must be able to (i) limit the damage to a repairable level, (ii) maintain physiological integrity in the dried state, and (iii) mobilise mechanisms upon rehydration that repair damage suffered during desiccation and subsequent rehydration (Bewley, 1979; Oliver & Bewley, 1997; Oliver et al., 2000). In a first conclusion, desiccation tolerance and prolonged longevity in the desiccated state seems to depend on the ability to prevent light damage (as discussed in the previous section) and activate the biochemical mechanism to scavenge free radicals produced during dehydration-rehydration cycles, using the two described pathways: antioxidant molecules such as glutathione or ascorbate and antioxidant enzymes capable of scavenging free radicals (Kranner & Birtic, 2005). Particularly, ROS-antioxidant interactions has been described in lichens and it is well known that an enhancement of antioxidant status occurs in the symbiotic partnership being more resistant to environmental stress than either partner alone (Kranner et al., 2005). However, within the few studies carried out with lichens, there is not a clear relationship between desiccation tolerance and antioxidant levels. Cellular activities of the antioxidant enzymes ascorbate peroxidase, catalase, and superoxide dismutase as well as the auxiliary enzyme glutathione reductase and the pentose-phospahte pathway key enzyme glucose-6-phosphate dehydrogenase were shown to increase, decrease, or remain unchanged in response to desiccation and rehydration, depending on the species and the experimental conditions (Weissman et al., 2005). For instance, Mayaba & Beckett (2001) observed that activities of SOD, CAT, and ascorbate peroxidase (AP) were similar during wetting and drying cycles in *Peltigera polydactyla*, *Ramalina celastri* and *Teloschistes capensis*, which grow in moist, xeric and extremely xeric habitats, respectively. Kranner (2002) neither observed a correlation between GR activity and the different degrees of desiccation-tolerance of three lichens, *Lobaria pulmonaria*, *Peltigera polydactyla* and *Pseudevernia furfuracea*. Weissman et al. (2005) even reported that after rehydration *Ramalina lacera* loss almost all CAT activity and SOD decreases by 50-70%. In our laboratories we have observed that during both desiccation period and recovery of *Asterochloris erici*, the levels of superoxide dismutase and peroxidase decreased under both slow and rapid dehydration. Thus, in *A. erici* a longer dehydration time does not lead to a higher accumulation or preservation of classical antioxidants during dehydration/desiccation (Gasulla et al., 2009). Therefore, as a general conclusion, enzymatic antioxidants are perhaps more likely to be involved in removing ROS produced during normal metabolism or by other stresses rather than during rehydration following severe desiccation (Kranner et al., 2008).

In many organisms, the major water-soluble low-molecular weight antioxidants are the tripeptide GSH (glutatione, γ-glutamyl-cysteinyl-glycine) and ascorbate (Noctor & Foyer, 1998). Ascorbate is known to be a non-enzymatic antioxidant of major importance to the assimilatory and photoprotective processes, its function being central to the defence system. Ascorbate acts as an antioxidant by removing hydrogen peroxide generated during photosynthetic processes in a group of reactions termed the "Mehler peroxidase reaction sequence" (Asada, 1994). Also, ascorbic acid may act as a direct electron donor in photosynthetic and mitochondrial electron transport (Miyake & Asada, 1992) in the ascorbate–glutathione cycle (Foyer & Halliwell, 1976). In addition, it is a cofactor for violaxanthin de-epoxidase in chloroplasts. In lichens, it has been reported that the ascorbate play an important antioxidant role against oxidative stress, such as excessive light or atmospheric pollution (Calatayud et al., 1999). In desiccation-tolerant higher plants, ascorbate forms the first line of defense against oxidative damage (Kranner et al., 2002). However, Kranner et al. (2005) did not find any relationship between ascorbate and the desiccation tolerance of the lichen *Cladonia vulcanii* nor its photobiont. Furthermore, ascorbate levels were undetectable in *A. erici* (Gasulla et al., 2009).

On the other hand, GSH not only can scavenge ROS reacting with OH$^{•}$ to form GS$^{•}$; it can also react with another GS$^{•}$, forming glutathione disulphide (GSSG) (Kranner, 2002). In addition, the redox couple of glutathione (GSH-GSSG) is involved in protecting protein thiol-groups by forming protein-bound glutathione (PSSG) (Kranner & Grill, 1996). In plants, accumulation of GSSG is often correlated with increased stress. Indeed, GSSG can be recycled by the NADPH dependent enzyme GR and in desiccation-tolerant organisms GSSG accumulates during desiccation and is re-reduced to GSH during rehydration (Kranner et al. 2006). Going back to the work of Kranner (2002), it is has been reported that desiccation caused oxidation of almost all GSH in the lichens *Lobaria pulmonaria*, *Peltigera polydactyla* and *Pseudevernia furfuracea*, and rehydration caused the inverse effect. However, after a long desiccation period, in *P. furfuracea* the recovery of the initial concentration of GSH was very rapid, while *P. polydactyla* did not re-establish the GSH pool initial level. It has been demonstrated that NADPH dependent enzyme GR activity is high during rehydration process and therefore it is not a limiting factor to explain the differences between different lichen species. It is more likely that the capacity to reduce GSSG was correlated with the reactivation or synthesis "de novo" of glucose-6-phosphate dehydrogenase, an enzyme of the oxidative pentose phosphate pathway (Kranner, 2002).

7. Conclusion

Two different ways to protect the photosynthetic machinary of lichen chlorobionts were considered in this review. The first one, based in the dissipation of light energy and the second based in the presence of antioxidant activities. Dissipation of light energy at the level of the chlorobiont´s PSII can be achieved by the classical mechanisms based in the xanthophyll cycle, but proof is increasing in favour of the presence of new sinks for conversion of light energy into heat. The activation of these sinks is independent of zeaxanthin accumulation and probably requires an additional pigment and conformational changes in some protein(s) associated to the reaction center or the antenna. Further research is needed to determine the chemical nature and action mechanism of this (or these) alternative energy sink(s).

The lichen is more resistant to oxidative stress than the photobiont or the mycobiont alone. The main antioxidant substance in lichens seems to be glutathione. However, there is not a clear relationship between desiccation tolerance and antioxidant levels. Glutathione, ascorbic acid and antioxidant activities such as SOD or POX can increase, decrease, or not change, depending on the desiccation tolerance of the organism as well as the mode and duration of the treatment. Probably, constitutive levels of these antioxidants are high enough to protect cells against abrupt changes in the environmental conditions, mainly light intensity and humidity.

8. Acknowledgment

This work was supported by grants from the MEC (BFU2006-11577/BFI and BFU2009-08151)-FEDER and CARM (08610/PI/08), as well as grants funded by the Spanish Ministry of Education and Science (CGL2006-12917-C02-01/02), the Spanish Ministry of Science and Innovation (CGL2009-13429-C02-01/02), the AECID PCI_A/024755/09) and the Generalitat Valenciana (PROMETEO 174/2008 GVA). Joaquín Herrero is a fellow of the spanish FPU program.

9. References

Abele D. (2002). Toxic oxygen: the radical life-giver. *Nature,* Vol. 420, (November 2002), pp. 27, ISSN: 0028-0836.

Adams, WW III & Demmig-Adams, B. (1992). Operation of the xanthophyll cycle in higher plants in response to diurnal changes in incident sunlight. *Planta* Vol. 186, No. 3, (February 1992), pp. 390–398, ISSN: 0032-0935

Adams, WW III; Demmig-Adams, B; Rosenstiel, TN; Ebbert, V. (2001). Dependence of photosynthesis and energy dissipation activity upon growth form and light environment during the winter. *Photosynthesis Research* Vol. 67, No. 1-2, (February 2001), pp. 51–62, ISSN: 0166-8595

Ahmadjian, V. (1967a). *The lichen symbiosis.* Blaisdell Publishing Company, Massachussetts.

Ahmadjian, V. (1967b). A guide to the algae occurring as lichen symbionts: isolation, culture, cultural physiology and identification. *Phycologia* Vol. 6, No. 2 and 3, (April 1967), pp. 127–160

Ahmadjian, V. (1993). *The lichen symbiosis.* John Wiley & Sons, Inc., ISBN:0-471-57885-1, New York.

Almagro, L; Gómez-Ros, LV; Belchí-Navarro, S; Bru, R; Ros Barceló, A; Pedreño, MA. (2009). Class III peroxidases in plant defence reactions. *Journal of Experimental Botany,* Vol. 60, No. 2, (February 2009), pp. 377–390, ISSN 0022-0957.

Alpert, P. (2006). Constrains of tolerance: why are desiccation-tolerance organisms so small or rare? *The Journal of Experimental Botany* Vol. 209, (May 2006), pp. 1575-1584, ISSN: 0022-0949

Anderson, B & Barber, J. (1996). Mechanism of photodamage and protein degradation during photoinhibition of photosystem II. In: *Photosynthesis and the enviroment.* Baker NR., (Ed.), 101-121, Kluwer Academic Publishers, ISBN 1-7923-4316-6, Dordrecht, Netherlands.

Aruoma, O.I; Halliwell, B; Gajewski, E & Dizdaroglu, M. (1989). Damage to the bases in DNA induced by hydrogen peroxide and ferric ion chelates. *The Journal of Biological Chemistry.* Vol. 264, No. 34, (December 1989), pp. 20509-20512.

Asada, K. (1994). Production and action of active oxygen species in photosynthetic tissues. In: *Causes of Photooxidative Stress and Amelioration of Defense Systems in Plants.* CH Foyer, CH & Mullineaux, P., (Ed.), 77-103, CRC Press, ISBN 0-8493-5443-9, Boca Raton, Florida, USA.

Ascaso, C. (1980). A rapid method for the quantitative isolation of green-algae from lichens. *Annals of Botany,* Vol. 45, No. 4, (April 1980), pp. 483–483, ISSN: 0305-7364.

Baker, NR. (2008). Chlorophyll fluorescence: A probe of photosynthesis in vivo. *Annual Review of Plant Biology,* Vol. 59, (June 2008), pp. 89–113.

Backor, M. & Hudak, J. (1999). The effect of cytokinins on the growth of lichen photobiont *Trebouxia irregularis* cultures. *The Lichenologist,* Vol. 31, No. 2, pp. 207-210.

Bewley, DJ. (1979). Physiological aspects of desiccation tolerance. *Annual Reviews of Plant Physiology,* Vol. 30, No.1, (June 1979), pp. 195–238, ISSN 0066-4294.

Bilger, W; Rimke, S; Schreiber, U. & Lange, OL. (1989). Inhibition of Energy-Transfer to Photosystem II in Lichens by Dehydration: Different Properties of Reversibility with Green and Blue-green Phycobionts. *Journal of Plant Physiology,* Vol.134, No. 3, pp. 261-268, ISSN 0176-1617.

Bilger, W. & Bjorkman, O. (1990). Role of the Xanthophyll Cycle in Photoprotection Elucidated by Measurements of Light-Induced Absorbency Changes, Fluorescence and Photosynthesis in Leaves of *Hedera canariensis. Photosynthesis Research,* Vol. 25, No. 3, (May 1990), pp. 173-185. ISSN 0166-8595.

Bischoff, HW. & Bold, HC. (1963). Phycologycal Studies IV. Some soil algae from Enchanted Rock and related algal species. University of Texas, Publication 6318, pp: 1-95.

Bowler, C.; Montagu, MV. & Inze, D. (1992) Superoxide dismutases and stress tolerance. *Annual Review in Plant Physiology and Plant Molecular Biology,* Vol. 43, (June 1992), pp. 83–116, ISSN 1040-2519.

Büdel, B. (1992). Taxonomy of lichenized procarytic blue-green algae. In : *Algae and Symbioses: Plants, Animals, Fungi, Viruses. Interactions Explored,* W. Reisser, (Ed.), 301-324, Biopress Ltd, Bristol, United Kingdom.

Büdel, B. & Lange, O.L. (1994). The role of cortical and epicortical layers in the lichen genus *Peltula. Cryptogamic Botany,* Vol. 4, pp. 262-269.

Bukhov, N.G ; Kopecky, J ; Pfundel, E.E ; Klughammer, C. & Heber, U. (2001). A few molecules of zeaxanthin per reaction centre of photosystem II permit effective thermal dissipation of light energy in photosystem II of a poikilohydric moss. *Planta,* Vol. 212, No. 5, (April 2001), pp. 739-748, ISSN 0032-0935.

Calatayud, A ; Deltoro, VI ; Barreno, E ; del Valle Tascon, S. (1997). Changes in *in vivo* chlorophyll fluorescence quenching in lichen thalli as a function of water content and suggestion of zeaxanthin associated photoprotection. *Physiologia Plantarum,* Vol. 101, No. 1, (November 1997), pp. 93–102, ISSN 1399-3054.

Calatayud, A.; Deltoro, VI.; Abadía, A.; Abadía, J. & Barreno, E. (1999). Effects of ascorbate feeding on chlorophyll fluorescence and xanthophyll cycle components in the lichen *Parmelia quercina* (Willd.) Vainio exposed to atmospheric pollutants. *Physiologia Plantarum,* Vol. 105, No. 4, pp. 679-684, ISSN 1399-3054.

Calatayud, A ; Guéra, A ; Fos, S ; Barreno, E. (2001). A new method to isolate lichen algae by using Percoll® gradient centrifugation. *The Lichenologist,* Vol. 33, No. 4, (July 2001), pp. 361–366, ISSN: 0024-2829.

Casano, LM ; Gómez, LD; Lascano, HR; González, CA; & Trippi, VS. (1997). Inactivation and degradation of CuZn-SOD by active oxygen species in wheat chloroplasts exposed to photooxidative stress. *Plant & Cell Physiology,* Vol. 38, No. 4, (January 1997), pp. 433-440, ISSN 0032-0781.

Casano, LM; Martín, M; Zapata, JM. & Sabater, B. (1999). Leaf age and paraquat concentration-dependent effects on the levels of enzymes protecting against photooxidative stress. *Plant Science,* Vol. 149, No. 1, (November 1999), pp. 13-22. ISSN 0168-9452.

Casano, LM.; del Campo, EM.; García-Breijo, FJ.; Reig-Armiñana, J.; Gasulla, F.; del Hoyo, A.; Guéra, A. & Barreno, E. (2010). Two Trebouxia algae with different physiological performances are ever-present in lichen thalli of Ramalina farinacea. Coexistence versus Competition? *Environmental Microbiology,* Vol. 13, No. 3, (December 2010), pp. 806-818, ISSN 1462-2920.

Catalá, M; Gasulla, F; Pradas del Real, A; García-Breijo, F; Reig-Armiñana, J. & Barreno, E. (2010). Fungal-associated NO is involved in the regulation of oxidative stress during rehydration in lichen symbiosis. *BMC Microbiology,* Vol. 10, No. 1, (November 2010), pp. 1-13, ISSN 1471-2180.

Chakir, S., & M. Jensen. (1999). How does *Lobaria pulmonaria* regulate photosystem II during progressive desiccation and osmotic water stress? A chlorophyll fluorescence study at room temperature and at 77 K. *Physiologia Plantarum,* Vol. 105, No. 9, pp. 257–265.

Chen, K; Feng, H; Zhang, M; Wang, X. (2003). Nitric oxide alleviates oxidative damage in the green alga *Chlorella pyrenoidosa* caused by UV-B radiation. *Folia Microbiologica,* Vol. 48, No. 3, (May 2003), pp. 389-393, ISSN 0015-5632.

Courtois, C; Besson, A; Dahan, J; Bourque, S; Dobrowolska, G; Pugin, A. & Wendehenne, D. (2008). Nitric oxide signalling in plants: interplays with Ca2+ and protein kinases. *Journal of Experimental Botany*, Vol. 59, No. 2, (January 2008), pp. 155-163, ISSN 0022-0957.

Crittenden, P.D; Katucka, I. & Oliver, E. (1994). Does nitrogen supply limit the growth of lichens? *Cryptogamic Botany*, Vol. 4, pp. 143-155.

Crouchman, S; Ruban, A. & Horton, P. (2006). PsbS enhances nonphotochemical fluorescence quenching in the absence of zeaxanthin. *FEBS Letters*, Vol. 580, No. 8, (March 2006) pp. 2053-2058, ISSN 0014-5793.

Darley-Usmar, VM; Pate, RP; O'Donell, VB; Freeman, BA. (2000). Antioxidant actions of nitric oxide. In: *Nitric Oxide: Biology and Pathology*. Ignarro L., (Ed.), 256-276, Academic Press, ISBN 0-12-370420-0, Los Ángeles, USA.

de los Rios, A; Ascaso, C. & Wierzchos, J. (1999). Study of lichens with different state of hydration by the combination of low temperature scanning electron and confocal laser scanning microscopies. *International Microbiology*, Vol. 2, No. 4, (December 1999), pp. 251–257.

del Hoyo, A.; Álvarez, R.; del Campo, E.M.; Gasulla, F.; Barreno, E. & Casano, M.(2011). Oxidative stress induces distinct physiological responses in the two Trebouxia phycobionts of the lichen Ramalina farinacea. Annals of Botany 107: 109–118. doi:10.1093/aob/mcq206

del Prado, R. & Sancho, L.G. (2007). Dew as a key factor for the distribution pattern of the lichen species *Teloschistes lacunosus* in the Tabernas Desert (Spain). *Flora*, Vol. 202, No. 5, (July 2007), pp. 417-428, ISSN 0367-2530.

Demmig-Adams, B; Adams, III WW.; Czygan, F-C.; Schreiber, U. & Lange, OL. (1990a). Differences in the capacity for radiationless energy dissipation in the photochemical apparatus of green and blue-green algal lichens associated with differences in carotenoid composition. *Planta*, Vol. 180, No. 4, (March 1990), pp.:582-589, ISSN 0032-0935.

Demmig-Adams, B; Adams, III WW.; Heber, U.; Neimanis, S.; Winter, K.; Krfiger, A.; Czygan, F-C.; Bilger, W. & Björkman, O. (1990b). Inhibition of zeaxanthin formation and of rapid changes in radiationless energy dissipation by dithiothreitol in spinach leaves and chloroplasts. *Plant Physiology*, Vol. 92, No. 2, (February 1990), pp. 293-301, ISSN 0032-0889.

Demmig-Adams, B. & Adams, WW. III. (1993). The xanthophyll cycle, protein turnover, and the high light tolerance of sun acclimated leaves. *Plant Physiology*, Vol. 103, No. 4, (December 1993), pp. 1413-1420, ISSN 0032-0889.

Demmig-Adams, B; Adams, WW.III; Barker, D.H; Logan, B; Bowling, D.R. & Verhoeven, A.S. (1996). Using chlorophyll fluorescence to asses the fraction of observed light allocated to thermal dissipation of excess excitation. *Physiologia Plantarum*, Vol. 98, No. 2, (October 1996), pp. 253-264, ISSN 1399-3054.

Demmig-Adams, B. & Adams, WW. III. (2000). Photosynthesis: Harvesting sunlight safely. Nature, Vol. 403, No. 6768, pp. 371-374, ISSN 0028-0836.

Dietz, S. & Hartung, W. (1999). The effect of abscisic acid on chlorophyll fluorescence in lichens under extreme water regimes. *New Phytologist*, Vol. 143, No. 3, (September 1999), pp. 495-501, ISSN 1469-8137.

Dietz, S.; Büdel, B.; Lange, O.L. & Bilger, W. (2000). Transmittance of light through the cortex of lichens from contrasting habitats. In: *Aspects in Cryptogamic Research. Contributions in Honour of Ludger Kappen*. B. Schroeter, M. Schlensog & T. G. A. Green, (Ed.), 171-182, Gebrüder Borntraeger Verlagsbuchhandlung, ISBN 978-3-443-58054-4, Berlin-Stuttgart, Germany.

Durner, J. & Klessig, D F. (1999). Nitric oxide as a signal in plants. *Current Opinion in Plant Biology*, Vol. 2, No. 5, (October 1999), pp. 369-374, ISSN 1369-5266.

Elstner, E.F. (1982). Oxygen activation and oxygen toxicity. *Annual Review of Plant Physiology*, Vol. 33, No. 1, (June 1982), pp. 73-96, ISSN 0066-4294.

Elstner, E.F. & Osswald, W.F. (1994). Mechanisms of oxygen activation during water stress. *Proceedings of the Royal Society of Edinburgh*, Vol. 102, pp. 131-154.

Ertl, L. (1951). Über die Lichtverhälnisse in Laubflecten. *Planta*, Vol. 39, No. 3, (june 1951), pp. 245-270, ISSN 0032-0935.

Esteban-Carrasco, A; Zapata, JM ; Casano, L ; Sabater, B. & Martin, M. (2000). Peroxidase activity in *Aloe barbadensis* commercial gel: probable role in skin protection. *Planta Medica*, Vol. 6, No. 8, (April, 2000), pp. 724–727, ISSN 0032-0943.

Esteban-Carrasco. A; López-Serrano, M; Zapata, JM; Sabater, B. & Martín, M. (2001). Oxidation of phenolic compounds from *Aloe barbadensis* by peroxidase activity: Possible involvement in defence reactions. *Plant Physiology and Biochemistry*, Vol. 39, No. 6, (June 2001), pp. 521−527, ISSN 0981-9428.

Esterbauer, H; Schaur, R.J. & Zollner H. (1991). Chemistry and biochemistry of 4-hydroxynonenal, malonaldehyde and related aldehydes. *Free Radical Biolology and Medicine*, Vol. 11, No. (December 1991), pp. 81-128, ISSN 0891-5849.

Farrar, J.F. (1976a). Ecological Physiology of the lichen *Hypogymnia physodes*. II. Effects of wtting and drying cycles and the concept of "physiological buffering". *New Phytologist*, Vol. 77, No. 1, (July 1976), pp. 105-113.

Farrar, J.F. (1976b). Ecological Physiology of the lichen *Hypogymnia physodes*. I. Some effects of constant water saturation. *New Phytologist*, Vol. 77, No. 1, (July 1976), pp. 93-103.

Feelisch, M. & Martin, JF. (1995). The early role of nitric-oxide in evolution. *Trends in Ecology & Evolution*, Vol. 10, No. 12, (December 1995), pp. 496-499, ISSN 0169-5347.

Fernandez-Marin, B; Becerril, JM. & Garcia-Plazaola, JI. (2010). Unravelling the roles of desiccation-induced xanthophyll cycle activity in darkness: a case study in *Lobaria pulmonaria*. *Planta*, Vol. 231, No. 6, (May 2010), pp.1335–1342, ISSN 0032-0935.

Ferrer, MA. & Ros Barceló, A. (1999). Differential effects of nitric oxide on peroxidase and H_2O_2 production by the xylem of *Zinnia elegans*. *Plant, Cell and Environment*, Vol. 22, No. 7, (July 1999), pp. 891-897, ISSN 1365-3040.

Finkel, T. & Holbrook, N.J. (2000). Oxidants, oxidative stress and the biology of ageing. *Nature*, Vol. 408, (November 2000), pp. 239-247, ISSN: 0028-0836.

Fos, S; Deltoro, V.I; Calatayud, A. & Barreno, E. (1999). Changes in water economy in relation to anatomical and morphological characteristics during thallus development in *Parmelia acetabulum*. *The Lichenologist*, Vol. 31, No. 4, (July 1999), pp.375-387, ISSN 0024-2829.

Foyer, C. & Halliwell, B. (1976). The presence of glutathione and glutathione reductase in chloroplasts: A proposed role in ascorbic acid metabolism. *Planta*, Vol. 133, No. 1, (January 1976), pp. 21–25, ISSN 0032-0935.

Foyer, C. & Noctor, G. (2003). Redox sensing and signalling associated with reactive oxygen in chloroplasts, peroxisomes and mitochondria. *Physiologia Plantarum*, Vol. 119, No. 3, (November 2003), pp. 355–364, ISSN 0031-9317.

Friedl, T & Büdel, B. (2008). Photobionts. In *Lichen Biology* (second ed.), T.H. Nash, (Ed.), 9-26, Cambridge University Press, ISBN 0-521-45368-2, Cambridge, UK.

Gärtner, G. (1992). Taxonomy of symbiotic eukaryotic algae. In *Algae and Symbioses: Plants, Animals, Fungi, Viruses. Interactions Explored*, W. Reisser, (Ed.), 325-338, Biopress Ltd, Bristol, UK.

Gasulla, F.; de Nova, P.G.; Esteban-Carrasco, A.; Zapata, J.M.; Barreno, E. & Guéra, A. (2009). Dehydration rate and time of desiccation affect recovery of the lichen alga [corrected] *Trebouxia erici*: Alternative and classical protective mechanisms. *Planta*, Vol. 231, No. 1, (December 2009), pp. 195–208, ISSN 0032-0935.

Gasulla, F; Guéra, A. & Barreno, E. (2010). A simple and rapid method for isolating lichen photobionts. *Symbiosis*, Vol. 51, No. 2, (May 2002), pp. 175–179, ISSN 0334-5114.

Gauslaa, Y. & Solhaug, KA. (1999). High-light damage in air-dry thalli of the old forest lichen *Lobaria pulmonaria*: interactions of irradiance, exposure duration and high temperature. *Journal of Experimental Botany*, Vol. 50, No. 334, (May 1999), pp.697–705, ISSN 0022-0957.

Guéra, A.; Calatayud, A.; Sabater, B. & Barreno, E. (2005). Involvement of the thylakoidal NADH-plastoquinone-oxidoreductase complex in the early responses to ozone exposure of barley (*Hordeum vulgare* L.) seedlings. *Journal of Experimental Botany*, Vol. 56, No. 409, (January 2005), pp. 205-218, ISSN 0022-0957.

Gülçin, I.; Oktay, M.; Küfrevioğlu, OI. & Aslan, A. (2002). Determination of antioxidant activity of lichen *Cetraria islandica* (L) Ach. *Journal of Ethnopharmacology*, Vol. 79, No. 3, (March 2002), pp. 325–329, ISSN 0378-8741.

Green, TGA ; Snelgar, W.P. ; Wilkins, A.L. (1985). Photosynthesis, water relations and thallus structure of *Stictaceae* lichens. In *Lichen physiology and cell biology*, Brown, D.H. (Ed.), 57-75, Plenum Press, ISBN 03-064-2200X, New York, USA.

Green, TGA.; Nash, TH.; Lange, OL. (2008). Physiological ecology of carbon dioxide exchange. In *Lichen biology*, Nash, TH., (Ed.), 152–181, Cambridge University Press, ISBN 0-521-45368-2, Cambridge, UK.

Hale, M.E. (1973). Growth. In *The Lichens*, V. Ahmadjian & M. E. Hale, (Ed.), 473-492. Academic Press.New York, USA.

Halliwell, B. (1984). *Chloroplast Metabolism*. Clarendon Press, Oxford, UK.

Halliwell, B. & Gutteridge J.M. (1999). *Free radicals in biology and medicine*. Oxford University Press, Oxford, UK.

Halliwell, B. (2006). Reactive species and antioxidants. Redox biology is a fundamental theme of aerobic life. *Plant Physiololgy*, Vol. 141, No. 2, (June 2006), pp. 312-322, ISSN 0032-0889.

Harris, G.P. & Kershaw, K.A. (1971). Thallus growth and the distribution of stored metabolites in the phycobionts of the lichens *Parmelia sulcata* and *P. physodes*. *Canadian Journal of Botany*, Vol. 49, No. 8 (August 1971), pp. 1367-1372, ISSN 0008-4026.

Heber, U.; Bilger, W.; Bligny, R. & Lange, OL. (2000). Photoreactions in two lichens, a poikilohydric moss and higher plants in relation to phototolerance of alpine plants: a comparison. *Planta*, Vol. 211, pp. 770–780.

Heber, U.; Bukhov, NG.; Shuvalov, VA.; Kobayashi, Y. & Lange, OL. (2001). Protection of the photosynthetic apparatus against damage by excessive illumination in homoiohydric leaves and poikilohydric mosses and lichens. *Journal of Experimental Botany*, Vol. 52, No. 363, pp. 1999-2006, ISSN 0022-0957.

Heber, U. & Shuvalov, VA. (2005). Photochemical reactions of chlorophyll in dehydrated photosystem II: two chlorophyll forms (680 and 700 nm). *Photosynthesis Research*, Vol. 84, No. 1, (June 2005), pp. 85–91, ISSN 0166-8595.

Heber, U.; Bilger, W. & Shuvalov, VA (2006a). Thermal energy dissipation in reaction centres and in the antenna of photosystem II protects desiccated poikilohydric mosses against photo-oxidation. *Journal of Experimental Botany*, Vol. 57, No. 12, (August 2006), pp. 2993–3006, ISSN 0022-0957.

Heber, U.; Lange, OL. & Shuvalov, VA. (2006b). Conservation and dissipation of light energy as complementary processes: homoiohydric and poikilohydric autotrophs. *Journal of Experimental Botany*, Vol. 57, No. 6, (March 2006), pp.1211–1223, ISSN 0022-0957.

Heber, U.; Azarkovich, M. & Shuvalov, V. (2007). Activation of photoprotection by desiccation and by light: poikilohydric photoautotrophs. *Journal of Experimental Botany*, Vol. 58, No. 11, (July 2007), pp. 2745–2759, ISSN 0022-0957.

Heber, U. (2008). Photoprotection of green plants: a mechanism of ultra-fast thermal energy dissipation in desiccated lichens. *Planta*, Vol. 228, No. 4, (September 2008), pp.641–650, ISSN 0032-0935.

Heber, U.; Bilger, W.; Türk, R. & Lange, O.L. (2010) Photoprotection of reaction centres in photosynthetic organisms: mechanisms of thermal energy dissipation in desiccated thalli of the lichen Lobaria pulmonaria. *New Phytologist*, Vol. 185, No 2, (January 2010), pp. 459–470, ISSN 1469-8137.

Heber, U. & Lüttge, U. (2011). Lichens and Bryophytes: Light Stress and Photoinhibition in Desiccation/Rehydration Cycles – Mechanisms of Photoprotection. In *Plant Desiccation Tolerance. Ecological Studies*, Lüttge, Ulrich; Beck, Erwin & Bartels, Dorothea, 121-137, Springer, ISBN 978-3-642-19106-0, Berlin Heidelberg, Germany.

Heber, U.; Sonib, V. & Strasser, R.J. (2011) Photoprotection of reaction centers: thermal dissipation of absorbed light energy vs charge separation in lichens. *Physiologia Plantarum*, Vol. 142, No 1 (May 2011), pp. 65–78. 2011 ISSN 0031-9317

Holder, J.M.; Wynn-Williams, D.D.; Rull Perez, F. & Edwards, H.G.M. (2000). Raman spectroscopy of pigments and oxalates in situ within epilithic lichens: Acarospora from the Antarctic and Mediterranean. *New Phytologist*, Vol. 145, No. 2, (February 200), pp. 271–280.

Holt, N.E.; Fleming, G.R. & Niyogi, K.K. (2004). Toward an understanding of the mechanism of non-photochemical quenching in green plants. *Biochemistry*, Vol. 43, No. 26, (May 2004), pp. 8281-8289.

Honegger, R. (1987). Questions about pattern formation in the algal layer of lichens with stratified (heteromerous) thalli. In *Progress and Problems in Lichenology in the Eighties*, Peveling, E., (Ed.), 59-71. Cramer, J, ISBN 3-443-58004-1, Berlin-Stuttgart, Germany.

Horton, P.; Wentworth. M. & Ruban, A. (2005). Control of the light harvesting function of chloroplast membranes: The LHCII-aggregation model for non-photochemical quenching. *FEBS Letters*, Vol. 579, No. 20, (August 2005), pp. 4201-4206, ISSN 0014-5793.

Jensen, M. & Feige, G. B. (1991). Quantum efficiency and chlorophyll fluorescence in the lichens *Hypogymnia physodes* and Parmelia sulcata. *Symbiosis*, Vol. 11, pp. 179–191.

Jensen, M.; Chakir, S. & Feige, G.B. (1999). Osmotic and atmospheric dehydration effects in the lichens *Hypogymnia physodes*, *Lobaria pulmonaria*, and *Peltigera aphthosa*: an in vivo study of the chlorophyll fluorescence induction. *Photosynthetica*, Vol. 37, No. 3, (November 1999), pp. 393-404, ISSN 0300-3604.

Kappen, L. (1974). Response to extreme environments. In *The lichens*, V. Ahmadjian & M. E. Hale, (Ed.), 311-380. Academic Press, New York, USA.

Kappen, L. & Valladares, F (2007). Opportunistic Growth and Desiccation Tolerance: The Ecological Success of Poikilohydrous Autotrophs. In functional Plant ecology, F. Pugnaire & F. Valladares (Ed.). pp.7-65. Taylor & Francis, New York, USA.

Kershaw, K.A. (1985). *Physiological Ecology of Lichens*. Cambridge University Press, ISBN 978-0-521-23925-7, Cambridge, UK.

Komura, M.; Yamagishia, A.; Shibataa, Y.; Iwasakib, I & Itoh, S. (2010). Mechanism of strong quenching of photosystem II chlorophyll fluorescence under drought stress in a lichen, *Physciella melanchla*, studied by subpicosecond fluorescence spectroscopy. *Biochimica et Biophysica Acta (BBA) – Bioenergetics*, Vol. 1797, No. 3, (March 2010), pp.331-338, ISSN 0005-2728.

Kopecky, J.; Azarkovich, M.; Pfundel, E.E.; Shuvalov, V.A. & Heber, U. (2005). Thermal dissipation of light energy is regulated differently and by different mechanisms in lichens and higher plants. *Plant Biology*, Vol. 7, No. 2, (March 2005), pp. 156-167, ISSN 0005-2728.

Kosugi, M.; Arita, M.; Shizuma, R.; Moriyama, Y.; Kashino, Y.; Koike, H. & Satoh, K. (2009). Responses to desiccation stress in lichens are different from those in their symbionts. *Plant and Cell Physiology*, Vol. 50, pp. 879–888.

Kramer, DM.; Johnson, G.; Kiirats, O. & Edwards, GE. (2004). New fluorescence parameters for the determination of Q(A) redox state and excitation energy fluxes. *Photosynthesis Research*, Vol. 79, No. 2, (February 2004), pp.209 - 218, ISSN 0166-8595.

Kranner, I. & Grill, D. (1996). Determination of Glutathione and Glutathione Disulphide in Lichens: a Comparison of Frequently Used Methods. *Phytochemical Analysis*, Vol. 7, No. 1, (January 1996), pp. 24–28, ISSN 1099-1565.

Kranner, I. & Lutzoni, F. (1999). Evolutionary consequences of transition to a lichen symbiotic state and physiological adaptation to oxidative damage associated with poikilohydry. In *Plant Response to Environmental Stress: From Phytohormones to Genome Reorganization*, H. R. Lerner, (Ed.), 591-628, Marcel Dekker Inc., NY, USA.

Kranner, I. (2002). Glutathione status correlates with different degrees of desiccation tolerance in three lichens. *New Phytologist*, Vol. 154, No. 2, (May 2002), pp. 451–460, ISSN 1469-8137.

Kranner, I.; Beckett, R.P.; Wornik, S.; Zorn, M. & Pfeifhofer, H.W. (2002). Revival of a resurrection plant correlates with its antioxidant status. *The Plant Journal*, Vol. 31, No. 1, (July 2002), pp. 13-24.

Kranner, I.; Zorn, M.; Turk ,B.; Wornik, S.; Beckett, RR. & Batic, F. (2003). Biochemical traits of lichens differing in relative desiccation tolerance. *New Phytologist*, Vol. 160, No. 1, (October 2003), pp. 167–176, ISSN 1469-8137.

Kranner, I.; Cram, WJ.; Zorn, M.; Wornik, S.; Yoshimura, I.; Stabentheiner, E. & Pfeifhofer, HW. (2005). Antioxidants and photoprotection in a lichen as compared with its isolated symbiotic partners. *Proceedings of the National Academy of Sciences of the United States of America*, Vol. 102, No. 8, (February 2005), pp. 3141–3146, ISSN 0027-8424.

Kranner, I. & Birtic, F. (2005). A modulation role for antioxidants in desiccation tolerance. *Integrative and Comparative Biology*, Vol. 45, No. 5, (November 2005), pp. 734–740, ISSN 1540-7063.

Kranner, I.; Birtic, S.; Anderson, K.M. & Pritchard, H.W. (2006). Glutathione half-cell reduction potential: a universal stress marker and modulator of programmed cell death? *Free Radical Biology and Medicine*. Vol. 40, No. 12, (June 2006), pp. 2155-2165, ISSN 0891-5849.

Kranner, I.; Beckett, R.; Hochman, A. & Nash, T.H. (2008). Desiccation-Tolerance in Lichens: A Review. *The Bryologist*, Vol. 111, No. 4, pp. 576-593, ISSN 0007-2745.

Krause, G.H. & Weis, E. (1991). Chlorophyll Fluorescence and Photosynthesis - the Basics. *Annual Review of Plant Physiology and Plant Molecular Biology*, Vol. 42, (June 1991), pp. 313-349, ISSN: 0066-4294.

Krause, GH. & Jahns, P. (2004). Non-photochemical energy dissipation determined by chlorophyll fluorescence quenching: characterization and function. In *Chlorophyll a fluorescence: a signature of photosynthesis. Advances in photosynthesis and respiration*, Papageorgiou, CG. & Govindjee (Ed.), vol 19, 463–495, Springer, ISBN 4020 3217 X, The Netherlands.

Kroncke, KD.; Fehsel, K. & Kolb-Bachofen, V. (1997). Nitric oxide: cytotoxicity versus cytoprotection–how, why, when, and where? *Nitric Oxide*, Vol. 1, No. 2, (April 1997), pp. 107-120, ISSN 1089-8603.

Lange, O.L. (1953). Hitze und Trockenresistenz der Flechten in Beziehung zu ihrer Verbreitung. *Flora*, Vol. 140, pp. 39, ISSN 0367-2530

Lange, O.L. (1970). Experimentell-ökologische Untersuchungen an Flechten der Negev-Wüste. I. CO$_2$-Gaswechsel von *Ramalina maciformis* (Del.) Bory unter kontrollierten Bedingungen im Laboratorium. *Flora* (Abt B), Vol. 158, pp. 324-359, ISSN 0367-2530.

Lange, O.L. & Matthes, U. (1981). Moisture-dependent CO$_2$ exchange in lichens. *Photosynthetica*, Vol. 15, pp. 555-574, ISSN:0300-3604.

Lange, O.L. & Tenhunen J.D. (1981). Moisture content and CO_2 exchange of lichens. II. Depression of net photsynthesis in *Ramalina mcaciformis* at high water content is caused by increased thallus carbon dioxide diffusion resistance. *Oecologia*, Vol. 51, pp. 426-429, ISSN 0029-8549.

Lange, O.L.; Green, T.G.A.; Melzer A. & Zellner H. (2006). Water relations and CO_2 exchange of the terrestrial lichen *Teloschistes capensis* in the Namib fog desert: Measurements during two seasons in the field and under controlled conditions. *Flora*, Vol. 201, No. 4, (June 2006), pp. 268-280, ISSN 0367-2530.

Lascano, HR.; Gómez, LD.; Casano, LM. & Trippi, VS. (1999). Wheat Chloroplastic Glutathione Reductase Activity is Regulated by the Combined Effect of pH, NADPH and GSSG. *Plant and Cell Physiology*, Vol, 40, No, 7, (April 1999), pp. 683-690, ISSN 0032-0781.

Li, X.P.; Muller-Moule, P.; Gilmore, A.M. & Niyogi, K.K. (2002). PsbS-dependent enhancement of feedback de-excitation protects photosystem II from photoinhibition. *Proceedings of the National Academy of Sciences of the United States of America*, Vol. 99, No. 23, (November 2002), pp. 15222-15227, ISSN 0027-8424.

Mayaba, N. & Beckett, R.P. (2001). The effect of desiccation on the activities of antioxidant enzymes in lichens from habitats of contrasting water status. *Symbiosis*, Vol. 31, No. 1, pp. 113-121, ISSN 0334-5114.

Minibayeva, F. & Beckett, RP. (2001). High rates of extracellular superoxide production in bryophytes and lichens, and an oxidative burst in response to rehydration following desiccation. *New Phytologist*, Vol. 152, No. 2, (November 2001), pp. 333-341, ISSN 1469-8137.

Miranda, KM.; Espey, MG.; Jourd'heuil, D.; Grisham, MB.; Fukuto, JM. & Feelish, M. (2000). The chemical biology of NO. In *Nitric Oxide. Biology and Pathology*, Ignarro L, (Ed.), , 41-55, CA: Academic Press, ISBN 0-12-370420-0, Los Angeles, USA.

Mittler, R. (2002). Oxidative stress, antioxidants and stress tolerance. *Trends in Plant Science*, Vol. 7, No. 9, (September 2002), pp. 405-410, ISSN 1360-1385.

Miyake, C. & Asada, K. (1992). Thylakoid-Bound Ascorbate Peroxidase in Spinach Chloroplasts and Photoreduction of Its Primary Oxidation Product Monodehydroascorbate Radicals in Thylakoids . *Plant & Cell Physiology*, Vol. 33, No. 5, (July 1992), pp. 541-553, ISSN 0032-0781.

Møller, I. M. (2001). Plant mitochondria and oxidative stress. Electron transport, NADPH turnover and metabolism of reactive oxygen species. *Annual Review of Plant Biology*, Vol. 52, (June 2001), pp. 561-591, ISSN 1543-5008.

Müller, P.; Li, X-P. & Niyogi, KK. (2001). Non-photochemical quenching: a response to excess light energy. *Plant Physiology*, Vol. 125, No. 4, (April 2001), pp. 1558-1566, ISSN 0032-0889.

Munekage, Y.; Hojo, M.; Meurer, J.; Endo, T.; Tasaka, M. & Shikanai, T. (2002). PGR5 is involved in cyclic electron flow around photosystem I and is essential for photoprotection in *Arabidopsis*. *Cell*, Vol. 110, No. 3, (August 2002), pp. 361-371, ISSN 0092-8674.

Niyogi, KK.; Li, XP.; Rosenberg, V. & Jung, HS. (2004). Is psbS the site of non-photochemical quenching in photosynthesis? *Journal of Experimental Botany*, Vol. 56, No. 411, (January 2005), pp. 375–382, ISSN 0022-0957.

Noctor, G. & Foyer, CH. (1998). Ascorbate and glutathione: keeping active oxygen under control. *Annual Review of Plant Physiology and Plant Molecular Biology*, Vol. 49, (June 1998), pp. 249-279, ISSN 1040-2519.

Oliver, MJ. & Bewley, JD. (1997). Desiccation-tolerance of plant tissues. A mechanistic overview. *Horticultural reviews* 18, Janick, J. (Ed.), 171–213, Wiley & Sons Inc., ISBN 9780470650608, New York, USA.

Oliver, M.J. & Wood, A J. (1997). Desiccation tolerance in mosses. In *Stress induced processes in higher eukaryotic cells*, Koval, T. M. (Ed.), 1-26, Plenum Press, New York, USA.

Oliver, MJ.; Velten, J. & Wood, AJ. (2000). Bryophytes as experimental models for the study of environmental stress: desiccation-tolerance in mosses. *Plant Ecology*, Vol. 151, No. 1, (November 2000), pp. 73–84, ISSN 1385-0237.

Ort, DR. (2001). When there is too much light. *Plant Physiology*, Vol. 125, No. 1, (january 2001), pp. 29–32, ISSN 0032-0889.

Palmieri, MC.; Sell, S.; Huang, X.; Scherf, M.; Werner, T.; Durner, J. & Lindermayr, C. (2008). Nitric oxide-responsive genes and promoters in *Arabidopsis thaliana*: a bioinformatics approach. *Journal of Experimental Botany*, Vol. 59, No. 2, (February 2008), pp. 177-186, ISSN 0022-0957.

Peltier, G.; Tolleter, D.; Billon, E. & Cournac, L. (2010). Auxiliary electron transport pathways in chloroplasts of microalgae. Photosynthesis Research, Vol. 106, No. 1-2, (November 2010), pp. 19-31.

Piccotto, M. & Tretiach, M. (2010).Photosynthesis in chlorolichens: the influence of the habitat light regime. *Journal of Plant Research*, Vol. 123, No. 6, (November 2010), pp. 763-765, ISSN 0918-9440.

Polle, A. (2001). Dissecting the superoxide dismutase–ascorbate peroxidase–glutathione pathway in chloroplasts by metabolic modeling. Computer simulations as a step towards flux analysis. *Plant Physiology*, Vol. 126, No. 1, (May 2001), pp. 445–462, ISSN 0032-0889.

Proctor, M.C. & Smirnoff, N. (2000). Rapid recovery of photosystems on rewetting desiccation-tolerant mosses: chlorophyll fluorescence and inhibitor experiments. *Journal of Experimental Botany*, Vol. 51, No, 351, (October 2000), pp. 1695-1704, ISSN 0022-0957.

Proctor, M.C.F. & Tuba, Z. (2002). Poikilohydry and homoiohydric: antithesis or spectrum of possibilities. *New Phytologist*, Vol. 156, pp. 327-349, ISSN 1469-8137.

Richardson, DH. (1971). Lichens. In: *Methods in Microbiology*, Booth, C. (Ed.), 267–293, Academic Press, ISBN 19722702089, New York, USA.

Ros Barceló, A. (1998). The generation of H_2O_2 in the xylem of *Zinnia elegans* is mediated by an NADPH-oxidase-like enzyme. *Planta*, Vol. 207, No. 2, (November 1998), pp. 207-216, ISSN 0032-0935.

Ros Barceló, A.; Gómez Ros, LV. & Esteban Carrasco, A. (2007). Looking for syringyl peroxidases. *Trends in Plant Science*, Vol.12 No.11, (November 2007), pp. 486-491, ISSN 1360-1385.

Ruban, A.V.; Pascal, A.A.; Robert, B. & Horton, P. (2002). Activation of zeaxanthin is an obligatory event in the regulation of photosynthetic light harvesting. *The Journal of Biological Chemistry*, Vol. 277 (March 2002), pp. 7785-7789, ISSN 0021-9258.

Rundel, PW. (1988). Water relations. In: *Handbook of lichenology, vol II*. Galun, M. (Ed.), 17–36. CRC Press, ISBN 0-8493-3583-3, Boca Raton, Florida, USA.

Scheidegger, C.; Schroeter B. & Frey, B. (1995). Structural and functional processes during water vapour uptake and desiccationin selected lichens with green algal photobionts. *Planta*, Vol. 197, No. 2, (September 1995), pp. 375-389, ISSN 0032-0935.

Schofield, S.C.; Campbell, D.A.; Funk, C. & MacKenzie, T.D.B. (2003). Changes in macromolecular allocation in nondividing algal symbionts allow for photosynthetic acclimation in the lichen *Lobaria pulmonaria*. *New Phytologist*, Vol. 159, No. 3, (September 2003), pp. 709-718, ISSN 1469-8137.

Scott, G.D. (1969). *Plant Symbiosis*. Edward Arnold, ISBN 0713122366, London, UK.

Schreiber, U.: Schliwa, U. & Bilger, W. (1986). Continuous Recording of Photochemical and Nonphotochemical Chlorophyll Fluorescence Quenching with A New Type of Modulation Fluorometer. *Photosynthesis Research*, Vol. 10, No. 1-2, (January 1986), pp. 51-62, ISSN 0166-8595.

Sharma, SS. & Dietz, KJ. (2009). The relationship between metal toxicity and cellular redox imbalance. *Trends in Plant Science*, Vol.14, No.1, (January 2009), pp. 43-50, ISSN 1360-1385.

Smirnoff, N. (1993). The role of active oxygen in the response of plants to water deficit and desiccation. *New Phytologist*, Vol. 125, No. 1(February 1993), pp. 27-58, ISSN 0028-646X.

Spinall-O'Dea, M.; Wentworth, M.; Pascal, A.; Robert, B.; Ruban, A. & Horton, P. (2002). In vitro reconstitution of the activated zeaxanthin state associated with energy dissipation in plants. *Proceedings of the National Academy of Sciences of the United States of America*, Vol. 99, No. 25, (December 2002), pp. 16331-16335, ISSN 0027-8424.

Tschermak-Woess, E. (1989). Developmental studies in trebouxioid algae and taxonomical consequences. *Plant Systematics and Evolution*, Vol. 164, No. 1, (March 1989), pp. 161-195, ISSN 0378-2697.

Tuba, Z.; Cintalan, Z. & Proctor, M.C.F. (1996). Photosysnthetic response of a moss, *Tortula ruralis, ssp. ruralis*, and the lichens *Cladonia convoluta* and *C. furcata* to water deficit and short periods of desiccation, and their ecophysiological significance: a baseline study at present-day CO_2 concentration. *New Phytologist*, Vol. 133:, No. 2, (June 1996), pp. 353-361, ISSN 0028-646X.

Valenzuela, A. (1991). The biological significance of malondialdehyde determination in the assessment of tissue oxidative stress. *Life Sciences*, Vol. 48, No. 4, pp. 301-309, ISSN 0024-3205.

Veerman, J.; Vasilev, S.; Paton, G.D.; Ramanauskas, J. & Bruce, D. (2007). Photoprotection in the lichen *Parmelia sulcata*: the origins of desiccation-induced fluorescence quenching. *Plant Physiology*, Vol. 145, No. 3, (November 2007), pp. 997-1005, ISSN 0032-0889.

Weissman, L.; Garty, J. & Hochman, A. (2005). Characterization of enzymatic antioxidants in the lichen *Ramalina lacera* and their response to rehydration. *Applied and Environmental Microbiology*, Vol. 71, No. 11, (November 2005), pp. 6508 - 6514.

Yruela, I.; Gatzen, G.; Picorel, R. & Holzwarth, AR. (1996). Cu(II)-inhibitory efect on Photosystem II from higher plants. A picosecond time-esolved fluorescence study. *Biochemistry*, Vol. 35, pp. 9469-9474, ISSN 0001527X.

Zapata, JM.; Sabater, B. & Martin, M. (1998). Identification of a thylakoid peroxidase of barley which oxidizes hydroquinone. *Phytochemistry*, Vol. 48, No. 7, (August 1998), pp. 1119–1123, ISSN 0031-9422.

Zapata, JM.; Guéra, A.; Esteban-Carrasco, A.; Martín, M. & Sabater, B. (2005). Chloroplasts regulate leaf senescence: delayed senescence in transgenic ndhF-defective tobacco. *Cell Death and Differentiation*, Vol. 12, (May 2005), pp. 1277–1284, ISSN 1350-9047.

Zapata, JM.; Gasulla, F.; Esteban-Carrasco, A.; Barreno, E. & Guéra, A. (2007). Inactivation of a plastid evolutionary conserved gene affects PSII electron transport, life span and fitness of tobacco plants. *New Phytologist*, Vol. 174, No. 2, (April 2007), pp. 357–366, ISSN 0028-646X.

Zarter, CR.; Adams, WW III.; Ebbert, V.; Adamska, I.; Jansson, S. & Demmig-Adams, B. (2006a). Winter acclimation of PsbS and related proteins in the evergreen *Arctostaphylos uva-ursi* as influenced by altitude and light environment. *Plant, Cell & Environment*, Vol. 29, No. 5, (May 2006), pp. 869–878, ISSN 0140-7791.

Zarter, CR.; Adams, WW III.; Ebbert, V.; Cuthbertson, D.; Adamska, I. & Demmig-Adams, B. (2006b). Winter downregulation of intrinsic photosynthetic capacity coupled with upregulation of Elip-like proteins and persistent energy dissipation in a subalpine forest. *New Phytologist*, Vol. 172, No. 2, (October 2006), pp. 272-282, ISSN 0028-646X.

Zorn, M.; Pfeifhofer, H.W.; Grill, D. & Kranner, I. (2001). Responses of platid pigments to desiccation and rehydration in the desert lichen Ramalina maciformis. *Symbiosis* 31, No. 1, pp. 201-211, ISSN 0334-5114.

9

The Photomorphogenic Signal: An Essential Component of Photoautotrophic Life

Sabrina Iñigo, Mariana R. Barber, Maximiliano Sánchez-Lamas,
Francisco M. Iglesias and Pablo D. Cerdán
Fundación Instituto Leloir, IIBBA-CONICET and FCEN-UBA
Argentina

1. Introduction

In a thermosolar plant, the engineers locate movable mirrors that concentrate solar radiation. These plants are designed to maximize energy capture. In green plants, their morphology changes to maximize energy capture as well, but also to avoid light in excess, which can damage plant tissues. Contrary to mirrors in thermosolar plants, located in desert land and organized in regular arrays, most green plants grow surrounded of vegetation and their own tissues are not regularly spaced, new leaves tend to shade older leaves. Hence, plants need to use light as a source of information in order to properly locate their "sunlight collectors" and be able to efficiently use light as an energy source for photosynthesis.

To monitor environmental light conditions, plants are equipped with several families of photoreceptors: the cryptochromes, the phytochromes and the LOV-domain bearing photoreceptors. While phytochromes perceive light most effectively in the red/far-red region of the spectrum, cryptochromes and LOV photoreceptors detect blue and UV-A light. Downstream these photoreceptors, plants have evolved sophisticated transcriptional networks that mediate metabolic and developmental changes in response to light. These light-regulated processes include seed germination, seedling photomophogenesis, greening, shade avoidance, photoperiodic responses and senescence.

Greening and chloroplast biogenesis are promoted after light exposure. Phytochromes and cryptochromes trigger to initiate this biogenesis, which includes the induction of photosynthesis-related genes at the transcriptional level, the import of nuclear-encoded proteins and the establishment of a thylakoid network fully assembled with photosynthetic electron transport complexes. Furthermore, these photoreceptors affect the synthesis of chlorophyll and other photosynthetic accessory pigments; modifying the photosynthetic apparatus properties as a result of light quality perception. On the other hand, phytochromes are also involved in the induction of Rubisco, a key enzyme of the Calvin Cycle. Light quality plays an important role in modulating the photosynthetic characteristics. It regulates chlorophyll degradation, modulates photosystem stoichiometry and the activity of the ROS scavenging system.

Besides the role in the formation of the photosynthetic apparatus, the photoreceptors play significant roles in establishing how the photosynthetic pigments will be exposed to light to harvest its energy content. Under weak light, chloroplasts move towards light, in a blue light dependent way, to optimize the light absorption and photosynthesis. However, under strong light, chloroplasts show an opposite response to avoid photodamage (Kodama *et al.*, 2011). Besides chloroplast movement, photoreceptors modulate plant architecture to maximize the photosynthetic surface exposed to light. When plants perceive the presence of plant neighbours, phytochromes trigger the elongation of the stem and petioles, a series of changes known as Shade Avoidance Syndrome (SAS). The manipulation of phytochrome levels has been used to improve the harvest index of tobacco plants (Robson *et al.*, 1996) by avoiding the diversion of resources to the SAS. However, the phytochromes are still very important to position the leaves in the canopy. Thus, manipulation of phytochrome activity must be precise, if used to improve crop performance (Maddonni *et al.*, 2002).

In this chapter, we focus on the role of the photomorphogenic signal to trigger the synthesis of photosynthetic genes and pigments during the greening process and later on, during photosynthetic plant development, with emphasis on the regulation of gene expression.

2. Photomorphogenesis

Plants are sessile organisms, and as such, have evolved a great deal of developmental plasticity to optimally respond to the immediate environment. Light is one of the most important cues for plant growth; plants respond to its intensity, wavelength, direction and periodicity (Franklin & Quail, 2009). The first physiological consequence of light perception is the reprogramming of the seedling development in a process termed deetiolation, or photomorphogenesis. In darkness, seedlings display a skotomorphogenesis development characterised by the following phenotypes: elongated hypocotyl, closed, pale and unexpanded cotyledons; the apical hook remains closed to protect the apical meristem before emerging from the soil and chlorophyll and anthocyanin biosynthesis do not take place. All these features allow the seedlings to grow through a layer of soil and eventually emerge into the light.   Once the seedlings perceive sufficient light, they exhibit a photomorphogenic development. They undergo deetiolation that includes inhibition of hypocotyl elongation, unfolding and greening of cotyledons, opening of the apical hook, chlorophyll and anthocyanin biosynthesis and differentiation of chloroplasts; processes aimed to achieve full autotrophy. This transition from skotomorphogenic to photomorphogenic development is steered by a complex molecular network that includes upstream signalling components (photoreceptors) and intermediate factors transducing the signal to downstream regulators. These downstream components integrate the light signals from the various photoreceptors and bring about changes in metabolism and gene expression that eventually lead to photomorphogenesis (Casal *et al.*, 2003).

2.1 Photomorphogenic photoreceptors

Light is directly perceived by protein molecules known as photoreceptors. Photoreceptors are considered as such if upon photon absorption they are able to deliver a signal to downstream components. Because membranes are transparent to light, most photoreceptors are cytoplasmic and water soluble, contrary to other type of receptors whose ligands are not able to move through membranes.

The solar spectrum at Earth's surface extends from UV (about 280 nm) through the blue to red/far red (about 750 nm). Because the polypeptide backbone and amino acid side chains do not absorb in most of this range, most photoreceptors contain an organic, non-protein component, known as the chromophore. Chromophores can be attached either covalently or non-covalently to the apoprotein (Moglich *et al.*, 2010). As explained above, plants possess several classes of photoreceptors whose absorption properties match the spectrum of the incoming light: the red/far-red photoreversible phytochromes, the UV-A/blue-light absorbing cryptochromes, the phototropins, the members of the Zeitlupe family (Moglich *et al.*, 2010) and, more recently, a UV-B specific photoreceptor, UVR8, has been added to the list (Rizzini *et al.*, 2011).

2.2 Phytochromes
2.2.1 Generalities about phytochromes

Phytochromes are the only red and far-red light photoreceptors in plants (Strasser *et al.*, 2010; Takano *et al.*, 2009) and, together with cryptochromes and phototropins, constitute one of the three mayor regulators of photomorphogenesis (Rockwell *et al.*, 2006). Phytochromes are synthesised in the cytosol as soluble dimers composed of two 125-kDa polypeptides. Each polypeptide folds into two main domains. The amino-terminal domain covalently binds phytochromobilin (PφB), a tetrapyrrole chromophore that confers the spectral properties characteristic of phytochromes. PφB is synthesised from haeme in plastids, haeme oxygenase encoded by *HY1* converts haeme into biliverdin IXα, which is reduced to 3Z-PφB by the PφB synthase (*HY2*). Then 3Z-PφB isomerises to 3E-PφB and attaches covalently to phytochrome (Tanaka & Tanaka, 2007).

The carboxy-terminal part of the phytochrome molecule is involved in dimerisation and transfer of the signal to downstream components (Rockwell *et al.*, 2006). Phytochromes are synthesised in the dark in a biologically inactive red-light absorbing form (known as Pr). Biological activity is acquired upon red-light triggered photoconversion to the far-red light absorbing form (known as Pfr). Photoconversion of Pfr back to Pr is triggered by far-red light. Both reactions are fully reversible, and eventually results in a dynamic photoequilibrium of Pr and Pfr in natural light conditions that depends on the proportion of red to far-red light (Franklin & Quail, 2009). The conversion is due to a single photochemical isomerisation of the chromophore about a specific double bond between the rings C and D of the phytochromobilin (Rockwell *et al.*, 2006). Following conversion, Pfr translocates into the nucleus (Fankhauser & Chen, 2008).

The phytochromes are encoded by a small gene family in angiosperms. The rice genome, for example, encodes three members, phyA, phyB and phyC, each representing one of the lineages found in plants (Sharrock, 2008). In Arabidopsis, the phytochrome family consists of five members, designated phytochrome A (phyA) to phytochrome E (phyE).

Classical photobiological experiments established three phytochromes modes of acting, the Very Low Fluence Response mode (VLFR), where responses to phytochromes are already saturated at very low fluencies of light, the Low Fluence Response (LFR) showing the classical red and far-red light reversibility and the High Irradiance responses (HIR) that require prolonged exposures to light of relative high intensity (Casal *et al.*, 2003). Now that we know about each phytochrome species, the phytochrome action modes can be explained by the different phytochrome species and different signal transduction pathways. phyA is the most specialized of the phytochromes; it is responsible for the VLFR and the HIR. The extraordinary sensitivity of this photoreceptor to light allows phyA to control germination of buried seeds in the soil and to induce germination when seeds are exposed briefly to light. phyA importance is evident when plants germinate under a dense canopy (Yanovsky, 1995) or for example, when weeds are induced to germinate after soil tillage (Ballaré, 1992). The other phytochromes control the R/FR reversible LFR and the responses to continuous red light. phyB is involved in seed germination, deetiolation, stem elongation, the SAS, stomatal development and several other aspects of plant development. phyD and phyE act in SAS by controlling internode elongation and flowering time, and phyE is also involved in far-red HIR- mediated seed germination (Franklin & Quail, 2009).

Before even knowing of the existence of multiple phytochromes, they were classified in type I, the light-labile pool and type II, the light stable pool. Now we know that type I is represented by phyA and type II by the other phytochromes. As illustrated above, type I and II phytochromes play distinct roles. The rapid proteolytic degradation of phyA is believed to be responsible for the termination of signalling. The light stable phytochromes are not totally resistant to proteolytic degradation (Jang *et al.*, 2010), but dark reversion also emerges as a switch-off mechanism. Dark reversion is a thermal process in which the Pfr form is slowly converted to the Pr form in the dark. Although dark reversion is not yet well characterised, it makes an important contribution to the balance between Pr and Pfr and hence, to determine the output state for a given phytochrome (Rockwell *et al.*, 2006).

2.2.2 Phytochrome structure and nuclear translocation

The two phytochrome mayor domains mentioned above are separated by a flexible hinge region. The N-terminal photosensory region (70 kDa) contains an N-terminal extension (ATE) and three conserved subdomains: PAS, GAF and PHY. The ATE is poorly conserved, possibly accounting for some functional differences among phytochromes and it might be implicated in stabilization of the Pfr form of photoreceptors. The GAF domain is associated with the bilin chromophore and possesses bilin lyase activity. The PAS and PHY domains are important for tuning the spectroscopic properties of the bound bilin.

A flexible hinge region separates the N-terminal domains from the C-terminal regulatory region (55 kDa), which is composed of two PAS subdomains, called PAS 1 and PAS 2, and a histidine kinase related domain (HKRD) (Figure 1). The PAS and HKRD domains contribute to the high-affinity subunit-subunit interaction between the phytochrome monomers to form dimers, and both domains are required for the formation of nuclear speckles. Besides, the PAS domains contain the nuclear localization signal (NLS) responsible for the relocalisation to the nucleus after phytochrome photoconversion (Rockwell *et al.*, 2006). Finally, at least one domain must be responsible for the serine/threonine kinase activity that governs phytochrome autophosphorylation and phytochrome–directed phosphorylation of other proteins, such as PHYTOCHROME-INTERACTING FACTOR 3 (PIF3). The functional significance of this kinase activity remains unknown. HKRD domain was initially suggested to be a kinase because of its relatedness to bacterial histidin kinases (Figure 1). However, it was shown that the kinase activity resides in the N-terminal domain (Bae & Choi, 2008). Further, it was recently shown that a Casein Kinase II is involved in phosphorylating phytochrome interacting factor 1 (PIF1), one or the downstream effectors of phytochrome signalling (see below) (Bu *et al.*, 2011).

2.3 Cryptochromes

Cryptochromes are receptors for blue and ultraviolet light. Arabidopsis contains two cryptochromes, cry1 and cry2. They are composed of two domains, an N-terminal photolyase related region (PHR), without photolyase activity, and a C-terminal extension domain (CCT), more variable among family members (Figure 1). The PHR region binds two chromophores, flavin adenine dinucleotide (FAD) and 5,10-methenyltetrahydrofolate

(MTFG). The CCT domain appears to be important for cryptochrome function, it interacts with downstream effectors and promotes photomorphogenic development in the dark by itself (Li & Yang, 2007). cry1 and cry2 form homodimers; dimerisation is mediated by the PRH domain and appears to be essential for signalling (Moglich *et al.*, 2010). cry1 and cry2 are predominantly nuclear. However, cry1 is also found in the cytoplasm. They mediate the regulation of gene expression and together are responsible for blue-light dependent changes in gene expression of up to 10-20% of the Arabidopsis genome (Lin & Todo, 2005). cry1 and cry2 participate in many blue-light responses including inhibition of hypocotyl elongation, anthocyanin accumulation, regulation of flowering time, stem and internode elongation, blue-light regulated gene expression, and entrainment of circadian rhythms. The function of cry1 and cry2 partially overlap, but differences are evident at different light intensities or at different developmental stages. Under high intensities of blue light, cry2 is rapidly degraded, leaving cry1 as the predominant photoreceptor, so the role of cry2 during seedling deetiolation is more evident under low blue light intensities. In contrast, cry2 role is predominant in the regulation of flowering time (Li & Yang, 2007).

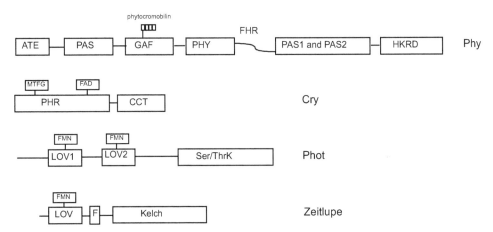

Fig. 1. Schematic representation of domain structure and chromophores of the main photoreceptors: phytochromes (phy), cryptochromes (cry), phototropins (phot) and the zeitlupe family. Domain abbreviation are ATE (N-terminal extension); PAS: domain acronym derived from period clock (PER) protein, aromatic hydrocarbon receptor nuclear translocator (ARNT), and single minded (SIM); GAF: (domain acronym derived from vertebrate cGMP-specific phosphodiesterase, cyanobacterial adenylate cyclase and formate hydrogen lyase transcription activator FhlA); PHY (phytochrome); FHR (flexible hinge region); HKRD (Histidine kinase related domain), PHR (photolyase related region), CCT (C-terminal extension domain), FAD (flavin adenine dinucleotide), MTFG (5,10-methenyltetrahydrofolate), LOV (Light-oxygen-voltage domains), Ser/Thrk (serine threonine kinase domain), FMN (Flavin mononucleotide), F (F box), Kelch (Kelch repeat).

2.4 LOV domain photoreceptors: The phototropins and the ztl family
The phototropic response of plants has been known at least since Darwin times. The photoreceptors involved were identified after finding mutants impaired in the phototropic response (Huala *et al.*, 1997), and were later named phot1 and phot2. The sequence revealed

the presence of two domains showing homology to domains that are involved in sensing Light, Oxygen or Voltage, the LOV domains (Figure 1). These domains bind FMN, the chromophore for phototropism. Phototropins were important in the identification of other LOV-domain containing photoreceptors.

The second family of LOV photoreceptors is comprised by Zeitlupe/Adagio/LOV KELCH Protein 1 (ztl/ado/lkp1), fkf1 and lkp2. Contrary to phototropins, the ztl family contains a single LOV domain, an F-box and a C-terminal Kelch domain. F-box proteins play a role in recruiting specific substrates for ubiquitination and protein degradation, whereas the Kelch domain might aid in this function by mediating protein-protein interactions (Moglich *et al.*, 2010). These photoreceptors have important functions in flowering time and circadian clock function, as we will explain in the following sections, mainly by controlling the stability of important clock associated proteins (Harmer, 2009; Mas, 2005).

2.5 The UV-B specific photoreceptor: The UVR8 protein

UV-B radiation (280-315 nm) is an integral part of the sunlight reaching the surface of the Earth and induces a broad range of physiological responses that are mediated by a recently identified UV-B specific photoreceptor, UVR8 (Rizzini *et al.*, 2011). The most extensively studied examples of photomorphogenic responses are the suppression of hypocotyl extension by low fluences of UV-B and the induction of genes involved in flavonoid biosynthesis (Jenkins, 2009).

UVR8 is a β-propeller protein with similar sequence to the eukaryotic RCC1, a guanine nucleotide exchange factor (GEF) for the small GTP-binding protein Ran (Gruber *et al.*, 2010). Aromatic amino acids absorb UV-B radiation. Tryptophan, with an absorption maximum in solution at around 280 nm (which extends to 300 nm and is likely to be further shifted in a protein environment), is particularly suited as UV-B chromophore. Structure modelling according to structurally related RRC1, identified 14 tryptophans of UVR8 all located at the top of the predicted β-propeller cluster in the centre of the protein structure. Evidence suggests UV-B perception is based on a tryptophan-based mechanism, an important difference with the other Photoreceptors that bear chromophores suited for visible light perception (Rizzini *et al.*, 2011).

3. Transducers of light signalling

3.1 COP1 is a general repressor of photomorphogenesis

Most of the photoreceptors mentioned above were identified after genetic screenings in Arabidopsis and led to the isolation of mutants defective in deetiolation. Other type of genetic screenings led to the isolation of mutants with constitutive photomorphogenic phenotypes in the dark (*cop*) or deetiolated (*det*). The phenotype of one of such mutants, *cop1* (Deng *et al.*, 1992), suggested that it was a negative regulator of photomorphogenesis. COP1 is an essential protein, null alleles are not viable. The overlap between the light-responsive transcriptome and the *cop1*-responsive transcriptome in dark grown seedlings clearly shows that COP1 is a general repressor of photomorphogenesis (Ma *et al.*, 2002). We now know that COP1 is a single unit E3 ubiquitin ligase, bearing both the substrate and E2 binding motifs. Ubiquitin ligation is the last step in the chain of events that leads to protein ubiquitination that marks proteins for degradation by the 26S proteasome. The COP1 protein bears three domains: a RING-finger motif, a coiled-coil domain and seven WD40 repeats. The RING domain is essential to recruit E2s and the other domains to recognize substrates. Several of the COP1 substrates have been characterised and they are transcription factors that act positively on photomorphogenesis.

3.2 COP1 targets positive regulators of photomorphogenesis for degradation
3.2.1 COP1 in phyA signalling
Genetic and molecular approaches have identified several transcription factors acting positively on photomorphogenesis downstream of photoreceptors. As phyA is the main photoreceptor perceiving continuous far-red light (acting in the HIR mode), mutants with long hypocotyls under far-red light were isolated, leading to phyA signalling components. Two of the genes identified, *long after far-red light 1* (*laf1*) and *long hypocotyl in far-red* (*hfr*) encode an R2/R3 MYB and a bHLH transcription factor respectively. Other mutants helped to identify other phyA signalling components; among them, two small plant-specific proteins involved in light-regulated phyA import to the nuclei, FAR-RED ELONGATED HYPOCOTYL1 (FHY1) and its homolog FHY1-LIKE (FHL) (Fankhauser & Chen, 2008), and two transposase-derived transcription factors, FHY3 and its homolog FAR-RED IMPAIRED RESPONSE1 (FAR1), which are direct activators of *FHY1* and *FHL* transcription, promoting phyA signalling (Lin *et al.*, 2007).

Genetic screenings for enhancers of phyA signalling led to the identification of SUPPRESSOR OF PHYTOCHROME A-105 1 (SPA1), which belongs to a small family of four proteins (SPA1-4). The quadruple mutant defective for the four *SPA* genes shows a constitutive photomorphogenesis phenotype in the dark, similar to *cop1* mutants (Laubinger *et al.*, 2004). SPA proteins and COP1 form complexes and, as mentioned above, show E3 ligase activity (Zhu *et al.*, 2008). This SPA-COP1 complex targets HFR and LAF1 for degradation, explaining part of its negative role in photomorphogenesis (Henriques *et al.*, 2009).

The SPA1-COP1 E3 ligase complex targets other important transcription factors for degradation, like elongated hypocotyl 5 (hy5) and hy5 homolog (hyh), two bZIP transcription factors. These transcription factors promote photomorphogenesis under various wavelengths and will be explained in the following sections.

3.2.2 COP1 in cryptochrome signalling
hy5 mutants display a long hypocotyl phenotype under diverse wavelengths of light, suggesting HY5 is a common promotor of photomorphogenesis downstream several photoreceptor signalling pathways. Interestingly, the association between HY5 and COP1 can be deduced from the overlapping set of differentially expressed genes in the respective mutants (Ma *et al.*, 2002). At the biochemical level, it was shown that both HY5 and HYH are targeted for degradation by COP1 (Holm *et al.*, 2002; Osterlund *et al.*, 2000). On the other hand, cryptochromes are known to interact with COP1 through its CCT domain and negatively regulate COP1 activity (Li & Yang, 2007). However, the precise light-mediated mechanism that controls COP1 activity remained unknown until recently. Three simultaneous publications addressed this issue (Lian *et al.*, 2011; Liu *et al.*, 2011; Zuo *et al.*, 2011). They showed that CRY1 interacts with the SPA proteins in a blue-light dependent manner and inhibit the interaction between COP1 and SPA proteins. This mechanism disrupts the complex E3 ligase activity and avoids HY5 degradation, promoting photomorphogenesis. In the case of CRY2, a similar blue-light dependent interaction with SPA proteins inhibits the activity of the COP1-SPA complex. This inhibition leads to higher levels of CONSTANS, a transcription factor that promotes flowering in long-day conditions.

These facts also explain some of the differences between the roles of CRY1 and CRY2 in plant development that we mentioned before.

3.2.3 COP1 in UV-B signalling
UVR8 forms a dimer but rapidly dissociates as the result of direct perception of UV-B. This is followed by a rapid nuclear accumulation of UVR8 and UVR8 interaction with COP1 that depends on the C-terminal WD40-repeat domain. The UVR8-COP1 interaction mediates the activation of numerous genes, including HY5, inducing photomorphogenic responses (Favory *et al.*, 2009; Jenkins, 2009).

3.3 The PIF family of bHLH transcription factor represses photomorphogenesis downstream of phytochromes
The photoconversion of Pr to Pfr with red light leads to conformational changes that unmask the NLS to become accessible for the nuclear-transport machinery and also allow the interacting surfaces for partner proteins. Within the nucleus, phytochromes accumulate in subnuclear foci, the phytochrome Nuclear Bodies (NBs). The identification of HEMERA, a protein involved in the formation of NBs, supports the notion that NBs are the sites of phytochrome-induced protein degradation (Chen *et al.*, 2010). Phytochrome-induced protein degradation is important to control the activity of the Phytochrome interaction factor (PIF) family of bHLH transcription factors. The Pfr form is rapidly translocated into the nucleus, where it interacts with PIFs, more strongly with the Pfr form (Fankhauser & Chen, 2008). Upon binding Pfr, the PIFs are phosphorylated and degraded. This event initiates a gene expression cascade leading to photomorphogenesis (Bae & Choi, 2008; Leivar *et al.*, 2009; Shen *et al.*, 2008).

The PIFs belong to a transcription factor superfamily, which forms dimers to target specific DNA sites and are well characterised in nonplant eukaryotes as important regulatory components in diverse biological processes. In Arabidopsis, there are at least 133 bHLH protein–encoding genes. Phylogenetic analysis of the bHLH domain sequences allowed the classification of these genes into 21 subfamilies (Heim *et al.*, 2003; Toledo-Ortiz *et al.*, 2003). The PIFs subfamily, called PHYTOCHROME INTERACTING FACTORs (PIFs) is involved in the repression of seed germination, promotion of seedling skotomorphogenesis and SAS, by regulating the expression of over a thousand genes (Leivar & Quail, 2011). PIF3 was the first member identified in this subfamily, isolated by a two-hybrid assay as a PHYB interactor (Ni *et al.*, 1998). Afterwards, other members of the family were identified by computational analysis (Leivar & Quail, 2011). Unlike other bHLHs, this subfamily have a characteristic active phytochrome binding motif (APB) in its N-terminal, that make them able to interact with the photoactivated phytochrome (Leivar & Quail, 2011).

PIFs can form homodimers and heterodimers that bind specifically to the G-box motif CACGTG (Toledo-Ortiz *et al.*, 2003) and, in some cases, HFR1 and other bHLH closely related to PIFs can form non-DNA binding bHLH heterodimers with some PIFs, preventing excessive responses (Hornitschek *et al.*, 2009). In addition, different PIFs are regulated preferentially by different phytochromes (Shen *et al.*, 2008).

As mentioned above, PIF3 is the founding member of this family. PIF3 acts as a negative regulator in both phyA and phyB-mediated seedling deetiolation processes such as hook opening and hypocotyl elongation. Both phyA and phyB bind to PIF3. This interaction leads to the phosphorylation of PIF3, triggering its degradation by the 26S proteasome-dependent pathway, and thus relieving its negative regulation of photomorphogenesis. phyA is responsible for the rapid degradation of PIF3 in response to far-red light, whereas phyA, phyB and phyD are responsible for PIF3 degradation in response to red light (Bae & Choi, 2008).

PIF1 (also known as PIL5), PIF4, PIF5 (also known as PIL6) and PIF6 (also known as PIL2) also have important roles in photomorphogenic development. Although they have highly similar sequences, their roles do not overlap completely. For example, PIF1 negatively regulates seed germination by inhibiting gibberelin (GA) biosynthesis and GA signalling, and simultaneously activating abscisic acid biosynthesis. In addition, PIF1 activates the expression of two DELLA genes, which are key negative GA signalling components. Phytochromes promote seed germination by inhibiting PIF1 activity. Conversely, PIF4 and PIF5 have important roles in the regulation of the SAS (Leivar & Quail, 2011). We will describe the roles of PIFs in chloroplast biogenesis and c chlorophyll synthesis in a following section.

4. The role of the circadian clock in photomorphogenic development

The circadian clocks are endogenous mechanisms that allow organisms to time their physiological changes to day/night cycles. These mechanisms are present in a wide range of organisms, from cyanobacteria to mammals. Circadian clocks generate rhythms with a ~24 hr period, which include changes in gene expression or protein activity. They regulate diverse aspects of plant growth and development, such as the movement of leaves and flowers, the production of volatiles, the stomatal opening, the hypocotyl expansion, the photosynthetic activity and the photoperiodic control of flowering, allowing plants to anticipate daily environmental changes and to synchronise their endogenous physiological processes to external environmental cues. Circadian rhythms persist with a period close to 24 hours after an organism is transferred from an environment that varies according to the time of the day (entraining condition) to an unchanging condition (free-running condition) (Harmer, 2009).

In a simple way, the circadian system can be divided into three main components: the *input pathways*, involved in the perception and transmission of environmental signals to synchronise the *central oscillator* that generates and maintains rhythmicity through multiple *output pathways*, connecting the oscillator to physiology and metabolism. However, this is an oversimplified model of the clock. The circadian system has to be considered as a complex network. The central clock is composed of multiple interlocked feedback loops, where clock *outputs* may be regulated directly by clock *input* signalling pathways and can also feedback to clock components and *input* signalling pathways. Clock genes have multiple functions, they can act within the *central oscillator* and in clock *input* and *output* signalling pathways (Mas, 2005). A key observation is that circadian clock mutants show defective developmental responses to red light (Harmer, 2009), but the endogenous clock oscillates in the absence of phyA phyB cry1 and cry2 (Yanovsky *et al.*, 2000) or in a mutant devoid of all phytochromes (Strasser *et al.*, 2010). These observations imply that the photoreceptors modulate the clock but they are not themselves part of the *central oscillator*.

4.1 Molecular basis of the circadian clock

In *Arabidopsis thaliana*, the current model for the circadian oscillator is composed of several interlocking positive and negative feedback loops. The first loop that was identified involves the Myb-related transcription factors CIRCADIAN CLOCK ASSOCIATED 1 (CCA1) and LATE ELONGATED HYPOCOTYL (LHY) and the pseudo-response regulator TIMING OF CAB EXPRESSION 1 (TOC1/PRR1) (Loop 1, figure 2). CCA1 and LHY proteins have partially redundant functions, bind directly to the TOC1 promoter and inhibit its expression during the day (Alabadi *et al.*, 2001). In turn, TOC1 promotes the expression of *CCA1* and *LHY* indirectly via a hypothetical component X in the morning. The mechanism by which TOC1 induces *CCA1* and *LHY1* is not completely understood, but it includes CCA1 HIKING EXPEDITION (CHE), a TCP type transcription factor, which associates with TOC1 to regulate *CCA1* (Pruneda-Paz *et al.*, 2009). Eventually, CHE also forms an additional loop with CCA1 (Imaizumi, 2010).

Mathematical modelling suggests that an evening-phased negative loop is coupled to the first loop, with an unknown component Y that positively regulates *TOC1* whereas *Y* is negatively regulated by TOC1, CCA1 and LHY (Locke *et al.*, 2005) (Loop 2, figure 2). It was suggested that a portion of Y activity is provided by the protein GIGANTEA (GI) (Locke *et al.*, 2005), but this is still unclear (Ito *et al.*, 2009; Martin-Tryon *et al.*, 2007).

The Arabidopsis genome contains four genes encoding proteins with similarity to TOC1: PSEUDORESPONSE REGULATOR (PRR), PRR3, 5, 7 and 9. All these *PRR* genes play a role in the circadian system, although the effect of single mutations is subtle. Multiple mutants generally have stronger phenotypes, for example the triple *prr5 prr7 prr9* mutants are essentially arrhythmic (Nakamichi *et al.*, 2005a; Nakamichi *et al.*, 2005b). Experimental and modelling studies suggest that morning expression of *CCA1* and *LHY* activates the transcription of *PRR7* and *PRR9* (Farre *et al.*, 2005; Nakamichi *et al.*, 2005b; Zeilinger *et al.*, 2006). This loop is called morning loop (Loop 3, figure 2) and is closed when PRR7 and PRR9 feedback to inhibit *CCA1* and *LHY* expression. Together the three interlinked feedback loops form an important part of the clock regulatory mechanism and enhance the robustness of the network against environmental perturbations (Harmer, 2009).

Other components that function within or close to the circadian oscillator have recently been identified: FIONA 1, TIME FOR COFFEE, LIGHT REGULATED WD-1 (LWD1) and LWD2 (Ding *et al.*, 2007; Kim *et al.*, 2008; Wu *et al.*, 2008). However, it is not known whether these clock proteins are part of pre-existing loops or constitute unidentified regulatory loops. It has been recently reported that LWD1/2 regulate the expression of multiple oscillator genes and attenuate light signals to adjust period length. Further, it was also proposed that LWD1 and PRR9 form a positive feedback loop within the central oscillator which is also involved in regulating the light input pathway (Wang *et al.*, 2011) (Figure 2). These results underscore the difficulties in dissecting which signalling events are part of the circadian oscillator and which ones are input pathways.

4.2 Light signalling input to the circadian clock

Several different photoreceptors mediate light input to the clock, including the phytochromes and the cryptochromes (Somers *et al.*, 1998; Devlin & Kay, 2000; Yanovsky *et al.*, 2001). However, the molecular mechanisms are only partially known. The ztl family of photoreceptors interacts with clock components and regulates their turnover; hence they are potentially part of input mechanisms. ztl interacts with TOC1 and PRR5, leading to their degradation via the proteasome pathway in the dark (Kiba *et al.*, 2007; Mas *et al.*, 2003). The TOC1-ztl interaction does not depend on light, but an interaction between ztl and GI is blue-light dependent, stabilizes both ztl and GI, and contributes to the robust rhythms of TOC1

(Kim *et al.*, 2007), contributing to a faster degradation of ztl, GI, TOC1 and PRR5 in darkness than in light (Loop 4, figure 2) (Kiba *et al.*, 2007; Kim *et al.*, 2007; Mas *et al.*, 2003). Within this loop 4, TOC1 binds to PRR3, interfering with TOC1 binding to ztl (Para *et al.*, 2007). Thus, PRR3 seems to stabilize TOC1 avoiding its recruitment to the SCF complex and its degradation by the proteasome (Loop 4, figure 2).

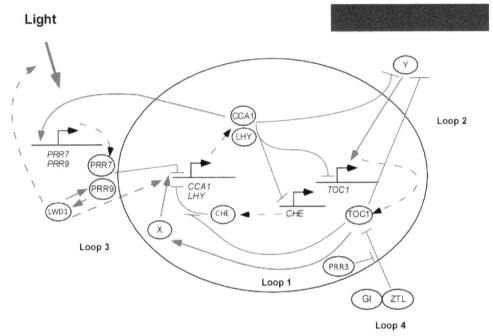

Fig. 2. A model for the *Arabidopsis* clock. The circadian clock is composed of several interlocking positive and negative loops.

The other members of the ztl family, fkf1 and lkp2, were also studied. Mutant combinations showed that fkf1 and lkp2 play similar roles to ztl in the circadian clock when ztl is absent, and that both of them interact with TOC1 and PRR5. These results indicate that ztl, fkf1 and lkp2 regulate TOC1 and PRR5 degradation and are important to determine the period of circadian oscillation (Baudry *et al.*, 2010).

Cryptochromes also signal to the circadian clock. However, the mechanisms are still unclear. One possibility is through the regulation of COP1 activity; COP1 directly interacts with ELF3 and with GI to promote GI degradation by the proteasome. This could be a mechanism by which cryptochromes regulate the activity of GI, a protein closely associated with circadian clock function (Yu *et al.*, 2008).

As mentioned above, one interesting aspect of phytochrome and circadian clock is that mutants affected in the clock are also affected in phytochrome responses (Ito *et al.*, 2007). However, how phytochromes contribute to the entrainment of the clock is still unclear. It was previously suggested that PIF3 could directly induce *CCA1* and *LHY* mRNA expression (Martinez-Garcia *et al.*, 2000). Later, it was shown that TOC1 interacts with PIF3 and PIL1

(Yamashino *et al.*, 2003). However, thorough analysis of PIF3 function has led to the conclusion that it does not play a significant role in controlling light input to the circadian clock (Viczian *et al.*, 2005).

Indeed, there is circumstantial evidence of phytochromes regulating CCA1 and LHY. Both genes are rapidly induced in a TOC1 dependent manner upon transfer of dark grown seedlings to red light. This induction requires *EARLY FLOWERING 4* (*ELF4*), which forms with CCA1 and LHY a negative feedback loop in an analogous manner to TOC1 (Kikis *et al.*, 2005) and ELF4 is itself a direct target of FHY3, FAR1 and HY5 (Li *et al.*, 2011). *ELF3*, is also necessary for light-induced expression of *CCA1* and *LHY* and this event seems to occur indirectly, through a direct repression of PRR9 by physically interacting with its promoter (Dixon *et al.*, 2011).

5. Downstream targets of light and clock signalling

5.1 The impact of the circadian clock in the expression of photosynthesis related genes

As presented above, the interconnections between the clock and light signalling are extremely complex. The regulation of outputs is not an exception. One unbiased measure of the impact of the circadian clock on plant development is the finding that at least one third of the Arabidopsis genome is circadian regulated (Covington *et al.*, 2008). The genes involved in photosynthesis are an important target group of the circadian clock, and tend to be expressed at the middle of the subjective day, together with genes involved in the phenylpropanoid pathway (Edwards *et al.*, 2006). In another global analysis it was shown that PRR5, PRR7 and PRR9 are negative regulators of the chlorophyll and carotenoid biosynthetic pathways (Fukushima *et al.*, 2009).

Despite what we know of the clock impact on photosynthetic gene expression, the mechanisms are still poorly understood. One such mechanism may involve CCA1. CCA1 was originally identified by its binding to an AA(CA)AATCT motif in the *lhcb1*3* promoter, and also shown to be required for phytochrome responsivity (Wang *et al.*, 1997). Hence, CCA1 can represent one of the mechanisms by which the clock regulates photosynthetic gene expression. Nevertheless, the reality is more complex. CCA1 binding site is similar to the Evening Element (AAAATATCT) found in promoters of clock regulated genes that peak toward the end of the subjective day (Harmer *et al.*, 2000), including TOC1, which is repressed by CCA1 (Alabadi *et al.*, 2001). However, *lhcb1*3* expression peaks earlier and is promoted by CCA1 (Wang *et al.*, 1997). These apparent contradictions can be reconciled by the finding that CCA1 effects depend on the context, showing also another level of complexity (Harmer & Kay, 2005).

5.2 Global expression analysis identifies the targets of photomorphogenesis master regulators

HY5, the bZIP targeted by COP1 for degradation, is necessary for responses to a broad spectrum of wavelengths of light and, as explained above, acts as a positive regulator in photomorphogenesis. Arabidopsis plants defective in HY5 show aberrant light mediated phenotypes, including an elongated hypocotyl, reduced chlorophyll/anthocyanin accumulation and reduced chloroplast development in greening hypocotyls (Lee *et al.*, 2007). HY5 regulates the transcription of multiple genes in response to light signals through binding to G-box elements in their promoters such as RBCS1A or CHS1 genes.

Genome-wide CHIP-chip analysis was used to identify HY5 binding regions and to compare this information to HY5-global expression data. This approach allowed the identification of more that 1100 direct targets where HY5 can either activate or repress transcription. However, not all the targets were light responsive genes, suggesting that HY5 must act in concert with other factors to confer light responsiveness (Zhang *et al.*, 2011).

5.3 The dissection of single light responsive promoters reveals another layer of complexity

Most of the photoreceptors, signalling components and transcription factors mentioned above were identified using genetic approaches, after Arabidopsis was established as the model plant. Another strategy to understand light signalling and photosynthetic gene expression has been underway since late mid 80s, after the first transgenic plants became available. This strategy was simple, the generation of transgenic plants bearing promoter:reporter gene fusions. With this approach, light responsive promoters were the subject of extensive research with the aim of finding the light responsive elements (LREs) and their cognate binding factors. The genes encoding the small subunits of the Rubisco (RbcS) and the light-harvesting chlorophyll a/b-binding proteins (Lhc; previously known as Cab), were considered a paradigm of light-regulated gene expression (Akhilesh & Gaur, 2003).

Several LREs were described, as GT-1-Boxes (core sequence GGTTAA), I-Boxes (GATAA), G-Boxes (CACGTG), H-Boxes (CCTACC), AT-rich sequences (consensus AATATTTTTATT) (Akhilesh & Gaur, 2003). Using complementary approaches as Gel Shift analysis and DNA footprinting, some of the cognate binding factors were identified. However, three difficulties hampered this approaches. First, the LREs identified were not always enough to sustain light regulation. Hence, it was proposed that combinations of different motifs but not multimerisation of single motifs could function as LREs, confirming the complex nature of these regulatory elements (Chattopadhyay *et al.*, 1998; Puente *et al.*, 1996). Second, when the cognate transcription factors were studied in Arabidopsis with available mutants, a direct role in light signal was not evident. This can be illustrated by the GT-element binding factors, a small family of plant trihelix DNA-binding proteins comprising Arabidopsis GT2 (AT1G76890), DF1L (AT1G76880), PTL (At5g03680), GT-2-LIKE1 (GTL1, AT1G33240), GT2L (At5g28300), EDA31 (AT3G10000) and GTL1L (AT5G47660). Some of these transcription factors have roles in the fusion of the polar nuclei, in the development of the embryo sac or even perianth development (Brewer *et al.*, 2004; Pagnussat *et al.*, 2005), but were not involved in responses to light. The third difficulty was the apparent "redundancy" of LREs in single promoters. This redundancy could be just the consequence of a single promoter responding to several different light inputs, as will be explained below.

In a few examples, thorough analysis of promoter sequences, combined with genetic approaches significantly advanced our understanding of light-regulated transcription, but also revealed the complex nature underneath this process. The Arabidopsis *Lhcb1*1* (*Cab 2*) promoter fused to luciferase reporters has been extensively used as a marker for light and circadian expression. Genetic screenings using this construct led to the isolation of *toc1* mutants (Strayer *et al.*, 2000). Promoter analysis of *Lhcb1*1* allowed the identification of a 78 bp fragment that was sufficient to confer phytochrome and circadian regulation to a minimal promoter (Anderson *et al.*, 1994). Further analysis of this promoter allowed the identification of HY5, CCA1 and a DET1 responsive elements (Maxwell *et al.*, 2003). Similarly, it has been shown that HY5 binds to the *Lhcb1*3* promoter and physically interacts with CCA1 to synergistically regulate expression (Andronis *et al.*, 2008).

Another promoter analysed in more detail was the tobacco *Lhcb1*2*. First, a 146 bp promoter fragment sufficed to confer VLFR (mediated by phyA), LFR (mediated by phyB) and HIR (mediated by phyA) to a minimal promoter (Cerdan *et al.*, 1997). Then, the motifs for VLFR and LFR were dissected from the HIR responsive motifs (Cerdan *et al.*, 2000) and finally, the TGGA motif was shown to bind Bell-like homeodomain 1 (BLH1) as part of the phyA mediated HIR (Staneloni *et al.*, 2009). This promoter is an example of how several different photoreceptors can regulate a single gene and integrate their signalling pathways at the promoter level; at least four different photoreceptors were shown to regulate this single promoter (Casal *et al.*, 1998; Cerdan *et al.*, 1999; Mazzella *et al.*, 2001) .

6. Light promotes chloroplast development

Proplastids are found in the embryo; they are undifferentiated plastids that are converted to other kind of plastids like chromoplasts, amyloplasts, chloroplasts and etioplasts. During skotomorphogenic development, proplastids turn into etioplasts, the chloroplast precursors. Etioplasts contain the prolamellar body, a structure rich in protochlorophyllide, the chlorophyll precursor, and the enzyme protochlorophyllide oxidoreductase (POR). During the development of etioplasts into chloroplasts, the POR is directly activated by light to convert protochlorophyllide into divinyl-chlorophyllide a, which is chlorophyll a and b precursor (Tanaka & Tanaka, 2007). This light-dependent step can be promoted by red-light in Arabidopsis, even in the absence of phytochromes (Strasser *et al.*, 2010). However, other events that occur during chloroplast biogenesis require the signals transduced by photoreceptors. These signals ensure proper coordination of synthesis and import of LHCB proteins, which are essential for the assembly of the photosynthetic complexes. These events are also coordinated with the synthesis of carotenoids, which are necessary for photoprotection (Cazzonelli & Pogson, 2010).

Phytochromes, through the action of PIFs, regulate the transition from amiloplasts to etioplasts and to chloroplasts. For example, the PIFs inhibit the conversion of endodermal amyloplasts to etioplasts, whereas the phytochromes antagonise this inhibition, promoting the formation of chloroplasts (Figure 3) (Kim *et al.*, 2011).

6.1 Chlorophyll biosynthesis is regulated by light

Chlorophyll biosynthesis and the synthesis of other components of the photosystems are tightly regulated by light and the circadian clock. This coordination is necessary because when the chlorophyll synthesis exceeds the accumulation of chlorophyll-binding apoproteins, reactive oxygen species are generated, ultimately leading to cell death. However, when the chlorophyll synthesis is not enough, the amount of fully functional chlorophyll-binding proteins is not sufficient to gain optimal photosynthetic activity. Another example highlighting the importance of proper coordination is that PIF deficient plants accumulate protochlorophyllide in the dark during skotomorphogenic development, but this accumulation leads to bleaching upon exposure to light (Stephenson *et al.*, 2009).

Plants have four classes of tetrapyrroles: chlorophyll, phytochromobilin, haeme and siroheme, all derived from the same biosynthetic pathway. The flow of the tetrapyrrole pathway is strictly regulated, keeping at low levels the potentially toxic intermediates

(Tanaka & Tanaka, 2007). Phytochrome and cryptochrome mutants contain lower levels of chlorophyll (Strasser *et al.*, 2010) stressing out the importance of the photomorphogenic signal for proper assembly of the photosynthetic machinery. In the next paragraphs we review how light signalling pathways regulate chlorophyll biosynthesis (Figure 4).

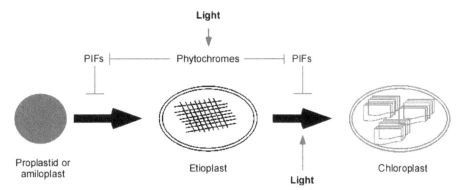

Fig. 3. Light interactions in plastid development. Phytochrome and PIFs roles during the transition from proplastid or amyloplast to chloroplast.

Chlorophyll synthesis occurs in plastids; in the first step glutamate is activated to Glutamyl-tRNA by the Glutamyl-tRNA synthetase, a step shared with plastid protein synthesis. The following step, the reduction of the Glutamyl-tRNA to produce glutamate-1-semialdehyde is subjected to tight regulation (Figure 4). In Arabidopsis, the Glutamyl-tRNA reductases are encoded by a little family of nuclear genes called *HEMA*. Of this family, the expression of *HEMA1* correlates with the expression of *Lhcb1* genes, which encode light-harvesting proteins of the photosystem II; in some way the expression of *HEMA1* reflects the demand of chlorophyll synthesis. On the other hand, *HEMA2* is not light regulated (Matsumoto *et al.*, 2004; McCormac *et al.*, 2001; McCormac & Terry, 2002a; McCormac & Terry, 2002b).

Glutamyl-tRNA reductase activity is regulated by negative feedback loops; the accumulation of Haeme, Mg-Protoporphyrin IX or Divinyl protochlorofilide a antagonise Glutamyl-tRNA reductase activity (Srivastava *et al.*, 2005). At the transcriptional level, *HEMA1* expression is induced by red and far-red light, implicating at least phyA and phyB, and blue light perceived by cry1 (McCormac *et al.*, 2001; McCormac & Terry, 2002a). *pif1* and *pif3* mutants contain higher levels of *HEMA1* mRNA, higher levels of protochlorophyllide and partially developed chloroplasts in the dark, a phenotype observed in *cop* mutants. The effects of *pif1* and *pif3* mutations are essentially additive, suggesting a model where phytochromes promote chloroplast biogenesis by antagonizing the activity of at least PIF1 and PIF3. As PIF1 and PIF3 are regulated by the circadian clock, but do not seem to affect central clock components (TOC1, CCA1, LHY), these PIFs seem to integrate chloroplast biogenesis with circadian and light signalling (Stephenson *et al.*, 2009).

The expression of photosynthetic nuclear genes is repressed by plastid signals if chloroplast biogenesis is blocked (retrograde signalling). This finding led to the isolation of mutants that disrupt chloroplast to nucleus communication, the genomes uncoupled mutants (*gun*) (Nott *et al.*, 2006). These mutants show high levels of *lhcb1* mRNA in the presence of norfluorzazon and were named *gun1* to *gun5*. *gun2* and *gun3* are allelic to *hy1* and *hy2* and disrupt phytochromobilin synthesis, leading to haeme accumulation and feedback inhibition of Glutamyl-tRNA reductase (Nott *et al.*, 2006). The product of the *GUN4* gene, a 22 kD protein localized to Chloroplasts, promotes Magnesium chelatase (MgCH) activity which catalyses the insertion of Mg2+ into protoporphyrin IX (Tanaka & Tanaka, 2007). The *GUN4* gene is also under circadian clock regulation and is repressed by PIF1 and PIF3 suggesting a similar regulatory mechanism to HEMA1 (Stephenson *et al.*, 2009). The expression of GUN4 is primarily under the control of phyA and phyB with some input from

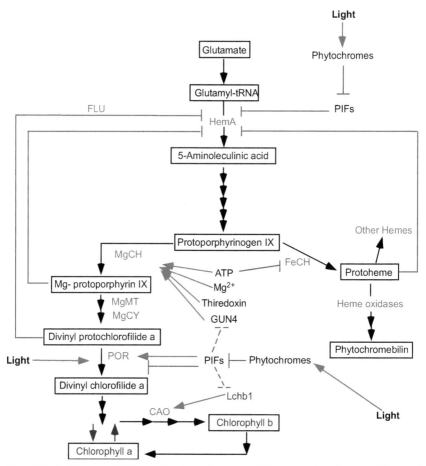

Fig. 4. Simplified chlorophyll biosynthesis pathway and light regulated steps. We emphasise how the light regulate directly the activity of NADPH:protochlorophyllide oxidoreductase (POR); or indirectly, through phytochrome and PIFs the expression of genes encoding the Glutamyl-tRNA reductases (HEMAs), Ferrum chelatase (FeCH), Magnesium chelatase (MgCH), NADPH:protochlorophyllide oxidoreductase (POR), Chlorophyllide a oxygenase (CAO), Mg-protoporphyrin IX methyltransferase (MgMT), and Mg-protoporphyrin IX monomethyl estercyclase (MgCy). The ATP/ADP ratio, the Mg2+ concentration and the thioredoxin levels also affect the MgCH activity, furthermore, these factors are light regulated (Tanaka & Tanaka, 2007). LHCs attach *chlorophyll a*, and CAO converts the *chlorophyll* a to b on the LHC apoprotein (Tanaka & Tanaka, 2007).

the cryptochromes, establishing GUN4 as a link between the phytochromes and the regulation of MgCH activity (Stephenson & Terry, 2008). *GUN5* encodes the H subunit of MgCH, known as CHLH (Nott *et al.*, 2006). The expression of *CHLH* is regulated at the mRNA level by light/dark cycles and by the circadian clock. Interestingly, this gene is co-regulated with *HEMA1, lhcb, Mg-protoporphyrin IX monomethyl estercyclase (MGCy)* and the gene encoding the chlorophyll(ide) a oxygenase *(CAO)* (Matsumoto *et al.*, 2004). On the other hand, *GUN1* encodes a pentatricopeptide repeat–containing protein that does not affect chlorophyll synthesis. GUN1 was proposed to generate a signal in chloroplast that represses nuclear photosynthetic gene expression; this repression on *lhcb* genes seems to be mediated by direct binding of *ABI4*, an AP2–type transcription factor (Koussevitzky *et al.*, 2007).

Another connection between light signalling and the retrograde signalling was recently established. A sensitive genetic screening for the *gun* phenotype uncovered new cry1 alleles. These results establish that cry1 is necessary for maximal repression of *lhcb* genes, when chloroplast biogenesis is blocked (Ruckle *et al.*, 2007).

One of the latest steps in chlorophyll synthesis is the reduction of 3,8-divinyl protochlorophyllide to 3,8-divinyl chlorophyllide. This protochlorophyllide to chlorophyllide conversion is catalysed by the POR enzyme. In angiosperms, POR is light-dependent and it is likely the source of red-light promoted chlorophyll synthesis in the absence of phytochromes (Strasser *et al.*, 2010). Angiosperms carry three POR-encoding genes, *PorA, PorB* and *PorC*, which are differentially regulated by both light and developmental stage. *PORA* expression is high in etiolated seedlings and rapidly becomes undetectable after illumination with FR, a HIR response mediated by phyA, whereas *PORB* expression persists throughout greening and in adult plants (Runge *et al.*, 1996). PORC is expressed during the adult life and together with PORB is responsible for bulk chlorophyll synthesis in green plants (Paddock *et al.*, 2010). It has been recently shown that *PORC* expression is directly activated by PIF1 binding to a G-box in *PORC* promoter, whereas *PORA* and *PORB* are also induced by PIF1, presumably in an indirect manner (Moon *et al.*, 2008).

7. Conclusion

During the last twenty years, plant biologists have witnessed major advances in our understanding of how plants use light as a source of information. These advances were possible thanks to the adoption of Arabidopsis as a model system. During these twenty years, 13 Arabidopsis photoreceptors were characterised In molecular terms and these findings extended to other species as well. A high number of signal transduction components were also characterised. With the advent of "omics" technologies, the networks that work downstream photoreceptors and their targets started to surface. However, with all these advances, we still do not know in detail how a single light responsive promoter works. How many transcription factors are sitting there? Which are their identities? How do they interact to fine tune expression under the diverse light conditions found in nature? If we multiply these questions by the number of light responsive promoters we can just have a hint of the enormous task ahead.

8. References

Akhilesh, K. T. &Gaur, T. (2003). Light regulation of nuclear photosynthetic genes in higher plants. *Critical Reviews in Plant Sciences* 22(5): 417-452.

Alabadi, D., Oyama, T., Yanovsky, M. J., Harmon, F. G., Mas, P. &Kay, S. A. (2001). Reciprocal regulation between TOC1 and LHY/CCA1 within the Arabidopsis circadian clock. *Science* 293(5531): 880-883.

Anderson, S. L., Teakle, G. R., Martino-Catt, S. J. &Kay, S. A. (1994). Circadian clock- and phytochrome-regulated transcription is conferred by a 78 bp cis-acting domain of the Arabidopsis CAB2 promoter. *Plant J* 6(4): 457-470.

Andronis, C., Barak, S., Knowles, S. M., Sugano, S. &Tobin, E. M. (2008). The clock protein CCA1 and the bZIP transcription factor HY5 physically interact to regulate gene expression in Arabidopsis. *Mol Plant* 1(1): 58-67.

Bae, G. &Choi, G. (2008). Decoding of light signals by plant phytochromes and their interacting proteins. *Annu Rev Plant Biol* 59: 281-311.

Ballaré, C. L. S., A.L.; Sánchez, R.A.; Radosevich, S. (1992). Photomorphogenic process in the agricultural environment. *Photochemistry and Photobiology* 56: 12.

Baudry, A., Ito, S., Song, Y. H., Strait, A. A., Kiba, T., Lu, S., Henriques, R., Pruneda-Paz, J. L., Chua, N. H., Tobin, E. M., Kay, S. A. &Imaizumi, T. (2010). F-box proteins FKF1 and LKP2 act in concert with ZEITLUPE to control Arabidopsis clock progression. *Plant Cell* 22(3): 606-622.

Brewer, P. B., Howles, P. A., Dorian, K., Griffith, M. E., Ishida, T., Kaplan-Levy, R. N., Kilinc, A. &Smyth, D. R. (2004). PETAL LOSS, a trihelix transcription factor gene, regulates perianth architecture in the Arabidopsis flower. *Development* 131(16): 4035-4045.

Bu, Q., Zhu, L., Dennis, M. D., Yu, L., Lu, S. X., Person, M. D., Tobin, E. M., Browning, K. S. &Huq, E. (2011). Phosphorylation by CK2 enhances the rapid light-induced degradation of phytochrome interacting factor 1 in Arabidopsis. *J Biol Chem* 286(14): 12066-12074.

Casal, J. J., Cerdan, P. D., Staneloni, R. J. &Cattaneo, L. (1998). Different phototransduction kinetics of phytochrome A and phytochrome B in Arabidopsis thaliana. *Plant Physiol* 116(4): 1533-1538.

Casal, J. J., Luccioni, L. G., Oliverio, K. A. &Boccalandro, H. E. (2003). Light, phytochrome signalling and photomorphogenesis in Arabidopsis. *Photochem Photobiol Sci* 2(6): 625-636.

Cazzonelli, C. I. &Pogson, B. J. (2010). Source to sink: regulation of carotenoid biosynthesis in plants. *Trends Plant Sci* 15(5): 266-274.

Cerdan, P. D., Staneloni, R. J., Casal, J. J. &Sanchez, R. A. (1997). A 146 bp fragment of the tobacco Lhcb1*2 promoter confers very-low-fluence, low-fluence and high-irradiance responses of phytochrome to a minimal CaMV 35S promoter. *Plant Mol Biol* 33(2): 245-255.

Cerdan, P. D., Staneloni, R. J., Ortega, J., Bunge, M. M., Rodriguez-Batiller, M. J., Sanchez, R. A. &Casal, J. J. (2000). Sustained but not transient phytochrome A signaling targets a region of an Lhcb1*2 promoter not necessary for phytochrome B action. *Plant Cell* 12(7): 1203-1211.

Cerdan, P. D., Yanovsky, M. J., Reymundo, F. C., Nagatani, A., Staneloni, R. J., Whitelam, G. C. &Casal, J. J. (1999). Regulation of phytochrome B signaling by phytochrome A and FHY1 in Arabidopsis thaliana. *Plant J* 18(5): 499-507.

Covington, M. F., Maloof, J. N., Straume, M., Kay, S. A. &Harmer, S. L. (2008). Global transcriptome analysis reveals circadian regulation of key pathways in plant growth and development. *Genome Biol* 9(8): R130.

Chattopadhyay, S., Puente, P., Deng, X. W. &Wei, N. (1998). Combinatorial interaction of light-responsive elements plays a critical role in determining the response characteristics of light-regulated promoters in Arabidopsis. *Plant J* 15(1): 69-77.

Chen, M., Galvao, R. M., Li, M., Burger, B., Bugea, J., Bolado, J. &Chory, J. (2010). Arabidopsis HEMERA/pTAC12 initiates photomorphogenesis by phytochromes. *Cell* 141(7): 1230-1240.

Deng, X. W., Matsui, M., Wei, N., Wagner, D., Chu, A. M., Feldmann, K. A. &Quail, P. H. (1992). COP1, an Arabidopsis regulatory gene, encodes a protein with both a zinc-binding motif and a G beta homologous domain. *Cell* 71(5): 791-801.

Devlin, P. F. &Kay, S. A. (2000). Cryptochromes are required for phytochrome signaling to the circadian clock but not for rhythmicity. *Plant Cell* 12(12): 2499-2510.

Ding, Z., Millar, A. J., Davis, A. M. &Davis, S. J. (2007). TIME FOR COFFEE encodes a nuclear regulator in the Arabidopsis thaliana circadian clock. *Plant Cell* 19(5): 1522-1536.

Dixon, L. E., Knox, K., Kozma-Bognar, L., Southern, M. M., Pokhilko, A. &Millar, A. J. (2011). Temporal repression of core circadian genes is mediated through EARLY FLOWERING 3 in Arabidopsis. *Curr Biol* 21(2): 120-125.

Edwards, K. D., Anderson, P. E., Hall, A., Salathia, N. S., Locke, J. C., Lynn, J. R., Straume, M., Smith, J. Q. &Millar, A. J. (2006). FLOWERING LOCUS C mediates natural variation in the high-temperature response of the Arabidopsis circadian clock. *Plant Cell* 18(3): 639-650.

Fankhauser, C. &Chen, M. (2008). Transposing phytochrome into the nucleus. *Trends Plant Sci* 13(11): 596-601.

Farre, E. M., Harmer, S. L., Harmon, F. G., Yanovsky, M. J. &Kay, S. A. (2005). Overlapping and distinct roles of PRR7 and PRR9 in the Arabidopsis circadian clock. *Curr Biol* 15(1): 47-54.

Favory, J. J., Stec, A., Gruber, H., Rizzini, L., Oravecz, A., Funk, M., Albert, A., Cloix, C., Jenkins, G. I., Oakeley, E. J., Seidlitz, H. K., Nagy, F. &Ulm, R. (2009). Interaction of COP1 and UVR8 regulates UV-B-induced photomorphogenesis and stress acclimation in Arabidopsis. *Embo J* 28(5): 591-601.

Franklin, K. A. &Quail, P. H. (2009). Phytochrome functions in Arabidopsis development. *J Exp Bot* 61(1): 11-24.

Fukushima, A., Kusano, M., Nakamichi, N., Kobayashi, M., Hayashi, N., Sakakibara, H., Mizuno, T. &Saito, K. (2009). Impact of clock-associated Arabidopsis pseudo-response regulators in metabolic coordination. *Proc Natl Acad Sci U S A* 106(17): 7251-7256.

Gruber, H., Heijde, M., Heller, W., Albert, A., Seidlitz, H. K. &Ulm, R. (2010). Negative feedback regulation of UV-B-induced photomorphogenesis and stress acclimation in Arabidopsis. *Proc Natl Acad Sci U S A* 107(46): 20132-20137.

Harmer, S. L. (2009). The circadian system in higher plants. *Annu Rev Plant Biol* 60: 357-377.

Harmer, S. L., Hogenesch, J. B., Straume, M., Chang, H. S., Han, B., Zhu, T., Wang, X., Kreps, J. A. &Kay, S. A. (2000). Orchestrated transcription of key pathways in Arabidopsis by the circadian clock. *Science* 290(5499): 2110-2113.

Harmer, S. L. &Kay, S. A. (2005). Positive and negative factors confer phase-specific circadian regulation of transcription in Arabidopsis. *Plant Cell* 17(7): 1926-1940.

Heim, M. A., Jakoby, M., Werber, M., Martin, C., Weisshaar, B. &Bailey, P. C. (2003). The basic helix-loop-helix transcription factor family in plants: a genome-wide study of protein structure and functional diversity. *Mol Biol Evol* 20(5): 735-747.

Henriques, R., Jang, I. C. &Chua, N. H. (2009). Regulated proteolysis in light-related signaling pathways. *Curr Opin Plant Biol* 12(1): 49-56.

Holm, M., Ma, L. G., Qu, L. J. &Deng, X. W. (2002). Two interacting bZIP proteins are direct targets of COP1-mediated control of light-dependent gene expression in Arabidopsis. *Genes Dev* 16(10): 1247-1259.

Hornitschek, P., Lorrain, S., Zoete, V., Michielin, O. &Fankhauser, C. (2009). Inhibition of the shade avoidance response by formation of non-DNA binding bHLH heterodimers. *EMBO J* 28(24): 3893-3902.

Huala, E., Oeller, P. W., Liscum, E., Han, I. S., Larsen, E. &Briggs, W. R. (1997). Arabidopsis NPH1: a protein kinase with a putative redox-sensing domain. *Science* 278(5346): 2120-2123.

Imaizumi, T. (2010). Arabidopsis circadian clock and photoperiodism: time to think about location. *Curr Opin Plant Biol* 13(1): 83-89.

Ito, S., Kawamura, H., Niwa, Y., Nakamichi, N., Yamashino, T. &Mizuno, T. (2009). A genetic study of the Arabidopsis circadian clock with reference to the TIMING OF CAB EXPRESSION 1 (TOC1) gene. *Plant Cell Physiol* 50(2): 290-303.

Ito, S., Nakamichi, N., Kiba, T., Yamashino, T. &Mizuno, T. (2007). Rhythmic and light-inducible appearance of clock-associated pseudo-response regulator protein PRR9 through programmed degradation in the dark in Arabidopsis thaliana. *Plant Cell Physiol* 48(11): 1644-1651.

Jang, I. C., Henriques, R., Seo, H. S., Nagatani, A. &Chua, N. H. (2010). Arabidopsis PHYTOCHROME INTERACTING FACTOR proteins promote phytochrome B polyubiquitination by COP1 E3 ligase in the nucleus. *Plant Cell* 22(7): 2370-2383.

Jenkins, G. I. (2009). Signal transduction in responses to UV-B radiation. *Annu Rev Plant Biol* 60: 407-431.

Kiba, T., Henriques, R., Sakakibara, H. &Chua, N. H. (2007). Targeted degradation of PSEUDO-RESPONSE REGULATOR5 by an SCFZTL complex regulates clock function and photomorphogenesis in Arabidopsis thaliana. *Plant Cell* 19(8): 2516-2530.

Kikis, E. A., Khanna, R. &Quail, P. H. (2005). ELF4 is a phytochrome-regulated component of a negative-feedback loop involving the central oscillator components CCA1 and LHY. *Plant J* 44(2): 300-313.

Kim, J., Kim, Y., Yeom, M., Kim, J. H. &Nam, H. G. (2008). FIONA1 is essential for regulating period length in the Arabidopsis circadian clock. *Plant Cell* 20(2): 307-319.

Kim, K., Shin, J., Lee, S. H., Kweon, H. S., Maloof, J. N. &Choi, G. (2011). Phytochromes inhibit hypocotyl negative gravitropism by regulating the development of endodermal amyloplasts through phytochrome-interacting factors. *Proc Natl Acad Sci U S A* 108(4): 1729-1734.

Kim, W. Y., Fujiwara, S., Suh, S. S., Kim, J., Kim, Y., Han, L., David, K., Putterill, J., Nam, H. G. &Somers, D. E. (2007). ZEITLUPE is a circadian photoreceptor stabilized by GIGANTEA in blue light. *Nature* 449(7160): 356-360.

Kodama, Y., Suetsugu, N. &Wada, M. (2011). Novel protein-protein interaction family proteins involved in chloroplast movement response. *Plant Signal Behav* 6(4): 483-490.

Koussevitzky, S., Nott, A., Mockler, T. C., Hong, F., Sachetto-Martins, G., Surpin, M., Lim, J., Mittler, R. &Chory, J. (2007). Signals from chloroplasts converge to regulate nuclear gene expression. *Science* 316(5825): 715-719.

Laubinger, S., Fittinghoff, K. &Hoecker, U. (2004). The SPA quartet: a family of WD-repeat proteins with a central role in suppression of photomorphogenesis in arabidopsis. *Plant Cell* 16(9): 2293-2306.

Lee, J., He, K., Stolc, V., Lee, H., Figueroa, P., Gao, Y., Tongprasit, W., Zhao, H., Lee, I. &Deng, X. W. (2007). Analysis of transcription factor HY5 genomic binding sites revealed its hierarchical role in light regulation of development. *Plant Cell* 19(3): 731-749.

Leivar, P. &Quail, P. H. (2011). PIFs: pivotal components in a cellular signaling hub. *Trends Plant Sci* 16(1): 19-28.

Leivar, P., Tepperman, J. M., Monte, E., Calderon, R. H., Liu, T. L. &Quail, P. H. (2009). Definition of Early Transcriptional Circuitry Involved in Light-Induced Reversal of PIF-Imposed Repression of Photomorphogenesis in Young Arabidopsis Seedlings. *Plant Cell.*

Li, Q. H. &Yang, H. Q. (2007). Cryptochrome signaling in plants. *Photochem Photobiol* 83(1): 94-101.

Li, G., Siddiqui, H., Teng, Y., Lin, R., Wan, X. Y., Li, J., Lau, O. S., Ouyang, X., Dai, M., Wan, J., Devlin, P. F., Deng, X. W. &Wang, H. (2011). Coordinated transcriptional regulation underlying the circadian clock in Arabidopsis. *Nat Cell Biol* 13(5): 616-622.

Lian, H. L., He, S. B., Zhang, Y. C., Zhu, D. M., Zhang, J. Y., Jia, K. P., Sun, S. X., Li, L. &Yang, H.Q. (2011). Blue-light-dependent interaction of cryptochrome 1 with SPA1 defines a dynamic signaling mechanism. *Genes Dev* 25(10): 1023-1028.

Lin, C. &Todo, T. (2005). The cryptochromes. *Genome Biol* 6(5): 220.

Lin, R., Ding, L., Casola, C., Ripoll, D. R., Feschotte, C. &Wang, H. (2007). Transposase-derived transcription factors regulate light signaling in Arabidopsis. *Science* 318(5854): 1302-1305.

Liu, B., Zuo, Z., Liu, H., Liu, X. &Lin, C. (2011). Arabidopsis cryptochrome1 interacts with SPA1 to suppress COP1 activity in response to blue light. *Genes Dev* 25(10): 1029-1034.

Locke, J. C., Southern, M. M., Kozma-Bognar, L., Hibberd, V., Brown, P. E., Turner, M. S. &Millar, A. J. (2005). Extension of a genetic network model by iterative experimentation and mathematical analysis. *Mol Syst Biol* 1: 2005 0013.

Ma, L., Gao, Y., Qu, L., Chen, Z., Li, J., Zhao, H. &Deng, X. W. (2002). Genomic evidence for COP1 as a repressor of light-regulated gene expression and development in Arabidopsis. *Plant Cell* 14(10): 2383-2398.

Maddonni, G. A., Otegui, M. E., Andrieu, B., Chelle, M. &Casal, J. J. (2002). Maize leaves turn away from neighbors. *Plant Physiol* 130(3): 1181-1189.

Martin-Tryon, E. L., Kreps, J. A. &Harmer, S. L. (2007). GIGANTEA acts in blue light signaling and has biochemically separable roles in circadian clock and flowering time regulation. *Plant Physiol* 143(1): 473-486.

Martinez-Garcia, J. F., Huq, E. &Quail, P. H. (2000). Direct targeting of light signals to a promoter element-bound transcription factor. *Science* 288(5467): 859-863.

Mas, P. (2005). Circadian clock signaling in Arabidopsis thaliana: from gene expression to physiology and development. *Int J Dev Biol* 49(5-6): 491-500.

Mas, P., Kim, W. Y., Somers, D. E. &Kay, S. A. (2003). Targeted degradation of TOC1 by ZTL modulates circadian function in Arabidopsis thaliana. *Nature* 426(6966): 567-570.

Matsumoto, F., Obayashi, T., Sasaki-Sekimoto, Y., Ohta, H., Takamiya, K. &Masuda, T. (2004). Gene expression profiling of the tetrapyrrole metabolic pathway in Arabidopsis with a mini-array system. *Plant Physiol* 135(4): 2379-2391.

Maxwell, B. B., Andersson, C. R., Poole, D. S., Kay, S. A. &Chory, J. (2003). HY5, Circadian Clock-Associated 1, and a cis-element, DET1 dark response element, mediate DET1 regulation of chlorophyll a/b-binding protein 2 expression. *Plant Physiol* 133(4): 1565-1577.

Mazzella, M. A., Cerdan, P. D., Staneloni, R. J. &Casal, J. J. (2001). Hierarchical coupling of phytochromes and cryptochromes reconciles stability and light modulation of Arabidopsis development. *Development* 128(12): 2291-2299.

McCormac, A. C., Fischer, A., Kumar, A. M., Soll, D. &Terry, M. J. (2001). Regulation of HEMA1 expression by phytochrome and a plastid signal during de-etiolation in Arabidopsis thaliana. *Plant J* 25(5): 549-561.

McCormac, A. C. &Terry, M. J. (2002a). Light-signalling pathways leading to the co-ordinated expression of HEMA1 and Lhcb during chloroplast development in Arabidopsis thaliana. *Plant J* 32(4): 549-559.

McCormac, A. C. &Terry, M. J. (2002b). Loss of nuclear gene expression during the phytochrome A-mediated far-red block of greening response. *Plant Physiol* 130(1): 402-414.

Moglich, A., Yang, X., Ayers, R. A. &Moffat, K. (2010). Structure and function of plant photoreceptors. *Annu Rev Plant Biol* 61: 21-47.

Moon, J., Zhu, L., Shen, H. &Huq, E. (2008). PIF1 directly and indirectly regulates chlorophyll biosynthesis to optimize the greening process in Arabidopsis. *Proc Natl Acad Sci U S A* 105(27): 9433-9438.

Nakamichi, N., Kita, M., Ito, S., Sato, E., Yamashino, T. &Mizuno, T. (2005a). The Arabidopsis pseudo-response regulators, PRR5 and PRR7, coordinately play essential roles for circadian clock function. *Plant Cell Physiol* 46(4): 609-619.

Nakamichi, N., Kita, M., Ito, S., Yamashino, T. &Mizuno, T. (2005b). PSEUDO-RESPONSE REGULATORS, PRR9, PRR7 and PRR5, together play essential roles close to the circadian clock of Arabidopsis thaliana. *Plant Cell Physiol* 46(5): 686-698.

Ni, M., Tepperman, J. M. &Quail, P. H. (1998). PIF3, a phytochrome-interacting factor necessary for normal photoinduced signal transduction, is a novel basic helix-loop-helix protein. *Cell* 95(5): 657-667.

Nott, A., Jung, H. S., Koussevitzky, S. &Chory, J. (2006). Plastid-to-nucleus retrograde signaling. *Annu Rev Plant Biol* 57: 739-759.

Osterlund, M. T., Hardtke, C. S., Wei, N. &Deng, X. W. (2000). Targeted destabilization of HY5 during light-regulated development of Arabidopsis. *Nature* 405(6785): 462-466.

Paddock, T. N., Mason, M. E., Lima, D. F. &Armstrong, G. A. (2010). Arabidopsis protochlorophyllide oxidoreductase A (PORA) restores bulk chlorophyll synthesis and normal development to a porB porC double mutant. *Plant Mol Biol* 72(4-5): 445-457.

Pagnussat, G. C., Yu, H. J., Ngo, Q. A., Rajani, S., Mayalagu, S., Johnson, C. S., Capron, A., Xie, L. F., Ye, D. &Sundaresan, V. (2005). Genetic and molecular identification of genes required for female gametophyte development and function in Arabidopsis. *Development* 132(3): 603-614.

Para, A., Farre, E. M., Imaizumi, T., Pruneda-Paz, J. L., Harmon, F. G. &Kay, S. A. (2007). PRR3 Is a vascular regulator of TOC1 stability in the Arabidopsis circadian clock. *Plant Cell* 19(11): 3462-3473.

Pruneda-Paz, J. L., Breton, G., Para, A. &Kay, S. A. (2009). A functional genomics approach reveals CHE as a component of the Arabidopsis circadian clock. *Science* 323(5920): 1481-1485.

Puente, P., Wei, N. &Deng, X. W. (1996). Combinatorial interplay of promoter elements constitutes the minimal determinants for light and developmental control of gene expression in Arabidopsis. *Embo J* 15(14): 3732-3743.

Rizzini, L., Favory, J. J., Cloix, C., Faggionato, D., O'Hara, A., Kaiserli, E., Baumeister, R., Schafer, E., Nagy, F., Jenkins, G. I. &Ulm, R. (2011). Perception of UV-B by the Arabidopsis UVR8 protein. *Science* 332(6025): 103-106.

Robson, P. R., McCormac, A. C., Irvine, A. S. &Smith, H. (1996). Genetic engineering of harvest index in tobacco through overexpression of a phytochrome gene. *Nat Biotechnol* 14(8): 995-998.

Rockwell, N. C., Su, Y. S. &Lagarias, J. C. (2006). Phytochrome structure and signaling mechanisms. *Annu Rev Plant Biol* 57: 837-858.

Ruckle, M. E., DeMarco, S. M. &Larkin, R. M. (2007). Plastid signals remodel light signaling networks and are essential for efficient chloroplast biogenesis in Arabidopsis. *Plant Cell* 19(12): 3944-3960.

Runge, S., Sperling, U., Frick, G., Apel, K. &Armstrong, G. A. (1996). Distinct roles for light-dependent NADPH:protochlorophyllide oxidoreductases (POR) A and B during greening in higher plants. *Plant J* 9(4): 513-523.

Sharrock, R. A. (2008). The phytochrome red/far-red photoreceptor superfamily. *Genome Biol* 9(8): 230.

Shen, H., Zhu, L., Castillon, A., Majee, M., Downie, B. &Huq, E. (2008). Light-induced phosphorylation and degradation of the negative regulator PHYTOCHROME-INTERACTING FACTOR1 from Arabidopsis depend upon its direct physical interactions with photoactivated phytochromes. *Plant Cell* 20(6): 1586-1602.

Somers, D. E., Devlin, P. F. &Kay, S. A. (1998). Phytochromes and cryptochromes in the entrainment of the Arabidopsis circadian clock. *Science* 282(5393): 1488-1490.

Srivastava, A., Lake, V., Nogaj, L. A., Mayer, S. M., Willows, R. D. &Beale, S. I. (2005). The Chlamydomonas reinhardtii gtr gene encoding the tetrapyrrole biosynthetic enzyme glutamyl-trna reductase: structure of the gene and properties of the expressed enzyme. *Plant Mol Biol* 58(5): 643-658.

Staneloni, R. J., Rodriguez-Batiller, M. J., Legisa, D., Scarpin, M. R., Agalou, A., Cerdan, P. D., Meijer, A. H., Ouwerkerk, P. B. &Casal, J. J. (2009). Bell-like homeodomain selectively regulates the high-irradiance response of phytochrome A. *Proc Natl Acad Sci U S A* 106(32): 13624-13629.

Stephenson, P. G., Fankhauser, C. &Terry, M. J. (2009). PIF3 is a repressor of chloroplast development. *Proc Natl Acad Sci U S A* 106(18): 7654-7659.

Stephenson, P. G. &Terry, M. J. (2008). Light signalling pathways regulating the Mg-chelatase branchpoint of chlorophyll synthesis during de-etiolation in Arabidopsis thaliana. *Photochem Photobiol Sci* 7(10): 1243-1252.

Strasser, B., Sanchez-Lamas, M., Yanovsky, M. J., Casal, J. J. &Cerdan, P. D. (2010). Arabidopsis thaliana life without phytochromes. *Proc Natl Acad Sci U S A* 107(10): 4776-4781.

Strayer, C., Oyama, T., Schultz, T. F., Raman, R., Somers, D. E., Mas, P., Panda, S., Kreps, J. A. &Kay, S. A. (2000). Cloning of the Arabidopsis clock gene TOC1, an autoregulatory response regulator homolog. *Science* 289(5480): 768-771.

Takano, M., Inagaki, N., Xie, X., Kiyota, S., Baba-Kasai, A., Tanabata, T. &Shinomura, T. (2009). Phytochromes are the sole photoreceptors for perceiving red/far-red light in rice. *Proc Natl Acad Sci U S A* 106(34): 14705-14710.

Tanaka, R. &Tanaka, A. (2007). Tetrapyrrole biosynthesis in higher plants. *Annu Rev Plant Biol* 58: 321-346.

Toledo-Ortiz, G., Huq, E. &Quail, P. H. (2003). The Arabidopsis basic/helix-loop-helix transcription factor family. *Plant Cell* 15(8): 1749-1770.

Viczian, A., Kircher, S., Fejes, E., Millar, A. J., Schafer, E., Kozma-Bognar, L. &Nagy, F. (2005). Functional characterization of phytochrome interacting factor 3 for the Arabidopsis thaliana circadian clockwork. *Plant Cell Physiol* 46(10): 1591-1602.

Wang, Y., Wu, J. F., Nakamichi, N., Sakakibara, H., Nam, H. G. &Wu, S. H. (2011). LIGHT-REGULATED WD1 and PSEUDO-RESPONSE REGULATOR9 form a positive feedback regulatory loop in the Arabidopsis circadian clock. *Plant Cell* 23(2): 486-498.

Wang, Z. Y., Kenigsbuch, D., Sun, L., Harel, E., Ong, M. S. &Tobin, E. M. (1997). A Myb-related transcription factor is involved in the phytochrome regulation of an Arabidopsis Lhcb gene. *Plant Cell* 9(4): 491-507.

Wu, J. F., Wang, Y. &Wu, S. H. (2008). Two new clock proteins, LWD1 and LWD2, regulate Arabidopsis photoperiodic flowering. *Plant Physiol* 148(2): 948-959.

Yamashino, T., Matsushika, A., Fujimori, T., Sato, S., Kato, T., Tabata, S. &Mizuno, T. (2003). A Link between circadian-controlled bHLH factors and the APRR1/TOC1 quintet in Arabidopsis thaliana. *Plant Cell Physiol* 44(6): 619-629.

Yanovsky, M. J., Mazzella, M. A. &Casal, J. J. (2000). A quadruple photoreceptor mutant still keeps track of time. *Curr Biol* 10(16): 1013-1015.

Yanovsky, M. J., Mazzella, M. A., Whitelam, G. C. &Casal, J. J. (2001). Resetting of the circadian clock by phytochromes and cryptochromes in Arabidopsis. *J Biol Rhythms* 16(6): 523-530.

Yanovsky, M. J. Casal., J.J & Whitelam, G.C. (1995). Phytochrome A, phytochrome B and HY4 are involved in hypocotyl growth responses to natural radiation in Arabidopsis: weak de-etiolation of the phyA mutant under dense canopies. *Plant Cell & Environment* 18: 788-794.

Yu, J. W., Rubio, V., Lee, N. Y., Bai, S., Lee, S. Y., Kim, S. S., Liu, L., Zhang, Y., Irigoyen, M. L., Sullivan, J. A., Lee, I., Xie, Q., Paek, N. C. &Deng, X. W. (2008). COP1 and ELF3 control circadian function and photoperiodic flowering by regulating GI stability. *Mol Cell* 32(5): 617-630.

Zeilinger, M. N., Farre, E. M., Taylor, S. R., Kay, S. A. &Doyle, F. J., 3rd (2006). A novel computational model of the circadian clock in Arabidopsis that incorporates PRR7 and PRR9. *Mol Syst Biol* 2: 58.

Zhang, H., He, H., Wang, X., Yang, X., Li, L. &Deng, X. W. (2011). Genome-wide mapping of the HY5-mediated gene networks in Arabidopsis that involve both transcriptional and post-transcriptional regulation. *Plant J* 65(3): 346-358.

Zhu, D., Maier, A., Lee, J. H., Laubinger, S., Saijo, Y., Wang, H., Qu, L. J., Hoecker, U. &Deng, X. W. (2008). Biochemical characterization of Arabidopsis complexes containing CONSTITUTIVELY PHOTOMORPHOGENIC1 and SUPPRESSOR OF PHYA proteins in light control of plant development. *Plant Cell* 20(9): 2307-2323.

Zuo, Z., Liu, H., Liu, B., Liu, X. &Lin, C. (2011). Blue-light dependent interaction of CRY2 with SPA1 regulates COP1 activity and floral initiation in Arabidopsis. *Curr Biol* 21(10): 841-847.

10

Energy Conductance from Thylakoid Complexes to Stromal Reducing Equivalents

Lea Vojta and Hrvoje Fulgosi
Rudjer Boskovic Institute
Croatia

1. Introduction

Oxygenic photosynthesis is the basis of heterotrophic life on Earth. It produces carbohydrates and oxygen and may be dived into two sets of reactions: light reactions taking place in the thylakoid membranes, and carbon fixation reactions in soluble stroma. The light reactions involve highly reactive species, and if not controlled properly, they can produce deleterious reactive oxygen species. The structure and function of photosynthetic machinery must be extremely dynamic to enable flawless primary production under a wide spectrum of environmental conditions. The molecular mechanisms behind these dynamic changes remain largely uncharacterized, in particular because various auxiliary proteins linking photosynthesis with physiological responses are still missing.

Cooperation of two photosystems in the chloroplast thylakoid membranes produces a linear electron flow (LEF) from H_2O to $NADP^+$. Efficient photosynthetic energy conversion requires a high degree of integration and regulation of various redox reactions in order to maximize the use of available light and to minimize damaging effects of excess light (Allen, 2002). The interplay between cyclic (CEF), linear, and pseudocyclic electron transport pathways is required for maintaining the poised state of the photosynthetic system (Allen, 2003). In the over-oxidized state there is no electron flow while in the over-reduced state photooxidation can cause damage to photosystems and eventually death. Common to all three pathways is the activity of PSI that transfers electrons from the plastocyanin located in the thylakoidal lumen to the stromal ferredoxin (Fd). This transfer is mediated by three subunits, C, D and E, of the so-called stromal ridge of PSI (Nelson and Yocum, 2006). In the reduced state Fd provides electrons for the ferredoxin:NADP+ oxidoreductase (FNR), which produces NADPH in a linear pathway (Carrillo & Ceccarelli, 2003), for the ferredoxin–thioredoxin reductase (FTR), which catalyses the reduction of chloroplast thioredoxins (Shaodong et al., 2007), for feeding of the CEF (Allen, 2003) or, alternatively, electrons can be transferred to superoxide, the terminal acceptor in pseudocyclic pathway (Allen, 2003). The generation of reducing power is crucial for all biosynthetic processes within chloroplasts. NAD(P)H and ATP may be considered cell's energetic equivalents and are the principal energetic links between membrane-associated redox reactions and metabolism in the cell soluble compartments. These two types of molecules are generated simultaneously in the chloroplast during light-dependent electron transport and photophosphorylation. They are utilized in the reductive assimilation of inorganic elements (carbon, nitrogen, sulphur) into cellular matter, from which ATP and reductant can be regenerated by oxidative phosphorylation in the mitochondria, which enables the reducing power of NAD(P)H to be converted into ATP. The synthesis of ATP and NADPH in linear electron flow is tightly coupled and if the substrates for the ATP synthase (ADP, inorganic phosphate) become limiting, then the proton motive force builds up, inhibiting electron transfer to NADP+. Likewise, if NADP+ is limiting, photosynthetic electron carriers become reduced, slowing electron transfer and associated proton translocation, thus limiting ATP synthesis (Kramer & Evans, 2006). Linear electron flow produces a fixed ATP/NADPH ratio, and each metabolic pathway directly powered by photosynthesis consumes different fixed ATP/NADPH ratios. Chloroplasts have very limited pools of ATP and NADPH and since mismatches in ATP/NADPH ratios rapidly (within seconds) inhibit photosynthesis (Kramer & Evans, 2006), chloroplasts must balance the production and consumption of both ATP and NADPH by augmenting production of the limiting intermediate (e.g. by CEF) or dissipating the intermediate in excess.

2. Ferredoxin: NADP⁺ oxidoreductase

Final electron transfer from ferredoxin to NADP+ is accomplished by the ferredoxin:NADP+ oxidoreductase (FNR), a key enzyme of photosynthetic electron transport required for generation of reduction equivalents. Reducing power derived this way may be further used for carbon assimilation (Calvin-Benson cycle), amino acid, lipid and chlorophyll biosynthesis or reduction of stromal redox-active components. FNR is a ubiquitous flavin adenine dinucleotide (FAD)-binding enzyme that has been identified in various organisms including heterotrophic and phototrophic bacteria, in mitochondria and plastids of higher plants and algae, as well as apicoplasts of some intracellular parasites (Ceccarelli et al., 2004). In higher plants FNR is encoded by a small nuclear gene family and has been found in various chloroplast compartments: at the thylakoid membrane, in the soluble stroma, and at the chloroplast inner envelope. Both the membrane-bound and the soluble FNR pools are photosynthetically active.

2.1 Structure and localization of chloroplast FNRs

FNR harbors one molecule of noncovalently bound FAD as a prosthetic group (Arakaki et al., 1997; Carillo & Ceccarelli, 2003) and it catalyzes reversible electron transfer between reduced Fd to NADP+ for production of NADPH according to the reaction $2Fd_{red} + NADP^+ + H^+ \leftrightarrow 2Fd_{ox} + NADPH$. The FAD cofactor of FNR functions as an one-to-two electron switch by reduction of FAD to a semiquinone form FADH·, followed by another round of reduction to FADH⁻, and hydride transfer from FADH⁻ to NADP+.

Chloroplast FNR proteins are hydrophilic proteins with a molecular weight of approximately 35 kDa. Sequence similarity of FNRs from various plant species varies from 40% to 97% (Arakaki et al., 1997), and especially regions involved in FAD and NADP+ binding share high degree of identity. The topology of all chloroplast FNRs is highly conserved, consisting of two distinct domains connected by a loop (Dorowski et al., 2001), which shows the biggest variance between the species. The N-terminal domain (ca. 150 amino acids) is made up of a β-barrel structure built of six antiparallel β-strands and capped by an α-helix and a long loop and is involved in FAD binding, while the C-terminal domain (ca. 150 amino acids) consists of a central five-stranded parallel β-sheet surrounded by six α-helices and is mainly responsible for binding of NADP+ (Karplus et al., 1991). Fd is bound to the large, shallow cleft between the two domains (Pascalis et al., 1993). The amino acids essential for the formation and activity of the Fd-FNR complex have been identified in detail and nuclear magnetic resonance and mutagenesis studies have further revealed that the flexible N-terminus of FNR is also involved in the interaction with Fd (Maeda et al., 2005). FNR is synthesized on cytosolic rybosomes as a precursor containing an amino-terminal transit peptide, which is responsible for targeting the protein to the chloroplasts (Newman & Gray, 1988). Upon import of FNR into chloroplasts cleavage of the transit peptide by a stromal processing peptidase occurs, followed by the interaction with stromal chaperones Hsp70 (heat shock protein of 70 kDa) and Cpn60 (chaperonin of 60 kDa), which assist in the proper folding of FNR (Tsugeki & Nishimura, 1993), and FAD incorporation, which is required for maintenance of the native structure. Binding of FAD is also a prerequisite for membrane binding of FNR (Onda & Hase, 2004). Regulation of the enzyme activity has been proposed to occur by binding of FNR to the thylakoid membrane (Nakatani & Shin, 1991). Although soluble and membrane-bound FNR form a complex with Fd with the same dissociation constant, the rate constant of $NADP^+$ photoreduction has been shown to be much higher in the membrane bound than in the soluble complex *in vitro* (Forti & Bracale, 1984). But, since the *Arabidopsis fnr1* knock out mutant does not contain any membrane-bound FNR and still possesses normal photosynthetic performance, it may be concluded that *in planta* the soluble FNR is photosynthetically competent (Lintala et al., 2007), and thus the solubility of FNR itself is not a crucial determinant of enzyme activity.

2.2 FNR gene families

In higher plants, chloroplast-targeted FNR is encoded by a small nuclear gene family with one to three *FNR* genes sharing approximately 80% homology with each other. The chloroplast FNR proteins seem to be at least partly redundant, but they also possess unique properties, which are probably required for adjustment of chloroplast metabolism according to changes in the ambient environment (Mulo, 2011). FNR gene families have been well studied in the dicot C_3 plant *Arabidopsis thaliana* (thale cress, Hanke et al., 2005), monocot C_3 plant *Triticum aestivum* (wheat, Gummadova et al., 2007; Grzyb et al., 2008), and monocot C_4 plant *Zea mays* (maize, Okutani et al., 2005).

In *Arabidopsis* two genes, *At5g66190* and *At1g20020*, encode the two distinct ~ 32 kDa leaf isoforms FNR1 and FNR2 (Hanke et al., 2005; Lintala et al., 2007). Both genes are predominantly expressed in the rosette leaves, whereas only minor amount of mRNA could be detected in the stems, flowers and siliques, and no FNR proteins could be detected in the root tissue (Hanke et al., 2005). Chloroplast FNR1 has been shown to be more abundant in the membrane fraction (Hanke et al., 2005), especially at the stroma thylakoids (Benz et al., 2009), whereas FNR2 accumulates in higher amounts in the soluble stroma (Hanke et al., 2005). Indeed, FNR1 serves as a membrane anchor to FNR2, since upon inactivation of *FNR1* all the chloroplast FNR (FNR2) exists as a soluble protein (Lintala et al., 2007). It is very likely that FNR *in vivo* exists as a (hetero)dimer (Hanke et al., 2005). Recently, also formation of large (~ 330 kDa) FNR oligomers, devoid of other proteins, has been documented (Grzyb et al., 2008). Inactivation of one chloroplast *FNR* isoform did not result in upregulation of the expression of the other, neither at the level of transcription nor translation (Lintala et al., 2007; Lintala et al., 2009). Inactivation of either *FNR* gene resulted in general down-regulation of the photosynthetic machinery, but neither of the isoforms showed any specific function in LEF or CEF of photosynthesis, or other alternative electron transfer reactions (Lintala et al., 2007; Lintala et al., 2009). Growth of the *fnr* knock-out plants under unfavorable conditions revealed a unique role for FNR2 in redistribution of electrons to various redox reactions (Lintala et al., 2009). Beside two leaf-type FNR, two root-type FNR isoenzymes, encoded by genes *At1g30510* and *At4g05390* are present in *Arabidopsis* as well (Hanke et al., 2005). The growth of *Arabidopsis* on different nitrogen regimes induced differential expression of the two chloroplast *FNR* genes showing that multiple FNR isoenzymes have variable metabolic roles and differentially contribute to nitrogen assimilation (Hanke et al., 2005). Studies *in vivo* have revealed that suppression of FNR expression leads to increased susceptibility to photo-oxidative damage, impaired plant growth and lowered photosynthetic activity of transgenic plants (Hajirezaei et al., 2002; Palatnik et al., 2003). On the other hand, overexpression of FNR results in slightly increased rates of electron transport from water to NADP+ and increased tolerance to oxidative stress (Rodriguez et al., 2007).

Based on isoelectric focusing and SDS-PAGE the proteome of wheat has revealed four distinct leaf FNR isoforms that can be divided into two groups, FNRI and FNRII. Both groups contain two proteins, which differ from each other by truncation of the N-terminus (Gummadova et al., 2007; Grzyb et al., 2008). It has been demonstrated that the presence of mature wheat FNR proteins with alternative N-terminal start points, differing by a three amino acid truncation in pFNRI and a two amino acid truncation in pFNRII, have statistically significant differences in response to the physiological parameters of chloroplast maturity, nitrogen regime, and oxidative stress (Moolna et al., 2009). It has been suggested that these isoforms may be crucial to the regulation of reductant partition between carbon fixation and other metabolic pathways. Also, differences in the N-terminus of the wheat FNR isoforms seem to result in changes of FNR activity, subchloroplast localization as well as affinity of FNR to different Fd isoforms (Moolna et al., 2009). Results of Moolna et al. (2009) also suggest that four pFNR protein isoforms are each present in the chloroplast in phosphorylated and nonphosphorylated states, probably as a response to physiological parameters.

The genome of maize codes for three distinct leaf FNR genes that share 83–92% homology with each other, and are present in the leaves at approximately equivalent concentration. FNR1 is found at the thylakoid membrane, FNR3 is an exclusively soluble stromal protein, while FNR2 has a dual location (Okutani et al., 2005). The activity of FNR2 is similar to FNR3, and higher than that of the FNR1, and the mode of interaction between Fd(s) and the FNR isoforms is dependent on both pH and redox status of the chloroplast (Okutani et al., 2005). It has also been proposed that leaf FNR isoenzymes 1 and 2 are relatively more abundant under conditions of high demand for NADPH (Okutani et al., 2005).

2.3 FNR and the cyclic electron flow

FNR functions in the crossing of various electron transfer pathways. Besides its confirmed involvment in the last step of the LEF from water via PSII, plastoquinone (PQ) pool, Cyt b_6f complex, PSI, Fd and FNR to NADP+, the role of FNR in cyclic electron transfer has not yet been defined. In CEF, electrons are transferred from PSI to Cyt b_6f complex via Fd, with associated formation of proton gradient. PQ is reduced by Fd or NADPH via one or more enzymes collectively called PQ reductase, rather than by PSII, as in LEF. From hidroplastoquinone (PQH_2), electrons return to PSI via the Cyt b_6f complex. Thus, CEF around PSI produces ATP without accumulation of NADPH. It is generally accepted that

CEF supplies the ATP needed for driving the CO_2 concentrating mechanism in the C_4 plants. Recently, the significance of CEF has been shown in C_3 plants as well, where under normal physiological conditions CEF might have a role in adjusting the stoichiometry of ATP/NADPH generated by photosynthesis (Munekage et al., 2004). FNR has been identified as a component of the Cyt b_6f complex (Clark et al., 1984).

Four possible routes of cyclic electron transfer have been proposed so far and may operate in parallel. **NAD(P)H dehydrogenase (NDH)-dependent route**, in which electrons are first transferred from NADPH to NAD(P)H dehydrogenase-1 complex, and secondly to the PQ pool (Kramer & Evans, 2006; Mulo, 2011). The redox reactions are coupled to proton translocation in two ways. First, protons are taken up on the negatively charged side of the membrane during quinone reduction and released on the positive side of the membrane during quinol oxidation. Four protons should be translocated for each electron transferred through the cycle, two via the reduction and oxidation of PQ and the Q cycle and two more via the NDH proton pump (Kramer & Evans, 2006). **Fd-dependent route**, in which electrons are funneled from Fd to Cyt b_6f complex via a partly uncharacterized route involving PGR5 and PGRL1 proteins, which work together to catalyze cyclic electron flow, and possibly FNR and hypothetical FQR (ferredoxin-plastoquinone oxidoreductase) (Munekage et al., 2002; Kramer & Evans, 2006). **Nda2, a type 2 NAD(P)H:PQ oxidoreductase route** is active in some green algae and conifers that lack the chloroplast NDH complex. In *Chlamydomonas*, PQ reduction in CEF has been proposed to occur via a Nda2 (Desplats et al., 2009). It is related to those found in bacteria and mitochondria and does not pump protons. Since type 2 complexes are structurally much simpler than complex I (one subunit with a single flavin cofactor compared with at least 11 protein subunits, nine FeS clusters, and a flavin), Nda2 may be less efficient in energy balancing (Kramer & Evans, 2006). **The Cyt b_6f complex and FNR route** uses the PQ reductase site of the Cyt b_6f complex to reduce PQ (Zhang et al., 2004). Electron transfer to Q_i probably involves the newly discovered heme c_i, which allows electrons to flow from Fd or FNR to the bound PQ (Zhang et al., 2004). This pathway probably involves the formation of a special cyclic electron flow supercomplex (Iwai et al., 2010).

2.4 FNR and oxidative stress tolerance

PSII and PSI in the chloroplasts of higher plants are potential sources of harmful reactive oxygen species (ROS). In *E. coli* FNR is involved in quenching of ROS (Krapp et al., 2002), and in methyl viologen resistant *Chlamydomonas reinhardtii* strains the steady state level of chloroplast FNR transcripts has been shown to be increased as compared to wild type (Kitiyama et al., 1995). Moreover, expression of plant FNR has been proven to restore the oxidative tolerance of a mutant *E. coli* (Krapp et al., 1997). The research on the participation of FNR in oxidative stress responses of higher plants has been performed on wheat and has shown that, in contrast to bacterial cells, the content of FNR mRNA as well as protein in higher plants rather decreases than increases in response to induction of oxidative stress (Palatnik et al., 1997). However, production of ROS results in marked release of FNR from the thylakoid membrane followed by reduction of NADP$^+$ photoreduction capacity, which might aim at maintaining the NADP$^+$/NADPH homeostasis of the stressed plants (Palatnik et al., 1997). Recently, it has been shown that FNR releases from the thylakoids in the plants suffering from drought stress (Lehtimaki et al., 2010), and the FNR containing thylakoid protein complexes disassemble upon high light illumination (Benz et al., 2009).

3. Supramolecular FNR complexes

It has been shown that FNR exists in soluble and membrane-bound forms (Palatnik et al., 1997; Lintala et al., 2007). Several potential FNR-binding partners have been discussed, which might be involved in membrane attachment of FNR. Various studies have shown interaction of FNR with the photosynthetic protein complexes Cyt b_6f (Clark et al., 1984; Zhang et al., 2001), PSI (Andersen et al., 1992) or NDH complex (Quiles & Cuello, 1998), but also interaction with glyceraldehyde-3-phosphate dehydrogenase, or direct membrane attachment have been suggested. However, until recently no exact protein partner responsible for FNR tethering has been identified. Two chloroplast proteins, Tic62 and TROL, were recently identified and shown to form high molecular weight protein complexes with FNR at the thylakoid membrane, and thus seem to act as molecular anchors of FNR to the thylakoid membrane. Tic62 and TROL have been shown to bind FNR by specific interaction via a conserved Ser/Pro-rich motif. In darkness, FNR forms large protein complexes at the thylakoids together with Tic62 and TROL. Similarly, Tic62 and presumably TROL bind FNR at the envelope. FNR is released from the membranes upon illumination.

3.1 Tic62

During its import into chloroplasts from the site of synthesis on cytosolic rybosomes, FNR has been found to interact with Tic62 protein, a 62 kDa component of the Translocon at the inner envelope of chloroplast (Küchler et al., 2002; Balsera et al., 2007; Stengel et al., 2008; Benz et al. 2009). Proteomics studies have identified Tic62 in the chloroplast envelope, stroma and thylakoid fraction (Benz et al., 2009). Furthermore, Tic62 at the thylakoid membrane was found in several high molecular mass protein complexes (250–500 kDa), and it was shown to be tightly associated with both chloroplast FNR isoforms (Benz et al., 2009). The N-terminus of Tic62 binds pyridine nucleotides, while the stroma exposed C-terminus contains repetitive, highly conserved FNR-binding domains (Küchler et al., 2002). Database searches have verified the presence of the FNR-binding domains only in the Tic62 protein of vascular plants (Balsera et al., 2007) and it occurs in different numbers dependent on the respective plants species.

The function of FNR in the Tic complex has been suggested to link redox regulation to chloroplast protein import. Indeed, *in vitro* experiments with compounds interfering either with NAD binding or NAD(P)/NAD(P)H ratio modulate the import characteristics of the leaf FNR isoforms: FNR1 is translocated preferentially at high NAD(P)/NAD(P)H ratio, while the translocation of FNR2 is not influenced by the redox status (Küchler et al., 2002). In maize, import of pre-FdI to chloroplast stroma is independent on illumination, while pre-FdIII and preFNR were efficiently targeted into stroma only in darkness (Hirohashi et al., 2001). These results imply that the diurnal changes in the chloroplast redox poise may control import characteristics of the organelle. It was recently shown that Tic62 shuttles between the soluble stroma and the chloroplast membranes, and that oxidation of stroma results in stronger association of Tic62 to the membrane fraction (Stengel et al., 2008). FNR shows similar shuttling behavior, and therefore the Tic62–FNR interaction is dependent on chloroplast redox state (Stengel et al., 2008). The lack of Tic62 and consequently the lack of Tic62–FNR complexes did not have any effects on the plant phenotype or photosynthetic properties, neither on LEF nor CEF (Benz et al., 2009), implying that the Tic62–FNR complexes serve for some other purpose(s) than photosynthesis. It has been observed that the membrane-bound Tic62–FNR protein complexes were most abundant in the dark, while increase in light intensity resulted in the disassembly of the complex. Similarly, *in vitro* alkalization of isolated thylakoids dissociated FNR and Tic62 (Benz et al., 2009). It is important to stess that the interaction of Tic62 with FNR stabilizes the activity of the FNR protein (Benz et al., 2009) and that FNR activity is lower in acidic than basic environment (Lee et al., 2007). These results indicate that Tic62 acts as a chaperone for FNR, and protects the flavoenzyme from inactivation and degradation during the photosynthetically inactive periods, e.g. in darkness (Benz et al., 2009).

3.2 TROL

By using antisense and gene inactivation strategies Jurić et al. (2009) identified a novel component of non-appressed thylakoid membranes which is responsible for anchoring of FNR. TROL (thylakoid rhodanese-like protein) is a 66 kDa nuclear encoded component of thylakoid membranes required for tethering of FNR and sustaining efficient LEF in vascular plants. TROL contains two transmembrane helices and a centrally positioned (inactive) rhodanese domain (Jurić et al., 2009). As an integral membrane protein, TROL is firmly attached to the thylakoid membrane and cannot be extracted from the membrane by high salt, urea or high pH treatments (Jurić et al., 2009). TROL possesses a unique fusion of two distinct modules: a centrally positioned rhodanese-like domain, RHO, which is found in all life forms, and a C-terminal single hydrophobic FNR-binding region, ITEP, which is ascribed to the vascular plants (Balsera et al., 2007). It is hypothesised that both N- and C-terminal parts of TROL face the stroma, while RHO faces the thylakoid lumen (Jurić et al., 2009). A closer investigation of the TROL protein sequence revealed an interesting region upstream of the ITEP domain. The Pro-Val-Pro repeat-rich region was designated PEPE. It consists of two identical repeats, followed by a possible PVP hinge. In membrane proteins, prolines are known to have a structural role in transmembrane helices, where they distort the alpha-helix due to the loss of at least one stabilising backbone hydrogen bond. Thus, PEPE region, which is presumed to be exposed into the stroma, is proposed to introduce flexibility in the helix that may result in kink and swivel motions of FNR-binding region.

Localization in non-appressed regions places TROL in the vicinity of the site of Fd reduction. TROL has been found in several complexes, indicating the presence of several TROL subpools in the thylakoid membrane. Only a 190 kDa complex appears to contain TROL in association with the FNR. Complexes at about 110 and 120 kDa indicate the existence of a small ligand which may be associated with other TROL domains, namely the large rhodanese-like domain which is predicted to be located in the thylakoid lumen.

The findings of Küchler et al. (2002) that Tic62 interacts with FNR prompted the analysis of TROL protein sequence in search for the similar binding module. Tic62 from *Pisum sativum* (PsTic62) contains three Pro/Ser-rich repetitive motives at the C-terminus, S-P-Y-x(2)-Y-x-D/E-L-K-P(2)-S/T/A-S/T-P-S/T-P, involved in the binding of FNR (Küchler et al., 2002). PsTic62 homolog in *A. thaliana*, encoded by a single-copy gene (*At3g18890*), shows approximately 60% identity for the deduced mature sequence and has a calculated molecular weight of 62.1 kDa (Küchler et al., 2002). AtTic62 contains four repetitive motives at the C-terminus, but it has been shown previously that only one repeat is sufficient for the binding of FNR (Küchler et al., 2002; Balsera et al., 2007). As TROL possesses almost identical domain to the Tic62 FNR-binding repeats, a modified yeast-two hybrid assay was used to confirm ITEP-FNR interaction (Jurić et al., 2009). Predictably, ITEP strongly binds to the FNR protein, even eight times stronger than Tic62. Established high-affinity interaction with FNR, together with the reports on TROL abundance at the thylakoid membranes (Peltier et al., 2004), implies that we are probably not dealing with highly dynamical and fast-responding interaction, but with more quantitative-based and rather inert interaction. The size of the complex is, however, smaller (190 kDa) than the Tic62–FNR complexes (250-500 kDa), which is in line with only one FNR binding motif found in the amino acid sequence of *Arabidopsis* TROL protein, as compared to four of such motifs present in the sequence of *Arabidopsis* Tic62. Although *in vitro* experiments indicate that the interaction between TROL and FNR is stronger than the interaction between FNR and Tic62 (Jurić et al., 2009), the exposure of plants to high light intensity results in faster dissociation of FNR from the TROL–FNR complexes than from the Tic62–FNR complexes (Jurić et al., 2009).

Using a synthetic peptide, representing the conserved binding motif, called the FNR-membrane-recruiting-motif (FNR-MRM) found in Tic62 and TROL, Alte et al. (2010) determined the crystal structure of the FNR:peptide complex and concluded that the FNR-MRM induces self-assembly of two FNR molecules. Although FNR is commonly distributed among all three domains of life, FNR-MRM of both Tic62 and TROL exclusively exists in vascular plants, thus, membrane tethering of FNR by this motif seems to be a recent evolutionary invention (Alte et al., 2010). Whereas most TROL proteins comprise a single FNR-binding domain, its number varies to a higher extent in Tic62 proteins. However, the binding affinity to FNR did not change significantly when constructs comprising one or three FNR-interacting motifs were analyzed (Alte et al., 2010). This indicates that binding to each domain occurs independently of the other motifs and excludes cooperative binding effects.

3.2.1 Redox regulation and tethering properties of TROL

Besides being involved in sulfur metabolism, rhodanese-like domains are implicated in redox regulation of various intracellular processes (Horowitz & Falksen, 1986; Horowitz et al., 1992; Nandi et al., 2000). It has been speculated that the rhodanese-like domain of TROL is involved in redox regulation of FNR binding and release (Jurić et al., 2009). It has been proposed that the redox regulation of FNR binding and release could be important for balancing the redox status of stroma with the membrane electron transfer chain. Such regulation could be important for prevention of over-reduction of any of these two compartments and maintenance of the redox poise (Jurić et al., 2009).

TROL, as FNR, is mainly located at the stroma thylakoids, but it can be also found embedded in the chloroplast inner envelope membrane in the non-processed form (70 kDa). Localization of the TROL precursor at the chloroplast envelope in its unprocessed form indicates its possible role in electron transfer chain specific for this membrane. This dual localization might also be dependent on the $NADP^+/NADPH$ ratio in the chloroplasts, similar to the shuttling of the Tic62 protein (Stengel et al., 2008).

FNR is supposed to be the key protein in transferring electrons to the final destination in LEF and *trol* plants indeed exhibit decreased LEF. On the basis of the investigation of membrane-bound pool of FNR and the 190 kDa-complex containing TROL and FNR in the wild-type plants grown under growth-light and high-light intensities (Jurić, 2010), a role for TROL as the FNR anchor could be proposed. TROL anchors and thus stabilises FNR during the night, possibly to prevent FNR from extracting electrons from NADPH molecules and compromising the downstream metabolic reactions. During the light period, under growth-light conditions, TROL anchors FNR and stabilises it, similar to association between Tic62 and FNR (Benz et al., 2009). FNR could be gradually released by binding/releasing of certain signalling molecules to the luminal RHO domain, and electrons are transferred from ferredoxin to $NADP^+$ at normal rates (Figure 1). Under the high-light/excess-light intensities TROL discharges FNR, possibly because of binding/releasing of certain signalling molecules to the luminal RHO domain, and FNR now catalyses the reverse reaction of transferring electrons from excess NADPH to potential electron-acceptor molecules (Figure 1) (Jurić et al., 2010). The PEPE swivel that preceeds the ITEP domain could maneuver the bound FNR protein due to the proline-introduced flexibility. For instance, the free-moving PEPE swivel could move FNR closer to the thylakoids to establish transient contacts with other transmembrane proteins, or it could move FNR away from the thylakoids. In addition, if the binding of FNR to the thylakoids is a precondition for efficient LEF (Forti & Bracale, 1984), then the TROL-bound FNR molecules could be easily displaced from TROL to the already discovered FNR-binding membrane proteins or the unknown ones. This is in accordance with the clearly visible TROL-FNR complex under dark conditions and with its disappearance during light periods (Benz et al., 2009).

The most interesting property of the FNR-Tic62/TROL interaction is the clear difference of the affinity at acidic (pH 6; KD $\approx$ 0.04 µM) compared to alkaline (pH 8; KD $\approx$ 3 µM) conditions (Alte et al., 2010). The pH variations reflect differences of the chloroplast stroma between light and dark cycles (Alte et al., 2010): During light phases, when photosynthetic activity is high, protons are transported into the thylakoid lumen, leading to an alkaline stromal pH. By contrast, when photosynthesis ceases during dark phases the stromal pH decreases again. Under these conditions, Tic62 and TROL are predominantly associated with the thylakoid membrane where they recruit FNR into stable high-molecular-weight complexes (Benz et al., 2009; Jurić et al., 2009). Light quantity can vary dramatically during the course of the day, therefore requiring constant adjustment of the light harvesting processes and the enzymatic reactions. Changes in light quantities alter stromal pH as well as the amount of FNR bound to the thylakoid membranes. Also, the membrane attachment of FNR is influenced by the stromal redox state ($NADP^+$/NADPH ratio), which mimics variations in environmental conditions (Stengel et al., 2008). Therefore, reversible attachment of FNR to the thylakoid membrane via Tic62/TROL provides an elegant way to store redundant molecules, not required when photosynthesis is less active or dormant.

4. FNR, TROL and Tic62 *Arabidopsis* mutants

When the amount of FNR was artificially reduced by antisense or silencing techniques (Hajirezaei et al., 2002; Lintala et al., 2009), or by interruption of a *FNR* gene by T-DNA (Lintala et al., 2007; Hanke & Hase, 2008), the plants suffered from chlorosis and reduced photosynthetic activity, which finally resulted in reduced growth. Although the level of total NADP(H) was not affected in the mutants, the NADPH/$NADP^+$ ratio was strongly reduced (Hajirezaei et al., 2002; Hald et al., 2008). These mutants are prone to photo-oxidative damage, and suffer from oxidative stress (Palatnik et al., 2003; Lintala et al., 2009). The redox poise of the NADP(H) pool is also likely to regulate photosynthetic electron transfer activity in order to balance production and consumption of reducing equivalents, and thereby to limit production of ROS in the chloroplasts (Hald et al., 2008). Over-expression of FNR, however, did not markedly up-regulate the rate of $NADP^+$ photoreduction or CO_2 assimilation, but showed increased tolerance to photodamage (Rodriguez et al., 2007).

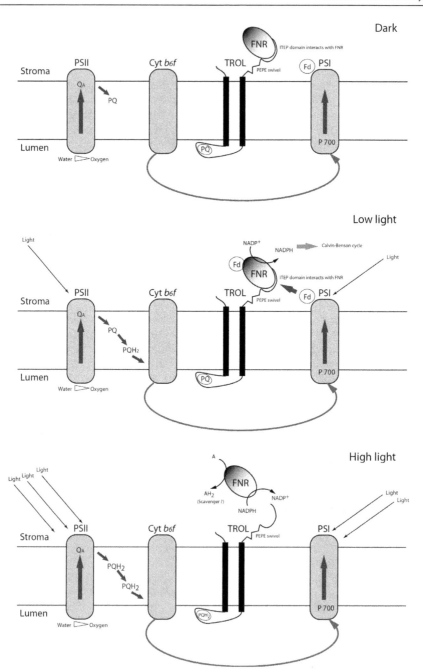

Fig. 1. The proposed mechanism of redox regulation of FNR binding and release. During the dark period, FNR is bound to the thylakoids *via* TROL. This stage could be sustained through the binding of small molecule, possibly oxidized PQ, to the RHO cavity. There is no

NADPH production. In conditions of growth-light, FNR is bound to the thylakoids *via* TROL and acts as an efficient NADPH producer. In conditions of saturating light, a molecule, possibly reduced PQ, competes for the RHO binding site and generates the signal for the FNR release, which, when soluble, acts as NADPH consumer. Protons are passed to an unknown scavenger A.

Overexpression of FNR in transgenic plants causes enhanced tolerance to photo-oxidative damage and herbicides that propagate reactive oxygen species (Rodriguez et al., 2007). On the other hand, antisense repression of FNR renders transgenic plants abnormally prone to photo-oxidative stress (Palatnik et al., 2003).

Analysis of *Arabidopsis* mutant lines indicates that, in the absence of TROL, relative electron transport rates at high-light intensities are severely lowered accompanied with significant increase in non-photochemical quenching (NPQ). If solubilization of FNR is necessary for the regulation of oxidative stress, then it is not surprising that, under high-light conditions, TROL-deficient plants exhibit increased rates of NPQ. This effect was also explained by recently proposed feedback redox regulation via the redox poise of the NADP(H) pool (Hald et al., 2008). Thus, TROL might represent a missing thylakoid membrane docking site for a complex between FNR, ferredoxin and $NADP^+$. Such association might be necessary for maintaining photosynthetic redox poise and enhancement of the NPQ (Jurić et al., 2009).

Inhibition of TROL accumulation by antisense expression results in quenching which is higher than that of the wild-type plants, but lower than that of the TROL knock-out plants (Jurić et al., 2009). This demonstrates dosage effect of TROL and indicates that FNR binding to the thylakoid membranes is dependent on the availability of tethering sites and that the amount of soluble FNR directly influences NPQ.

It has been proposed that the balance between $NADP^+$ and NADPH regulates the photosynthetic electron transport at the level of cyt b_6/f complex in a feedback manner (Hald et al., 2008). Interestingly, NADP-malic enzyme 2 that catalyses the oxidative decarboxylation of malate, producing pyruvate, carbon dioxide and NAD(P)H in cytosol (Wheeler et al., 2005) was significantly up-regulated in TROL-deficient plants grown under growth-light conditions, thus providing the possible pathway of maintaining $NADP^+$/NADPH balance through the malate valve (Jurić et al., 2009). In this case, *trol* plants would act as efficient NADPH producers, and, in an effort not to hyperreduce the thylakoids, they would export the reducing energy in a form of malate to the cytosol.

Also, in *trol* plants, genes encoding proteins involved in stress management are strongly up-regulated. As plant growth and development are driven by electron transfer reactions (Noctor, 2006), it is not surprising that leaf anatomy is altered in the knock-out. Furthermore, chloroplasts in the knock-out are small and have less developed thylakoids. These morphological changes reflect alterations in gene expression of a specific set of genes encoding chloroplast proteins. Many processes important for chloroplast morphogenesis could be influenced by NADPH production, or be dependent on metabolic retrograde signaling (Jurić et al., 2009).

When Tic62 was knocked out, formation of high molecular weight FNR protein complexes was hindered, while some free FNR still was detected at the thylakoids of *tic62* plants. The amount of FNR in the soluble pool, however, remained more or less constant. *Vice versa*, if either one of the chloroplast targeted FNR isoforms was missing the membrane binding of the Tic62 protein was prevented. Since no changes in the *FNR* gene expression or in the FNR pre-protein import could be detected in the *tic62* plants, reduction of FNR level most probably resulted from differences in the turnover of FNR isoforms inside the chloroplasts (Benz et al., 2009).

In contrast to *tic62* plants, *trol* knock-out mutants have a clear photosynthetic phenotype (Jurić et al., 2009). The appearance of the *trol* plants is slightly smaller than the WT, but with no distinct differences in pigment composition. However, the mutant chloroplasts are small, irregular in morphology and show deteriorated thylakoid structure. The abnormalities in chloroplast structure are reflected on the function with marked differences in electron transfer rate under high light intensity (500 to 800 μmol photons m^{-2} s^{-1}). Since nonphotochemical quenching increases and variable chlorophyll fluorescence decreases in the *trol* leaves upon increasing illumination, it seems that the absence of TROL results in increased ability to dissipate excess absorbed energy.

The absence of TROL disables FNR from being tethered to the membrane, therefore a substantial amount of FNR remains soluble. Forti & Bracale (1984) demonstrated that the soluble form of FNR is very inefficient in NADP$^+$ photoreduction by isolated thylakoids. It has been shown in TROL-deficient plants that linear photosynthetic flow can be sustained until light intensity exceeds 250 μmol photons m^{-2} s^{-1}. As soluble FNR is no longer able to reduce NADP$^+$ at high rates, this could lead to over-reduction of the entire electron transport chain. In this case, NPQ modulation could be particularly important to prevent photo-damage caused by build-up of reduced electron carriers which block LEF before the lumen could be significantly acidified (Kanazawa & Kramer, 2002). In conclusion, TROL is necessary for maintenance of efficient LEF, induction of NPQ, as well as for redox-regulation of key thylakoid signal-transduction pathways. Furthermore, discovery of TROL provides new information for linking leaf and chloroplast morphogenesis with photosynthetic cues.

5. Conclusion

The discovery of TROL protein and its important role in tethering of the flavoenzyme FNR not only answers the old dilemma about the position at which this crucial photosynthetic enzyme is located, but opens new approaches for the investigations of oxidative stress management in chloroplasts. FNR in bacteria acts as an important scavenger of free radicals and it will be interesting to see if it possesses similar function in plant cells. Linking FNR with cellular energetics and possibly with retrograde signaling will also be investigated. Is TROL the source element in signal-transduction cascade linking photosynthesis with plant growth and cellular responses? TROL possesses several elaborate elements of signal transduction, namely rhodanese-like domain located in lumen, proline-rich swivel involved in signal attenuation, and FNR membrane recruitment motif. Functions of each of these domains will be investigated *in planta*, by using genetic transformation techniques. Global gene expression analysis revealed that genes depending on NADPH synthesis and availability are up-regulated in TROL deficient plants. Among them are NADP-malic enzyme and protochlorophyllide oxidoreductase B. How is TROL linked with malate shuttle enzymes? What is the role of TROL and Tic62 in the inner envelope membrane? Finally, other proteins following the same expression pattern in correlation analyses will be investigated for their ability to interact with TROL and Tic62. Supramolecular complexes of TROL and membrane yeast-two-hybrid screens will likely reveal so far overlooked elements of thylakoid signal transduction. An exciting quest for auxiliary proteins involved in fine-tuning of photosynthetic energy conversion lies ahead.

6. Aknowledgement

The work of H.F. and L.V. has been funded by the Grant 098-0982913-2838 of the Ministry of Science, Education and Sports of the Republic of Croatia.

7. References

Allen, J. F. (2002). Photosynthesis of ATP-electrons, proton pumps, rotors, and poise. *Cell*, Vol. 110, No. 3, pp. 273–276.

Allen, J. F. (2003). Cyclic, pseudocyclic and noncyclic photophosphorylation: new links in the chain. *Trends in Plant Sciences*, Vol. 8, No. 1, pp. 15–19.

Alte, F.; Stengel, A.; Benz, J. P.; Petersen,   E.; Soll, J.; Groll, M. & Bölter,   B. (2010). Ferredoxin:NADPH oxidoreductase is recruited to thylakoids by binding to a polyproline type II helix in a pH-dependent manner. *Proceedings of the National Academy of Sciences of the United States of America*, Vol. 107, No. 45, pp. 19260-19265.

Andersen,   B.; Scheller,   H. V. & Moller,   B. L. (1992). The PSI E subunit of photosystem I binds ferredoxin:NADP$^+$ oxidoreductase. *FEBS Letters*, Vol. 311, No. 2, pp. 169–173.

Arakaki,   A. K.; Ceccarelli,   E. A. & Carrillo,   N. (1997). Plant-type ferredoxin-NADP$^+$ reductases: a basal structural framework and a multiplicity of functions. *FASEB Journal*, Vol. 11, No. 2, pp. 133–140.

Balsera, M.; Stengel, A.; Soll, J & Bölter, B. (2007). Tic62: a protein family from metabolism to protein translocation. BMC *Evolutionary Biology*, Vol.7, pp. 43.

Benz, J.P.; Stengel, A.; Lintala, M.; Lee, Y.H.; Weber, A.; Philippar, K.; Gugel, I.L.; Kaieda, S.; Ikegami, T.; Mulo, P.; Soll, J. & Bölter, B. (2009). *Arabidopsis* Tic62 and ferredoxin-NADP(H) oxidoreductase form light-regulated complexes that are integrated into the chloroplast redox poise. *Plant Cell*, Vol. 21, No.12, pp. 3965–3983.

Carrillo, N. & Ceccarelli, E. A. (2003). Open questions in ferredoxin-NADP$^+$ reductase catalytic mechanism. *European Journal of Biochemistry*, Vol. 270, No. 9, pp. 1900–1915.

Ceccarelli, E.A.; Arakaki, A.K.; Cortez, N. & Carrillo, N. (2004). Functional plasticity and catalytic efficiency in plant and bacterial ferredoxin-NADP(H) reductases. *Biochimica et Biophysica Acta - Proteins Proteomics*, Vol. 1698, No. 2, pp. 155–165.

Clark, R.D.; Hawkesford, M.J.; Coughlan, S.J.; Bennett, J. & Hind, G. (1984). Association of ferredoxin–NADP$^+$ oxidoreductase with the chloroplast cytochrome b–f complex. *FEBS Letters*, Vol. 174, No.1, pp. 137–142.

Desplats, C.; Mus, F.; Cuiné, S.; Billon, E.; Cournac, L. & Peltier, G. (2009). Characterization of Nda2, a plastoquinone-reducing type II NAD(P)H dehydrogenase in *Chlamydomonas* chloroplasts. *Journal of Biological Chemistry*, Vol. 284, No. 7, pp. 4148–4157.

Dorowski, A.; Hofmann, A.; Steegborn, C.; Boicu, M. & Huber, R. (2001). Crystal structure of paprika ferredoxin-NADP$^+$ reductase. Implications for the electron transfer pathway, *Journal of Biological Chemistry*, Vol. 276, No. 12, pp. 9253–9263.

Forti, G. & Bracale, M. (1984). Ferredoxin-ferredoxin NADP reductase interaction. *FEBS Letters*, Vol. 166, No. 1, pp. 81–84.

Grzyb, J.; Malec, P.; Rumak, I.; Garstka, M. & Strzalka, K. (2008). Two isoforms of ferredoxin:NADP$^{(+)}$ oxidoreductase from wheat leaves: purification and initial biochemical characterization. *Photosynthesis Research*, Vol. 96, No. 1, pp. 99–112.

Gummadova, J.O.; Fletcher, G.J.; Moolna, A.; Hanke, G.T.; Hase, T. & Bowsher, C.G. (2007). Expression of multiple forms of ferredoxin NADP$^+$ oxidoreductase in wheat leaves. *Journal of Experimental Botany*, Vol. 58, No. 14, pp. 3971–3985.

Hajirezaei, M.R.; Peisker, M.; Tschiersch, H.; Palatnik, J.F.; Valle, E.M.; Carrillo, N. & Sonnewald, U. (2002). Small changes in the activity of chloroplastic NADP(+)-dependent ferredoxin oxidoreductase lead to impaired plant growth and restrict photosynthetic activity of transgenic tobacco plants. *Plant Journal*, Vol. 29, No. 3, pp. 281-293.

Hald, S.; Pribil, M.; Leister, D.; Gallois, P. & Johnson, G.N. (2008). Competition between linear and cyclic electron flow in plants deficient in Photosystem I. *Biochimica et Biophysica Acta*, Vol. 1777, No. 9, pp. 1173-1183.

Hanke, G.T.; Okutani, S.; Satomi, Y.; Takao, T.; Suzuki, A. & Hase, T. (2005). Multiple iso-proteins of FNR in *Arabidopsis*: evidence for different contributions to chloroplast function and nitrogen assimilation. *Plant Cell and Environment*, Vol. 28, No. 9, pp. 1146-1157.

Hanke, G.T. & Hase, T. (2008). Variable photosynthetic roles of two leaf-type ferredoxins in arabidopsis, as revealed by RNA interference. *Photochemistry and Photobiology*, Vol. 84, No. 6, pp. 1302-1309.

Hirohashi, T.; Hase, T.; & Nakai, M. (2001). Maize non-photosynthetic ferredoxin precursor is mis-sorted to the intermembrane space of chloroplasts in the presence of light. *Plant Physiology*, Vol. 125, No. 4, pp. 2154-2163.

Horowitz, P.M. & Falksen, K. (1986). Oxidative inactivation of the enzyme rhodanese by reduced nicotinamide adenine dinucleotide. *Journal of Biological Chemistry*, Vol. 261, No. 36, pp. 16953-16956.

Horowitz, P.M.; Butler, M. & McClure, G.D. Jr (1992). Reducing sugars can induce the oxidative inactivation of rhodanese. *Journal of Biological Chemistry*, Vol. 267, No. 33, pp. 23596-23600.

Iwai, M.; Takizawa, K.; Tokutsu, R.; Okamuro, A.; Takahashi, Y. & Minagawa, J. (2010). Isolation of the elusive supercomplex that drives cyclic electron flow in photosynthesis. *Nature*, Vol. 464, No. 7292, pp. 1210-1213.

Juric, S.; Hazler-Pilepic, K.; Tomasic, A; Lepedus, H.; Jelicic, B.; Puthiyaveetil, S.; Bionda, T.; Vojta, L.; Allen, J.F.; Schleiff, E. & Fulgosi, H. (2009). Tethering of ferredoxin:NADP+ oxidoreductase to thylakoid membranes is mediated by novel chloroplast protein TROL. *The Plant Journal*, Vol. 60, No. 5, pp. 783-794.

Juric, S. (2010). The role of the gene product *At4g01050* in the regulation of photosynthesis in *Arabidopsis thaliana* (L.) Heynh. *Doctoral thesis*, Faculty of Science, University of Zagreb, Zagreb, Croatia.

Kanazawa, A. & Kramer, D.M. (2002). *In vivo* modulation of nonphotochemical exciton quenching (NPQ) by regulation of the chloroplast ATP synthase. *Proceedings of the National Academy of Sciences of the United States of America*, Vol. 99, No. 20, pp. 12789-12794.

Karplus, P.A.; Daniels, M.J. & Herriott, J. R. (1991). Atomic structure of ferredoxin-NADP+ reductase: prototype for a structurally novel flavoenzyme family. *Science*, Vol. 251, No. 4989, pp. 60-66.

Kitayama, K.; Kitayama, M. & Togasaki, R.K. (1995). Characterization of paraquat-resistant mutants of *Chlamydomonas reinhardtii*, In: *Photosynthesis, From Light to Biosphere* Vol. 3, P. Mathis, (Ed.), 595-598, Kluwer Academic Publishers, Amsterdam, Netherlands.

Kramer, D. M.& Evans, J. R. (2006). The importance of energy balance in improving photosynthetic productivity. *Plant Physiology*, Vol. 155, No. 1, pp. 70-78.

Krapp, A.R.; Tognetti, V.B.; Carrillo, N. & Acevedo A. (1997). The role of ferredoxin-NADP+ reductase in the concerted cell defense against oxidative damage. Studies using

Escherichia coli mutants and cloned plant genes. *Europaean Journal of Biochemistry*, Vol. 249, No. 2, pp. 556–563.

Krapp, A.R.; Rodriguez, R.E.; Poli, H.O.; Paladini, D. H.; Palatnik, J.F. & Carrillo, N. (2002). The flavoenzyme ferredoxin (flavodoxin)-NADP(H) reductase modulates NADP(H) homeostasis during the soxRS response of Escherichia coli. *Journal of Bacteriology*, Vol. 184, No. 5, pp. 1474–1480.

Küchler, M.; Decker, S.; Hörmann, F.; Soll, J. & Heins, L. (2002) Protein import into chloroplasts involves redox-regulated proteins. *EMBO Journal*, Vol. 21, No. 22, pp. 6136–6145.

Lee, Y.H.; Tamura, K.; Maeda, M.; Hoshino, M.; Sakurai, K.; Takahashi, S.; Ikegami, T.; Hase, T. & Goto, Y. (2007). Cores and pH-dependent dynamics of ferredoxin-NADP+ reductase revealed by hydrogen/deuterium exchange. *Journal of Biological Chemistry*, Vol. 282, No. 8, pp. 5959–5967.

Lehtimaki, N.; Lintala, M.; Allahverdiyeva, Y.; Aro, E.M. & Mulo, P. (2010). Drought stress-induced upregulation of components involved in ferredoxin-dependent cyclic electron transfer. *Journal of Plant Physiology*, Vol. 167; No.12, pp. 1018-1022.

Lintala, M.; Allahverdiyeva, Y; Kidron, H.; Piippo, M.; Battchikova, N; Suorsa, M; Rintamäki, E.; Salminen, T. A.; Aro, E. M. & Mulo, P. (2007). Structural and functional characterization of ferredoxin-NADP(+)-oxidoreductase using knock-out mutants of *Arabidopsis. The Plant Journal*, Vol. 49, No. 6, pp. 1041–1052.

Lintala, M.; Allahverdiyeva, Y; Kangasjärvi, S.; Lehtimäki, N.; Keränen, M.; Rintamäki, E.; Aro, E. M. & Mulo, P. (2009). Comparative analysis of leaf-type ferredoxin-NADP oxidoreductase isoforms in *Arabidopsis thaliana. The Plant Journal*, Vol. 57, No. 6, pp. 1103–1115.

Maeda, M.; Lee, Y. H.; Ikegami, T.; Tamura, K.; Hoshino, M.; Yamazaki, T.; Nakayama, M.; Hase T. & Goto, Y. (2005). Identification of the N- and C-terminal substrate binding segments of ferredoxin-NADP+ reductase by NMR. *Biochemistry*, Vol. 44, No. 31, pp. 10644–10653.

Moolna, A. & Bowsher, C. G. (2010) The physiological importance of photosynthetic ferredoxin NADP+ oxidoreductase (FNR) isoforms in wheat. *Journal of Experimental Botany*, Vol. 61, No. 10, pp. 2669–2681.

Mulo, P. (2011). Chloroplast-targeted ferredoxin-NADP(+) oxidoreductase (FNR): Structure, function and location. *Biochimica et Biophysica Acta*, Vol. 1807, No. 8, pp. 927–934.

Munekage, Y.; Hashimoto, M.; Miyake, C.; Tomizawa, K. I.; Endo, T.; Tasaka, M. & Shikanai, T. (2004). Cyclic electron flow around photosystem I is essential for photosynthesis. *Nature*, Vol. 429, No. 6991, pp. 579–582.

Nakatani, S. & Shin, M. (1991). The reconstituted NADP photoreducing system by rebinding of the large form of ferredoxin-NADP reductase to depleted thylakoid membranes. *Archives of Biochemistry and Biophysics*, Vol. 291, No. 2, pp. 390–394.

Nandi, D. L.; Horowitz, P. M. & Westley, J. (2000). Rhodanese as a thioredoxin oxidase. *The International Journal of Biochemistry and Cell Biology*, Vol. 32, No. 4, pp. 465–473.

Nelson, N. & Yocum, C.F. (2006). Structure and function of photosystem I and II. *Annual Reviews in Plant Biology*, Vol. 57, pp. 521–565.

Newman, B. J. & Gray, J. C. (1988). Characterization of a full-length cDNA clone for pea ferredoxin-NADP+ reductase. *Plant Molecular Biology*, Vol. 10, pp. 511–520.

Noctor, G. (2006). Metabolic signalling in defence and stress: the central roles of soluble redox couples. *Plant Cell and Environment*, Vol. 29, No. 3, pp. 409–425.

Okutani, S.; Hanke, G. T.; Satomi, Y.; Takao, T.; Kurisu, G.; Suzuki, A. & Hase, T. (2005). Three maize leaf ferredoxin: NADPH oxidoreductases vary in subchloroplast location, expression, and interaction with ferredoxin. *Plant Physiology*, Vol. 139, No. 3, pp. 1451–1459.

Onda, Y. & Hase, T. (2004). FAD assembly and thylakoid membrane binding of ferredoxin:NADP⁺ oxidoreductase in chloroplasts. *FEBS Letters*, Vol. 564, No. 1-2, pp. 116–120.

Palatnik, J. F.; Valle, E. M. & Carrillo, N. (1997). Oxidative stress causes ferredoxin NADP(+) reductase solubilization from the thylakoid membranes in methyl viologen treated plants, *Plant Physiology*, Vol. 115, No. 4, pp. 1721–1727.

Palatnik, J.F.; Tognetti, V. B.; Poli, H. O.; Rodriguez, R. E.; Blanco, N.; Gattuso, M.; Hajirezaei, M. R.; Sonnewald, U.; Valle, E. M. & Carrillo, N. (2003). Transgenic tobacco plants expressing antisense ferredoxin-NADP(H) reductase transcripts display increased susceptibility to photo-oxidative damage. *The Plant Journal*, Vol. 35, No. 3, pp. 332–341.

De Pascalis, A. R.; Jelesarov, I.; Ackermann, F.; Koppenol, W. H.; Hirasawa, M.; Knaff, D. B. & Bosshard, H. R. (1993). Binding of ferredoxin to ferredoxin:NADP⁺ oxidoreductase: the role of carboxyl groups, electrostatic surface potential, and molecular dipole moment. *Protein Science*, Vol. 2, No. 7, pp. 1126–1135.

Peltier , J. B.; Ytterberg, A. J.; Sun, Q. & van Wijk, K.J. (2004). New functions of the thylakoid membrane proteome of *Arabidopsis thaliana* revealed by a simple, fast, and versatile fractionation strategy. *Journal of Biological Chemistry*, Vol. 279, No. 47, pp. 49367–49383.

Quiles, M.J. & Cuello, J. (1998). Association of ferredoxin–NADP oxidoreductase with the chloroplastic pyridine nucleotide dehydrogenase complex in barley leaves. *Plant Physiology*, Vol. 117, No. 1, pp. 235–244.

Rodriguez, R. E.; Lodeyro, A.; Poli, H. O.; Zurbriggen, M.; Peisker, M.; Palatnik, J. F.; Tognetti, W. B.; Tschiersch, H.; Hajirezaei, M. R.; Valle E. M. & Carrillo, N. (2007). Transgenic tobacco plants overexpressing chloroplastic ferredoxin-NADP(H) reductase display normal rates of photosynthesis and increased tolerance to oxidative stress. *Plant Physiology*, Vol. 143, No. 2, pp. 639–649.

Shaodong, D.; Friemann, R.; Glauser, D. A.; Bourquin, F.; Manieri, W.; Schürmann, P. & Eklund, H. (2007). Structural snapshots along the reaction pathway of ferredoxin-thioredoxin reductase. *Nature*, Vol. 448, No. 7149, pp. 92–96.

Stengel, A.; Benz, P.; Balsera, M.; Soll, J. & Bölter, B. (2008). TIC62 redox-regulated translocon composition and dynamics. *Journal of Biological Chemistry*, Vol. 283, No. 11, pp. 6656–6667.

Tsugeki, R. & Nishimura, M. (1993). Interaction of homologues of Hsp70 and Cpn60 with ferredoxin-NADP⁺ reductase upon its import into chloroplasts. *FEBS Letters*, Vol. 320, No. 3, pp. 198–202.

Wheeler, M. C.; Tronconi, M. A.; Drincovich, M. F.; Andreo, C. S.; Flügge, U. I. & Maurino, V. G. (2005). A comprehensive analysis of the NADP-malic enzyme gene family of *Arabidopsis*. *Plant Physiology*, Vol. 139, No. 1, pp. 39-51.

Zhang, H.; Whitelegge, J. P. & Cramer, W.A. (2001). Ferredoxin:NADP⁺ oxidoreductase is a subunit of the chloroplast cytochrome b₆f complex. *Journal of Biological Chemistry*, Vol. 276, No. 41, pp. 38159–38165.

Zhang, H.; Primark, A.; Cape, J.; Bowman, M. K.; Kramer, D. M. & Cramer, W. A. (2004). Characterization of the high-spin heme x in the cytochrome b₆f complex of oxygenic photosynthesis. *Biochemistry*, Vol. 43, No. 51, pp. 16329–16336.

11

Chloroplast Photorelocation Movement: A Sophisticated Strategy for Chloroplasts to Perform Efficient Photosynthesis

Noriyuki Suetsugu and Masamitsu Wada
Kyushu University
Japan

1. Introduction

Chloroplasts move to weak light so that they can perceive light efficiently (the accumulation response), whereas they escape from strong light to avoid photodamage (the avoidance response) (Fig. 1).

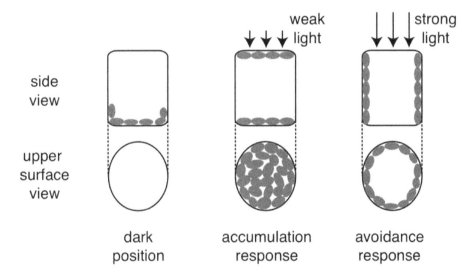

Fig. 1. Typical intracellular distribution pattern of chloroplasts by their photorelocation movement. In darkness, chloroplasts are located on the cell bottom in *Arabidopsis thaliana*. Note that the dark position varies among plant species. Weak light induces the chloroplast accumulation response along the peliclinal walls so that chloroplasts can perceive light efficiently. Strong light induces the chloroplast avoidance response toward the anticlinal walls to reduce photodamage.

The phototropin photoreceptor family of proteins, which includes phototropin (phot) and neochrome (neo), mediate chloroplast photorelocation movement in green plants (reviewed by Suetsugu & Wada, 2007b, 2009). Phot mediates blue-light-induced chloroplast movement in most green plant species, and neo mediates red-light-induced chloroplast movement in ferns and some green alga (reviewed by Suetsugu & Wada, 2005, 2007a). Like other plant organelle movement responses, chloroplast photorelocation movement depends on actin filaments (reviewed by Wada & Suetsugu, 2004). Detailed physiological and photobiological analyses revealed that chloroplasts could move in any direction without turning or rolling within a short lag time during both the accumulation and the avoidance responses (Tsuboi et al., 2009; Tsuboi & Wada, 2011a). This fact argued that chloroplasts move by utilizing preexisting actin filaments and myosins. However, recent detailed microscopic analyses in the flowering plant *A. thaliana* (Kadota et al., 2009), the fern *Adiantum capillus-veneris* (Tsuboi & Wada, 2011b) and the moss *Physcomitrella patens* (Yamashita et al., 2011) have revealed that short actin filaments around the periphery of chloroplasts (called as cp-actin filaments) but not cytoplasmic actin cables are involved in chloroplast photorelocation movement and in the attachment to the plasma membrane (Fig. 2). Furthermore, chloroplast photorelocation movement was normal in all of the examined multiple-myosin knockout plants and even in myosin mutant plants severely defective in movements of the mitochondria, Golgi bodies, peroxisomes, endoplasmic reticulum and cytoplasm (Suetsugu et al., 2010b). Molecular genetic analyses using *A. thaliana* are of a great benefit to the study of chloroplast movement. First, various molecular factors that regulate cp-actin filament generation and reorganization during chloroplast movement can be identified (Fig. 3). Two phototropins, phot1 and phot2, mediate chloroplast photorelocation movement (Jarillo et al., 2001; Kagawa et al., 2001; Sakai et al., 2001) by reorganizing cp-actin filaments (Kadota et al., 2009, Ichikawa et al., 2011). Two interacting coiled-coil proteins, WEB1 (weak chloroplast movement under blue light 1) and PMI2 (plastid movement impaired 2), regulate the velocity of chloroplast movement via light-induced cp-actin filament reorganization, possibly by suppressing JAC1 (J-domain protein required for chloroplast accumulation response 1) (Kodama et al., 2010, 2011; Luesse et al., 2006; Suetsugu et al., 2005a). A chloroplast outer envelope protein, CHUP1 (chloroplast unusual positioning 1), and two kinesin-like proteins, KAC1 (kinesin-like protein for actin-based chloroplast movement 1) and KAC2, are indispensable for cp-actin filament formation (Kadota et al., 2009; Oikawa et al., 2003; Suetsugu et al., 2010a). CHUP1 and KAC1 showed in vitro F-actin binding activity, and CHUP1 also interacted with G-actin and profilin in vitro (Oikawa et al., 2003; Schmidt von Braun & Schleiff, 2008; Suetsugu et al., 2010a), suggesting the direct involvement of these proteins in cp-actin filament generation and regulation. Most of these components are highly conserved in land plants from bryophytes to angiosperms (reviewed by Suetsugu et al., 2010b), suggesting that cp-actin filament-mediated chloroplast movement may facilitate the explosive evolution of land plants when in a fluctuating, ambient light environment. Second, the availability of mutants deficient in chloroplast movement encouraged us to verify a long-standing hypothesis that chloroplast movement is required for efficient photosynthesis in fluctuating light conditions. Experiments using mutants deficient in avoidance movement showed that the avoidance response is necessary for reducing photodamage under strong light conditions (Kasahara et al., 2002)(Fig. 4a). Some reports have suggested that the avoidance response (i.e. the distribution of chloroplasts on the anticlinal walls) affects CO_2 diffusion by changing the chloroplast surface that is exposed to intercellular air spaces and that the avoidance response in the upper part of the leaf could facilitate leaf photosynthesis by allowing greater light penetration to lower parts within the leaf (reviewed by Suetsugu & Wada, 2009) (Fig. 4b & 4c). However, these hypotheses are controversial and have not yet been clearly demonstrated experimentally.

In this chapter, we review three topics of chloroplast photorelocation movement: (i) the insights gained from physiological and photobiological analyses, (ii) the molecular mechanism and (iii) the contribution to photosynthesis.

2. From physiological analyses to molecular genetic analyses

Light-induced chloroplast movement (chloroplast photorelocation movement) has fascinated plant biologists since its discovery in the mid-nineteenth century (Böhm, 1856). Comprehensive analyses by Gustav Senn (1875-1945) of chloroplast movement in various plant species revealed the general responses of chloroplasts to light intensity and direction; chloroplasts are distributed at a position that ensures more efficient light absorption under weak light conditions, and they are positioned away from strong light, as if they had escaped (Senn, 1908). In land plant species, which generally bear multiple small chloroplasts in a cell, low light induces chloroplast movement and distribution toward the periclinal walls (the accumulation response), whereas strong light induces chloroplast avoidance toward the anticlinal walls (the avoidance response) (Fig. 1). Blue light is most effective at inducing chloroplast movement in most plant species, but in some cryptogam plant species, red light as well as blue light is effective. This red light effect exhibits red/far-red light reversibility, suggestive of the involvement of a red/far-red light receptor phytochrome. Detailed photobiological analyses, especially by the research groups of Wolfgang Haupt (1921-2005) and Jan Zurzycki (1925-1984), have provided many important insights on putative photoreceptor molecules that regulate chloroplast movement (Haupt, 1999; Zurzycki, 1980). They found that membrane-bound blue light photoreceptors other than phytochrome mediate blue-light-induced chloroplast movement in most plant species and that membrane-bound phytochromes mediate red-light induced chloroplast movement in some cryptogam plants. These predictions were demonstrated by the recent identification of photoreceptor genes in various plant species. The blue light receptor phot mediates blue-light-induced chloroplast movement in various plant species (Jarillo et al., 2001; Kagawa et al., 2001, 2004; Kasahara et al., 2004; Sakai et al., 2001). Neo, the chimeric photoreceptor that is a fusion of phytochrome and phototropin, regulates red-light-induced chloroplast movement in ferns and some green alga (Kawai et al., 2003; Suetsugu et al., 2005b). The photoreceptors are not discussed here; for a comprehensive review, see Suetsugu & Wada, 2005, 2007a, 2007b, 2009. First, we show our attempts to elucidate the mechanism of chloroplast photorelocation movement by detailed photobiological analyses. Second, we review recent molecular biological analyses of chloroplast photorelocation movement. Finally, the contribution of chloroplast movement and positioning to photosynthesis is discussed.

2.1 Elucidation of the mechanism of chloroplast photorelocation movement by detailed photobiological analyses

The underlying processes of chloroplast photorelocation movement can be categorized into three parts: photoperception, signal transduction and the motility system. Most of the photobiological analyses of chloroplast photorelocation movement were performed to identify the photoreceptor molecules (reviewed by Haupt, 1999; Zurzycki, 1980; Wada et al., 1993). Many pharmacological (i.e. treatment with chemicals and inhibitors) and microscopic (i.e. staining of the cytoskeleton) analyses have provided valuable insights, such as the possible involvement of calcium ions in the signal transduction pathway and the actin filament-dependency of the motility system (reviewed by Suetsugu & Wada, 2007b, 2009). However, the data from pharmacological treatment and microscopic observation of fixed samples should be carefully considered because of possible artifactual results. Thus, we decided to analyze chloroplast relocation movement using detailed physiological and photobiological analyses of the gametophytic cells of a fern *A. capillus-veneris* as a model system (reviewed by Wada, 2007). By changing light conditions, we can easily obtain two types of gametophytes: a filamentous protonemal cell or a two-dimensional prothallus, which is a cell sheet made of a one-cell layer. This fern regulates chloroplast movement by utilizing phot family proteins and actin filaments, like *A. thaliana* (Kadota & Wada, 1992; Kagawa et al., 2004; Tsuboi & Wada, 2011b). Using a microbeam irradiation system, we analyzed chloroplast photorelocation movement in protonemal and phothallial cells and elucidated several aspects of chloroplast movement, especially putative signaling molecules and movement.

2.1.1 Physiological properties of putative signals in chloroplast photorelocation movement

Blue light mediates the influx of calcium ions (Ca^{2+}) into the cytosol and this influx is dependent upon phots in *A. thaliana* and *P. patens* (reviewed by Harada & Shimazaki, 2007). Importantly, a Ca^{2+} chelator inhibited chloroplast movement and external Ca^{2+} ions and Ca^{2+} ionophores changed the distribution of chloroplasts when placed in darkness (reviewed by Suetsugu & Wada, 2009). However, plasma membrane Ca^{2+} channel blockers, which effectively inhibited phot-dependent blue light-mediated Ca^{2+} influx (reviewed by Harada & Shimazaki, 2007), were totally ineffective in suppressing chloroplast photorelocation movement in various plant species (reviewed by Suetsugu & Wada, 2009). Thus, the putative signals that control chloroplast movement remain to be determined. To characterize the properties of these putative signals, chloroplast photorelocation movement was induced by partial cell irradiation with a microbeam irradiator and analyzed in detail (Kagawa & Wada, 1999, 2000; Tsuboi & Wada, 2010a, b).

An open question was whether the signals were different between the accumulation and avoidance responses. When a dark-adapted cell of an *A. capillus-veneris* prothallus (Kagawa & Wada, 1999) and an *A. thaliana* leaf (Kagawa & Wada, 2000) (in this situation, a few chloroplasts were on the upper periclinal walls) were partially irradiated with strong blue light, chloroplasts moved to the irradiated area but could not enter the beam area. Immediately after the light was turned off, the chloroplasts moved into the formerly irradiated area. Similar responses were also found in filamentous protonemal cells in *A. capillus-veneris* (Yatsuhashi et al., 1985) and *P. patens* (Kadota et al., 2000; Sato et al., 2001). These results suggested several characteristics of putative signals (reviewed by Suetsugu & Wada, 2009): (i) Signals for both the accumulation and the avoidance responses are simultaneously generated by strong light. (ii) Signals for the avoidance response function only at the irradiated area whereas those for the accumulation response can be transferred toward chloroplasts when located far from the irradiated area. (iii) Signals for the avoidance response disappear immediately after the light is turned off, whereas those for the accumulation response are long-lived. (iv) Signals for the avoidance response can be override those for the accumulation response, at least in the irradiated area. Alternatively, it is possible that the signals for the avoidance response can be generated only when chloroplasts are directly irradiated with strong light. In this case, it is likely that the photoreceptor is localized on chloroplasts or that a plasma membrane-localized receptor generates the signals only when it exists in close proximity to the chloroplasts.

However, it is clear that the signals for the accumulation response are generated by the activation of photoreceptors in the irradiated area and are subsequently transferred toward chloroplasts. If measuring the speed of signal transfer for the accumulation response were possible, we could guess as to what is the putative signal by comparing the speeds between the putative signal and the known signaling molecules. When the chloroplast accumulation response was induced by microbeam irradiation in *A. capillus-veneris* protonemal cells, the onset of the accumulation movement lagged in proportion to the increase in the distance between the irradiated area and the chloroplasts, suggesting that the speed of the signal can be calculated as the lag time before the onset of movement (Tsuboi & Wada, 2010a, b). Similar calculations were also performed in *A. capillus-veneris* prothallial and *A. thaliana* mesophyll cells (Tsuboi & Wada, 2010a, b). These analyses revealed three interesting features of the putative signals. First, in protonemal cells, the speed of the signals in the basal-to-apical directions (about 2.3-2.4 μm min^{-1}) was about three times faster than that in the apical-to-basal direction (about 0.6-0.9 μm min^{-1}). However, the speed of the signals was almost equal in each cell type (about 0.9-1.1 μm min^{-1} in *A. capillus-veneris* prothallial cells and about 0.7 μm min^{-1} in *A. thaliana* mesophyll cells) (Tsuboi & Wada, 2010a, b). This difference in speed could result from the difference in the cell growth pattern of the cells, i.e. polarized (protonemal cells) and diffusive (prothallial and mesophyll cells). Second, in fern gametophytic cells, the speed of the signal and the maximum distance over which the signals could be transferred were almost equal irrespective of the intensity of the red or blue light microbeam, although in this case, more chloroplasts that were located farther away responded under continuous irradiation, compared to those submitted to a 1 min pulse of irradiation (Tsuboi & Wada, 2010a, b). Interestingly, the velocity of chloroplast accumulation movement was constant, regardless of the intensity of the microbeam placed on the prothallial cells (Kagawa & Wada, 1996; Tsuboi & Wada, 2011a). These results suggested that the properties of the signal, such as the speed, the amount and the activity, do not change in proportion to the change of light intensity. However, chloroplasts in the protonemal cells accumulated in the area that had been irradiated by a beam with a higher fluence rate, compared to adjacent areas that had been irradiated with beam of lower fluence rates of blue or red light. This result suggests that the amount or activity of the signal was increased when exposed to a beam with a higher fluence rate (Yatsuhashi et al., 1987; Yatsuhashi, 1996). Third and most importantly, the speed of signals (about 0.6-2.4 μm min^{-1}) was much slower than that caused by calcium ion spiking or waves known to occur in plant and animal systems (about several μm sec^{-1} to 100 μm sec^{-1}) (Tsuboi & Wada, 2010a, b). Furthermore, the signal transfer must not be actomyosin-dependent because the transfer of the signal still occurred when actin filaments were disrupted by treatment with inhibitor (Sato et al., 2001). Collectively, although the signals for chloroplast movement remained to be determined, our detailed physiological analyses will provide the clue to identify the actual signals.

2.1.2 An actin-based motility system deduced by detailed observation of chloroplast movement

For a long time, it was believed that the actomyosin system mediated chloroplast movement in various species. Many analyses using several kinds of techniques (such as inhibitor treatment, immunocytochemistry and observation of the in vivo dynamics of actin filaments) clarified the involvement of actin filaments in chloroplast movement in various plant species (reviewed by Suetsugu & Wada, 2009; Suetsugu et al., 2010b). However, the involvement of myosin motor proteins was still controversial (reviewed by Suetsugu et al., 2010b). If the actomyosin system is involved in chloroplast movement, it is expected that chloroplasts move along long actin cables that preexist or elongate in the direction of movement immediately after light exposure and, thus, that chloroplast movement should be polarized (i.e. parallel to actin cables). However, this was not the case with chloroplast photorelocation movement at least in *A. capillus-veneris* prothallial and *A. thaliana* mesophyll cells (Tsuboi & Wada, 2009, 2011a). Importantly, chloroplasts moved by sliding but not rolling during both the accumulation and the avoidance responses (Tsuboi & Wada, 2009, 2011a), suggesting that chloroplasts moved by attaching one side to the plasma membrane via actin filaments that spanned between the chloroplasts and the plasma membrane. When observed with a microscope, chloroplasts look elliptic (or dumbbell-shaped for dividing chloroplasts) but not completely round. Therefore, it is plausible that chloroplasts keep their long axis in parallel with the moving direction so that they can take the path of least resistance. If that is the case, then they should turn at an angle formed by an imaginary line spanning their long axis and a second imaginary line that connects the center of the chloroplast, at the original position, to the center of the irradiated area. However, chloroplasts were capable of moving in any direction even without turning. Even if chloroplasts turned immediately before or while they moved, the extent of their turning was so small (Tsuboi & Wada, 2009, 2011a). Exceptionally, chloroplasts of Arabidopsis mesophyll cells tended to adjust their short axis to be parallel with the moving direction during the avoidance movement, although they started to move without turning (Tsuboi & Wada, 2011a). Importantly, chloroplasts escaped from strong light by taking the shortest route, suggesting that they are capable of determining the location of the closest area that is out of the strong light (Tsuboi & Wada, 2011a). Moreover, when sequentially irradiated with weak or strong light, chloroplasts could change their moving direction according to the position of subsequent irradiated beam, with a short lag time (Tsuboi & Wada, 2009, 2011a). Collectively, these detailed microscopic analyses argued against the hypothesis that chloroplasts utilize of pre-existing actin filaments for photomovement and suggested that they move using actin filaments that dynamically reorganize in response to light irradiation.

2.2 Conserved molecular mechanism of chloroplast photorelocation movement in land plants

Generally, blue light is most effective in inducing chloroplast photorelocation movement, although red light can also induce the movement in some cryptogam plants (green algae, mosses and ferns). Phot is the blue light receptor for chloroplast movement and also mediates phototropism and stomatal opening (reviewed by Christie, 2007). Phototropins were identified in green plants, from green alga to seed plants, and were shown to regulate blue-light-induced chloroplast movement at least in *A. thaliana*, *A. capillus-veneris* and *P. patens* (reviewed by Suetsugu & Wada, 2005, 2007a, 2007b, 2009). Red-light-induced chloroplast movement is mediated by neo in several ferns and probably in some green algae (reviewed by Suetsugu & Wada, 2005, 2007a, 2007b, 2009). Regardless of significant advances in photoreceptor identification, the molecular mechanism of signal transduction and the identity of the motility system for chloroplast movement have been obscure. However, molecular genetic analyses using *A. thaliana* have identified several components that regulate chloroplast movement. Furthermore, recent imaging analyses have revealed that a novel actin-based mechanism governs chloroplast photorelocation and positioning. By combining these results, we could imagine the molecular framework of chloroplast photorelocation movement.

2.2.1 Unique actin-based mechanism for chloroplast movement in land plants

For many years, it was thought that chloroplasts moved along long cytosolic actin cables by myosin motor proteins, similar to the movements of other organelles. However, the aforementioned studies (Tsuboi & Wada, 2009, 2011a) suggested that chloroplasts could utilize an actin-based mechanism that is different from those of other organelles.

To find the actin-based mechanism for chloroplast movement, we utilized Arabidopsis transgenic lines in which actin filaments could be visualized by various fusions of fluorescent proteins and actin binding proteins (such as GFP-talin and tdTomato-fimbrin) and analyzed the behavior of the actin filaments during chloroplast movement using a custom-made microscope and a confocal microscope (Kadota et al., 2009). Although cytoplasmic actin cables and filaments were associated with chloroplasts, they did not change much in response to light irradiation, and their behavior did not associate with directional chloroplast movement. Instead, we found that short actin filaments found around chloroplasts dynamically changed their structure in response to light irradiation and that their dynamics correlated with the direction and speed of chloroplast movement. We have named these actin filaments chloroplast-actin filaments (abbreviated as cp-actin filaments) (Kadota et al., 2009)(Fig. 2). When chloroplasts were stationary, cp-actin filaments were distributed around the chloroplast periphery. In response to strong blue light, cp-actin filaments transiently disappeared within about 30 seconds and then reappeared at the one side of the chloroplasts, which would eventually be the front region of the moving chloroplasts (Fig. 2). We called this pattern of localization of cp-actin filaments at the front region "biased" (Fig. 2). After biased cp-actin filaments were fully formed, chloroplasts moved toward the side where the cp-actin filaments accumulated (Kadota et al., 2009). The generation of biased cp-actin filaments was also found during the accumulation response that had been induced by weak blue light, but this was not accompanied by a transient disappearance of cp-actin filaments, unlike what occurred during the avoidance response (Kadota et al., 2009). Thus, the light-induced generation of biased cp-actin filaments is a prerequisite for both the avoidance and the accumulation responses. Possibly, a transient disappearance of cp-actin filaments induced by strong blue light facilitated an acceleration of chloroplast avoidance movement. As more cp-actin filaments accumulated at the front halves of the chloroplasts, in relation to the rear halves, the velocity of chloroplast avoidance also increased. When irradiated with a higher fluence of blue light, even more cp-actin filaments accumulated at the front halves, and chloroplasts moved even faster (Kadota et al., 2009). Thus, strong light caused a greater difference in the amount of cp-actin filaments at certain locations on the chloroplasts because cp-actin filaments located at the rear halves of the chloroplasts did not increase after transient disappearance. Conversely, weak light could not induce transient disappearance of cp-actin filaments, so a greater difference in the amount of cp-actin filaments at certain locations was not made. Actually, the velocity of chloroplast accumulation movement was constant irrespective of light intensity (Kagawa & Wada, 1996; Tsuboi & Wada, 2011a). Cp-actin filaments localized at the interface between the chloroplast and the plasma membrane, elongated from the edge of the chloroplast and shortened toward the chloroplast periphery, suggesting that the nucleation site of cp-actin filaments might exist on the chloroplast edge and that the force for chloroplast movement by cp-actin filaments might be generated there (Kadota et al., 2009). Cp-actin filaments mediated the anchoring of chloroplasts to the plasma membrane as well as their directional movement (Kadota et al., 2009). The strong-light-induced disappearance of cp-actin filaments was accompanied by increased chloroplast motility in random directions before avoidance movement, suggestive of the detachment of chloroplasts from the plasma membrane. Conversely, weak blue light induced the increase of cp-actin filaments around the chloroplast periphery and accompanied a decrease in chloroplast motility, likely facilitating chloroplast anchoring to the plasma membrane. In summary, there are three types of blue-light-induced rearrangements of cp-actin filaments that mediate both directional movement and the anchoring of chloroplasts to the plasma membrane: (i) the formation of biased cp-actin filaments during both the accumulation and the avoidance responses; (ii) a strong-blue-light-induced transient disappearance of cp-actin filaments; (iii) a weak-blue-light-induced increase in cp-actin filaments. Importantly, cp-actin filament-mediated chloroplast movement is conserved in a fern, *A. capillus-veneris*, and in a moss, *P. patens* (Tsuboi & Wada, 2011b; Yamashita et al., 2011). Thus, the regulation of chloroplast movement by cp-actin filaments was likely to be utilized during the early stages of land plant evolution.

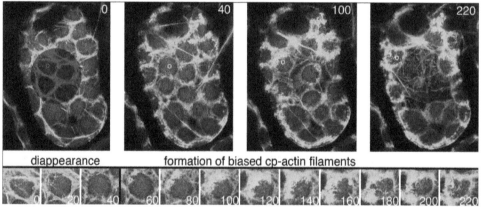

Fig. 2. Light-induced cp-actin filament reorganization during the chloroplast avoidance response. The chloroplast avoidance response was induced by scanning circular regions of interest (diameter 15 µm indicated by a shaded circle) with 2.8 mW of a 458 nm laser and using a confocal microscope (SP5, Leica). Time-lapse images of chloroplast movement and the associated cp-actin filament dynamics were captured at the indicated times (sec). Detailed dynamics of the cp-actin filaments of a chloroplast marked with a white circle are indicated below. After 40 sec of light irradiation, cp-actin filaments disappeared and then reappeared at the front region of the chloroplast, which had moved to the upper left side in this figure via the avoidance response.

2.2.2 Molecular components regulating the generation and/or reorganization of cp-actin filaments

We identified various Arabidopsis mutants deficient in chloroplast photorelocation movement (Kagawa et al., 2001; Kodama et al., 2010; Oikawa et al., 2003; Suetsugu et al., 2005, 2010a), and thus, the analyses of cp-actin filament behavior in these mutants have shed light on the molecular mechanism of cp-actin filament-mediated chloroplast movement (Kadota et al., 2009; Kodama et al., 2010; Suetsugu et al., 2010a; Ichikawa et al., 2011)(Fig.3).

Phototropin is a blue light receptor bearing two photosensory LOV (light, oxygen and voltage) domains at its N-terminus and a C-terminal serine/threonine kinase domain (reviewed by Christie, 2007). In *A. thaliana*, phot1 and phot2 redundantly mediated the chloroplast accumulation response (Sakai et al., 2001), and phot2 alone regulated the avoidance response (Kagawa et al., 2001; Jarillo et al., 2001). In *phot1phot2* double mutant plants, which are completely defective in chloroplast photorelocation movement (Sakai et al., 2001), blue-light-induced cp-actin filament reorganization did not occur, indicating that phototropins mediated chloroplast movement via the regulation of cp-actin filaments (Kadota et al., 2009; Ichikawa et al., 2011). The *phot1phot2* double mutant plants also did not change their amounts of cp-actin filaments in response to both weak and strong blue light and thus showed no light-induced motility changes (Kadota et al., 2009). This outcome indicated that phototropins mediated anchoring of the chloroplast to the plasma membrane via regulation of the amounts of cp-actin filaments. The strong-blue-light-induced transient disappearance of cp-actin filaments did not occurr at all in *phot2* mutant plants, which were impaired in the avoidance response. However, they showed normal biased cp-actin filament formation during the accumulation response (Kadota et al., 2009; Ichikawa et al., 2011), indicating that phot2 mediated the strong-blue-light-induced transient disappearance of cp-actin filaments (Fig.3) and that this reorganization of cp-actin filaments could be a prerequisite for the avoidance response. In *phot1* mutant plants, chloroplast photorelocation movement was only slightly impaired in the accumulation response (Kagawa & Wada, 2000), and therefore light-induced reorganization of cp-actin filaments in these plants was mostly normal (Kadota et al., 2009). However, in response to strong blue light, the onset of biased cp-actin formation and the avoidance movement in *phot1* mutants occurred earlier than in wild-type plants (Ichikawa et al., 2011), suggesting a small inhibition of cp-actin filament reorganization by phot1 during the avoidance movement.

JAC1 has a J-domain at the C-terminus and is similar to a clathrin uncoating factor, auxilin (Suetsugu et al., 2005). The J-domain of JAC1 is necessary for JAC1 function and the crystal structure showed high similarity between that domain and that of the bovine auxilin J-domain (Takano et al., 2010; Suetsugu et al., 2010c). *jac1* mutant plants were completely defective in the accumulation response but retain the avoidance response (Suetsugu et al., 2005). In response to weak blue light, the reorganization of cp-actin filaments did not occur in most chloroplasts of *jac1* mutant plants, but a few chloroplasts that avoided weak light formed biased cp-actin filaments (Ichikawa et al., 2011), indicating that JAC1 is essential for the reorganization of cp-actin filaments during the accumulation response but not for biased cp-actin filament formation (Fig. 3). Interestingly, in *jac1* mutant plants, whole cell irradiation with strong blue light did not induce the disappearance and subsequent biased localization of cp-actin filaments and thus the avoidance movement did not occur. However, when part of a cell was irradiated, chloroplasts that were close to the beam edge showed the avoidance movement with biased cp-actin filament formation, although cp-actin filaments on chloroplasts inside the beam did not disappear and their motility did not increase (Ichikawa et al., 2011). These results indicate that JAC1 is essential for an efficient chloroplast avoidance response by regulating the disappearance of cp-actin filaments (Fig. 3).

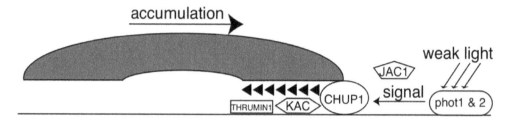

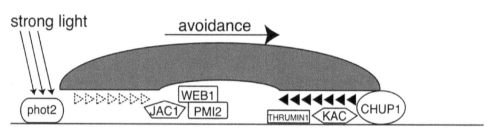

Fig. 3. A schematic model of cp-actin filament-mediated chloroplast movement. Weak light activates both phot1 and phot2, which are localized on the plasma membrane and subsequently generate an as yet unidentified signal that initiates the chloroplast accumulation response. JAC1 may be involved in signal generation, transport and/or perception. The signal activates a cp-actin filament nucleation complex, which is localized at the chloroplast edge, resulting in the polymerization of cp-actin filaments at the leading edge of the chloroplasts (black arrowheads indicate G-actins). CHUP1 could be the nucleation factor, and KAC proteins could be involved in cp-actin filament nucleation and/or maintenance. THRUMIN1 may interact with cp-actin filaments because THRUMIN1-YFP fusion protein decorated actin filaments in vivo. Strong-light-induced cp-actin filament disappearance (indicated by broken-lined arrowheads) is mediated by phot2, JAC1 and WEB1/PMI2. After disappearance, cp-actin filaments reappeared at the leading edge, and the chloroplasts escaped from the strong light.

Recently, we identified two coiled-coil proteins, WEB1 and PMI2, as factors that regulate light-induced cp-actin filament reorganization (Kodama et al., 2010; Luesse et al., 2006). WEB1 and PMI2 belong to a coiled-coil protein family that contains a DUF827 (Domain of Unknown Function 827) domain (Kodama et al., 2010, 2011). WEB1 and PMI2 interacted with each other in yeast and plant cells, and WEB1 showed self-interaction activity, forming large complexes in plant cells, indicating that both WEB1 and PMI2 have protein-protein interaction activity (Kodama et al., 2010). Both *web1* and *pmi2* mutant plants showed severe defects in the avoidance response and slight defects in the accumulation response. Because the phenotypes of *web1pmi2* double mutant plants were very similar to those of *web1* and *pmi2* single-mutant plants, it was concluded that WEB1 and PMI2 probably function in the same pathway, possibly as a complex (Kodama et al., 2010, 2011). Because these mutants

were severely impaired in the strong-light-induced disappearance and subsequent biased localization of cp-actin filaments, it was concluded that the mutant phenotypes observed were a result of the impairment in cp-actin filament reorganization (Kodama et al., 2010)(Fig. 3). The defective avoidance response phenotype in these mutants was suppressed by a *jac1* mutation, suggesting a role for WEB1 and PMI2 in suppressing JAC1 activity, which regulates the accumulation response under high light conditions (Kodama et al., 2010). Given that the strong-light-induced reorganization of cp-actin filaments was severely impaired in *web1*, *pmi2* and *jac1* mutant plants (Kodama et al., 2010; Ichikawa et al., 2011), it is possible that WEB1/PMI2 and JAC1 cooperatively mediate the strong-light-induced reorganization of cp-actin filaments, although the detailed molecular mechanism remains to be determined.

Currently, two types of proteins, CHUP1 and KAC (KAC1 and KAC2), were identified as the factors necessary for the existence of cp-actin filaments, possibly serving as nucleators and/or stabilizers of cp-actin filaments (reviewed by Suetsugu et al., 2010b). CHUP1 is a multi-domain protein that bears an N-terminal hydrophobic region, a coiled-coil region, an F-actin-binding domain, a proline-rich region and a highly conserved C-terminal region (Oikawa et al., 2003). The hydrophobic region is essential for the localization of chloroplast outer envelope (Oikawa et al., 2003, 2008; Schmidt von Braun & Schleiff, 2008) and the coiled-coil region confer the ability of the protein to dimerize in vitro (Lehmann et al., 2011). The actin-binding domain was capable of interacting with F-actin in vitro (Oikawa et al., 2003), and the proline-rich region might serve as the profilin-interacting domain (Schmidt von Braun & Schleiff, 2008). KAC proteins belong to a microtubule motor kinesin-14 subfamily, but their motor and microtubule-binding activities have not yet been detected (Suetsugu et al., 2010a). A subset of KAC proteins was associated with the plasma membrane and chloroplast envelope although the bulk of the KAC proteins were found as soluble proteins (Suetsugu et al., 2010a). Both *chup1* and *kac1kac2* double mutant plants completely lacked cp-actin filaments but retained the normal cytosolic actin filament structure, indicating that the CHUP1 and KAC proteins are essential for cp-actin filament formation and/or maintenance (Kadota et al., 2009; Suetsugu et al., 2010a)(Fig. 3). Importantly, both mutants showed no chloroplast photorelocation movement and defects in the anchoring of chloroplasts to the plasma membrane. This result reinforced the notion that cp-actin filaments mediate photorelocation and the anchoring of chloroplasts to the plasma membrane. In *kac1* single mutant plants, significantly fewer amounts of cp-actin filaments were observed, the accumulation response was severely impaired and the velocity of the avoidance movement was much slower compared to wild-type plants (Suetsugu et al., 2010a). Thus, the amount of KAC proteins is an important factor that determines the chloroplast velocity by regulating the amounts of cp-actin filaments.

Although PMI1 and THRUMIN1 were identified through the analyses of mutants deficient in chloroplast photorelocation movement (DeBlasio et al., 2005; Whippo et al., 2011), their involvement in cp-actin filament regulation is unknown. Because THRUMIN1 has an actin bundling activity (Whippo et al., 2011), analyses of mutants deficient in these factors will reveal a more detailed framework of cp-actin filament-dependent chloroplast movement.

Three essential genes for cp-actin filament-mediated chloroplast movement, *PHOT*, *CHUP1* and *KAC*, are found in the genome of a liverwort, *Marchantia polymorpha*, and a moss, *P. patens*. Because cp-actin filament-mediated chloroplast movement is found in *P. patens*

(Yamashita et al., 2011), the molecular mechanism of cp-actin filament-mediated chloroplast movement must have evolved early in land plant evolution.

2.3 Contribution of chloroplast photorelocation movement to photosynthesis

The intracellular distribution of chloroplasts is essential for the promotion of photosynthetic performance. For example, the chloroplast distribution in bundle sheath cells of C4 plants may be necessary for efficient C4 photosynthesis because it controlled CO_2 diffusion and/or facilitated the metabolite exchange between mesophyll and bundle sheath cells (reviewed by von Caemmerer & Furbank, 2003). The chloroplast accumulation response could play an important role in efficient light capture under weak light conditions, although it has not been demonstrated experimentally (Zurzycki, 1955). The avoidance response is required for chloroplats to escape from photodamage under excess light conditions (Fig. 4a); however, two other hypotheses exist that could also explain the ecological advantages of chloroplast distribution on anticlinal walls by the avoidance response (Fig. 4b & 4c).

2.3.1 Promotion of light penetration to deeper leaf cell layer by the avoidance response

In the leaves of *Oxalis*, *Marah* and *Cyrtomium*, changes in leaf absorptance due to chloroplast movement positively correlated with changes in fluorescence emission; in particular, changes in fluorescence emissions increased during the avoidance response induced by strong blue light, whereas they decreased during subsequent relaxation in red light (Brugnoli & Björkman, 1992). Considering that leaves consist of multiple layers of photosynthetic cells and that the efficiency of net leaf photosynthesis depends on the efficient light utilization of chloroplasts in all cell layers, it is reasonable to conclude that chloroplast distribution to anticlinal cell walls, as a result of the avoidance response in the upper cell layer (or palisade layer), could facilitate the penetration of the incident light to a deeper cell layer (or sponge layer)(Fig. 4b). Light transmittance through the palisade layer was greater in high light-irradiated *Alocasia* leaves than in dark-adapted leaves. However, the difference in light transmittance through the spongy layer was not significant between high light- and dark-adapted leaves, indicating that chloroplast positioning on the anticlinal walls by the avoidance response in the palisade layer could facilitate light penetration to a deeper layer (Gorton et al., 1999). In *Tradescantia* leaves, which consist of three mesophyll cell layers (the first is a palisade layer, and the second and third are sponge layers), the chloroplasts in the second layer did not move to the anticlinal walls when irradiated with strong light of 100 μmol m^{-2} s^{-1} from either the adaxial or the abaxial side. However, by abaxial-side irradiation, chloroplasts in the third layer were positioned on the anticlinal walls by way of the avoidance response (Terashima & Hikosaka, 1995). These results suggest that the avoidance response in the surface mesophyll layer facilitates light capture in the deeper cell layers, resulting in a net increase of whole leaf photosynthesis. However, this hypothetical role of the avoidance movement in the enhancement of light penetration to the deeper cell layers has not yet been demonstrated conclusively.

2.3.2 Influence of the chloroplast avoidance response on CO_2 diffusion between air spaces and mesophyll cells

The diffusion path length of CO_2 from intercellular air spaces to the chloroplast stroma must be short so that mesophyll chloroplasts can efficiently utilize CO_2 from those air spaces.

Thus, chloroplasts should be located on the cell wall facing air spaces. Mesophyll cell chloroplasts tended to be located along intercellular air spaces in various plant species (Senn, 1908; Psaras et al., 1996) (Fig. 4c). Senn (1908) hypothesized that this positioning might result from the chemotaxis of chloroplasts to the CO_2 in air spaces. Using different approaches and plant species, three groups examined whether chloroplast photorelocation movement, especially the avoidance response, influenced CO_2 diffusion in leaves (Gorton et al., 2003; Loreto et al., 2009; Tholen et al., 2008). One research group hypothesized that chloroplast distribution on the anticlinal walls by the avoidance response facilitated CO_2 utilization by shortening the CO_2 diffusion path length from air spaces to the mesophyll chloroplasts (Gorton et al., 2003). This group measured oxygen diffusion times using pulsed photoacoustics (as a substitution for CO_2 diffusion) between control and strong light-irradiated *Alocasia* leaf samples, but they could not find any differences in CO_2 diffusion rates between the two samples (Gorton et al., 2003). Additionally, they could found no difference in the distance between the centers of the chloroplasts and the closest air spaces between two samples (Gorton et al., 2003). Another group found that blue light rapidly reduced CO_2 diffusion from intercellular air spaces to the chloroplasts in both *Nicotiana* and *Platanus* leaves and that this reduction was completed before chloroplasts finished the avoidance movement to the aniticlinal walls (Loreto et al., 2009). Importantly, the blue-light-induced reduction of CO_2 diffusion was still normal in samples treated with an anti-actin inhibitor cytochalasin, which completely inhibits chloroplast movement (Loreto et al., 2009). Thus, results by two independent groups indicated that chloroplast movement did not significantly change the efficiency of CO_2 diffusion from intercellular air spaces to the chloroplasts. However, the results by a third group suggested that the avoidance response reduced the CO_2 diffusion rate rather than increased it (Tholen et al., 2008). In *Arabidopsis* wild-type plants, the surface area of chloroplasts facing air spaces was reduced after the induction of the chloroplast avoidance response and resulted in the reduction of CO_2 diffusion. However, these reductions were not found in *phot2* and *chup1* mutant plants or in cytochalasin-treated plants (Tholen et al., 2008). Compared to wild type, the surface area of chloroplasts that faced air spaces and the rate of CO_2 diffusion were constitutively lower in *chup1* mutant plants because of aberrant positioning of their chloroplasts (Tholen et al., 2008). Collectively, these results suggested that the chloroplast avoidance response was not involved in CO_2 diffusion or that it possibly decreased, rather than increased, the diffusion rate. However, these three groups examined the contribution of chloroplast movement to CO_2 diffusion using different plant species and techniques: pulsed photoacoustics (Gorton et al., 2003), a chlorophyll fluorescence-based method (Loreto et al., 2009) and a carbon isotope discrimination method (Tholen et al., 2008). The examination of one plant species using different techniques and/or that of various plant species by one technique are required to uncover whether chloroplast photorelocation movement influences CO_2 diffusion.

2.3.3 Chloroplast avoidance response is essential for protection against photodamage by strong light

The two aforementioned hypotheses on the roles of the avoidance response are applicable only in multilayered leaf tissue and not in gametophytic cells that have single cell layers or in the filamentous structures of fern, moss, liverwort and green alga. During the early period of land plant evolution, plants were exposed directly to sunlight until seed plants eventually dominated terrestrial ecosystems and formed a dense canopy. Thus, it is plausible that the main role of the chloroplast avoidance response is to prevent chloroplasts from photodamage caused by strong light (Zurzycki, 1957) (Fig. 4a). Shade plants (such as *Oxalis oregana* and *Tradesdantia albiflora*) showed a greater avoidance response than that of non-shade plants (such as sunflower and pea) (Brugnoli & Björkman, 1992; Park et al., 1996), partly explaining why *T. albiflora* was more tolerant of strong light stress than pea plants (Park et al., 1996). Furthermore, cytochalasin-treated *Platanus* leaves, whose chloroplast movements were inhibited by the drug, but not untreated leaves showed a strong inhibition of photochemical efficiency (Loreto et al., 2009). Clumping of chloroplasts in succulent plants (Kondo et al., 2004) and the aggregative movement of C4 mesophyll chloroplasts (Yamada et al., 2009) were induced by drought stress in a light-dependent fashion, and these

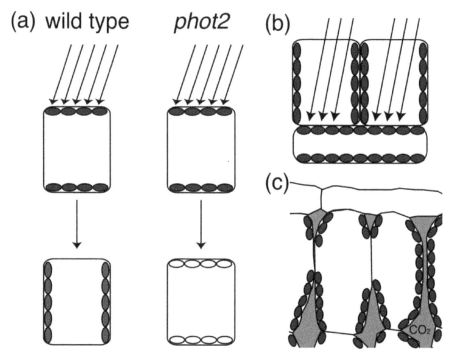

Fig. 4. Three hypotheses for the ecological significance of the chloroplast avoidance response: (a) Protection from photodamage. Chloroplasts escape from strong light and distribute along the anticlinal walls so that they do not directly perceive excess light energy, which could cause photodamage. As a result, plants can tolerate strong light stress. Mutants deficient in the avoidance response, such as Arabidopsis *phot2* mutants, cannot survive under the strong light conditions because their chloroplasts are directly exposed to extremely strong light and are therefore severely damaged and die. (b) Promotion of light penetration in leaves. Chloroplasts distributed along the anticlinal walls in the upper cell layer facilitate light penetration to deeper cell layers. Consequently, light perception and thus photosynthesis in the deeper cell layers increases. (c) Modulation of CO_2 diffusion from intercellular air spaces to the chloroplast. The chloroplast avoidance response can change the total chloroplast surface area facing airspace and thus the efficiency of CO_2 diffusion from intercellular air spaces to the chloroplast may increase.

responses were implicated in the protection from photodamage in plants that inhabit tropical and/or dry areas. Experiments using *Arabidopsis* wild-type and mutant plants definitely demonstrated that chloroplast avoidance movement is essential for the protection of plants from photodamage by strong light (Kasahara et al., 2002). Leaf transmittance in wild type and *phot1* mutant plants increased as light intensity was increased to about five-fold of the initial value (about 500 μmol m^{-2} s^{-1} of white light). However, little change in leaf transmittance occurred in *phot2* and *chup1* mutant leaves, even at 2000 μmol m^{-2} s^{-1} of white light, indicating that *phot2* and *chup1* mutant plants are defective in the avoidance response under a wide range of light intensity. When low-light-acclimated plants were shifted to a strong light condition, the leaves of *phot2* and *chup1* mutant plants were bleached after 10 h and were severely necrotic after 22 h. However, wild-type and *phot1* plants did not show leaf necrosis even after 31 h. When the change in the chlorophyll fluorescence parameter F_v/F_m (representing the maximal quantum yield of photosystem II photochemistry) was analyzed during strong light treatment, the F_v/F_m value in wild-type and *phot1* plants steeply declined to about 80% of the initial value within 1 h and then gradually decreased and finally reached about 70% of the initial value in 5 h. However, the F_v/F_m values in *phot2* and *chup1* mutant plants declined more rapidly than in wild type and consequently reached about 50% of the initial value in 5 h. In *phot2* mutants, the extent of the decrease of the F_v/F_m value after 1 h of light treatment was larger at all examined light intensities than that of wild-type plants. Furthermore, the F_v/F_m value in *phot2* mutant plants did not fully recover after 6 h, whereas the F_v/F_m values in wild-type plants almost fully recovered after 3 h under low light. Collectively, these results indicate that *phot2* and *chup1* mutant plants are highly susceptible to strong light stress and their photosystem IIs are much less tolerant of light stress resulting in leaf necrosis. Note that *phot2* and *chup1* mutant plants were normal in chlorophyll content, chlorophyll fluorescence parameters, antioxidant contents and the activities of reactive oxygen-scavenging enzymes. *phot2* mutant plants showed a slight defect in stomatal opening (Kinoshita et al., 2001), but this defect was less than that in *phot1* mutant plants and was negligible in the strong light conditions used by Kasahara et al. Recently, another research group confirmed that photosystem II of *phot2* mutant plants was more susceptible to strong light (Sztatelman et al., 2010). Overall, we conclude that chloroplast avoidance movement is indispensable for plant survival under strong light conditions.

3. Conclusion

Although chloroplast photorelocation movement has been extensively studied by many researchers, we still cannot accurately explain the molecular mechanism of chloroplast photorelocation movement. Some unanswered questions remain: what is the signal for the chloroplast accumulation response?; what protein(s) nucleate cp-actin filaments?; and how do cp-actin filaments generate the motive force for chloroplast movement? To answer these questions, chloroplast movement must be analyzed by combining various approaches: physiology, molecular biology, proteomics, crystallography and imaging techniques. Chloroplast movement under natural light conditions must also be examined because natural light is usually much more severe and always fluctuates, compared to laboratory conditions. Because various mutants deficient in chloroplast movement are available, the growth of these mutant plants under natural conditions must be analyzed, and the ecological significance of chloroplast photorelocation movement must be verified.

4. Acknowledgment

This work was supported in part by the Japanese Ministry of Education, Sports, Science, and Technology (MEXT 13139203 and 17084006 to M. W.) and the Japan Society of Promotion of Science (JSPS 13304061, 16107002, and 20227001 to M. W.; 20870030 to N. S.).

5. References

Böhm, J.A. (1856). Beiträge zur näheren Kenntnis des Chlorophylls. *S.B. Akad. Wiss. Wien, Math.-nat. Kl.,* Vol.22, (1856), pp. 479-498

Brugnoli, E. & Björkman, O. (1992). Chloroplast movements in leaves: Influence on chlorophyll fluorescence and measurements of light-induced absorbance changes related to ΔpH and zeaxanthin formation. *Photosynthesis Research,* Vol.32, No.1, (April 1992), pp. 23-35, ISSN 0300-3604

Christie, K. (2007). Phototropin blue-light receptors. *Annual Review of Plant Biology,* Vol.58, No.1, (June 2007), pp. 21-45, ISSN 1543-5008

DeBlasio, S.L.; Luesse, D.R. & Hangarter, R.P. (2005). A plant-specific protein essential for blue-light-induced chloroplast movements. *Plant Physiology,* Vol.139, No.1, (September 2005), pp. 101-114, ISSN 0032-0889

Gorton, H.L.; Herbert, S.K. & Vogelmann, T.C. (2003). Photoacoustic analysis indicates that chloroplast movement does not alter liquid-phase CO_2 diffusion in leaves of *Alocasia brisbanensis. Plant Physiology,* Vol.132, No.3, (July 2003), pp. 1529-1539, ISSN 0032-0889

Gorton, H.L.; Williams W.E. & Vogelmann, T.C. (1999). Chloroplast movement in *Alocasia macrorrhiza. Physiologia Plantarum,* Vol.106, No.4, (August 1999), pp. 421-428, ISSN 0031-9317

Harada, A. & Shimazaki, K. (2007). Phototropins and blue light-dependent calcium signalling in higher plants. *Photochemistry and Photobiology,* Vol.83, No.1, (January-February 2007), pp. 102-111, ISSN 0031-8655

Haupt, W. (1999). Chloroplast movement : from phenomenology to molecular biology, In: *Progress in Botany,* Vol.60, K. Esser, J.W. Kadereit, U. Lüttge, & M. Runge, (Eds.), 3-36, Springer-Verlag, ISBN 3-540-646892, Heidelberg, Germany

Ichikawa, S.; Yamada, N.; Suetsugu, N.; Wada, M. & Kadota, A. (2011). Red light, phot1 and JAC1 modulate phot2-dependent reorganization of chloroplast actin filaments and chloroplast avoidance movement. *Plant & Cell Physiology,* Vol.53, No.8, (July 2011), pp. 1422-1432, ISSN 0032-0781

Jarillo, J.A.; Gabrys, H.; Capel, J.; Alonso, J.M.; Ecker, J.R. & Cashmore, A.R. (2001). Phototropin-related NPL1 controls chloroplast relocation induced by blue light. *Nature,* Vol.410, No.6831, (April 2002), pp. 952-954, ISSN 0028-0836

Kadota, A.; Sato, Y. & Wada, M. (2000). Intracellular chloroplast photorelocation in the moss *Physcomitrella patens* is mediated by phytochrome as well as by a blue-light receptor. *Planta,* Vol.210, No.6, (May 2000), pp. 932-937, ISSN 0032-0935

Kadota, A.; Yamada, N.; Suetsugu, N.; Hirose, M.; Saito, C.; Shoda, K.; Ichikawa, S.; Kagawa, T.; Nakano, A. & Wada, M. (2009). Short actin-based mechanism for light-directed chloroplast movement in *Arabidopsis. Proceedings of the National Academy of Sciences of the United States of America,* Vol.106, No.31, (August 2009), pp. 13106-13111, ISSN 0027-8424

Kadota, A. & Wada, M. (1992). Photoorientaion of chloroplasts in protonemal cells of the fern *Adiantum* as analyzed by use of a video-tracking system. *The Botanical Magazine, Tokyo*, Vol.105, No.2, (June 1992), pp. 265-279, ISSN 0006-808X

Kagawa, T. & Wada, M. (1996). Phytochrome- and blue-light-absorbing pigment-mediated directional movement of chloroplasts in dark-adapted prothallial cells of fern *Adiantum* as analyzed by microbeam irradiation. *Planta*, Vol.198, No.3, (September 1996), pp. 488-493, ISSN 0032-0935

Kagawa, T. & Wada, M. (1999). Chloroplast-avoidance response induced by high-fluence blue light in prothallial cells of the fern *Adiantum capillus-veneris* as analyzed by microbeam irradiation. *Plant Physiology*, Vol.119, No.3, (March 1999), pp. 917-923, ISSN 0032-0889

Kagawa, T. & Wada, M. (2000). Blue light-induced chloroplast relocation in *Arabidopsis thaliana* as analyzed by microbeam irradiation. *Plant & Cell Physiology*, Vol.41, No.1, (January 2000), pp. 84-93, ISSN 0032-0781

Kagawa, T.; Sakai, T.; Suetsugu, N.; Oikawa, K.; Ishiguro, S.; Kato, T.; Tabata, S.; Okada, K. & Wada, M. (2001). *Arabidopsis* NPL1: a phototropin homolog controlling the chloroplast high-light avoidance response. *Science*, Vol.291, No.5511, (March 2001), pp. 829-832, ISSN 0036-8075

Kagawa, T.; Kasahara, M.; Abe, T.; Yoshida, S. & Wada, M. (2004). Functional analysis of phototropin2 using fern mutants deficient in blue light-induced chloroplast avoidance movement. *Plant & Cell Physiology*, Vol.45, No.4, (April 2002), pp. 416-426, ISSN 0032-0781

Kasahara, M.; Kagawa, T.; Oikawa, K.; Suetsugu, N.; Miyao, M. & Wada, M. (2002). Chloroplast avoidance movement reduces photodamage in plants. *Nature*, Vol.420, No.6917, (December 2002), pp. 829-832, ISSN 0028-0836

Kasahara, M.; Kagawa, T.; Sato, Y.; Kiyosue, T. & Wada, M. (2004). Phototropins mediate blue and red light-induced chloroplast movements in *Physcomitrella patens*. *Plant Physiology*, Vol.135, No.3, (July 2004), pp. 1388-1397, ISSN 0032-0889

Kawai, H.; Kanegae, T.; Christensen, S.; Kiyosue, T.; Sato, Y.; Imaizumi, T.; Kadota, A. & Wada, M. (2003). Responses of ferns to red light are mediated by an unconventional photoreceptor. *Nature*, Vol.421, No.6920, (January 2003), pp. 287-290, ISSN 0028-0836

Kinoshita, T.; Doi, M.; Suetsugu, N.; Kagawa, T.; Wada, M. & Shimazaki, K. (2001). phot1 and phot2 mediate blue light regulation of stomatal opening. *Nature*, Vol.414, No.6864, (December 2001), pp. 656-660, ISSN 0028-0836

Kodama, Y.; Suetsugu, N.; Kong, S.G. & Wada, M. (2010). Two interacting coiled-coil proteins, WEB1 and PMI2, maintain the chloroplast photorelocation movement velocity in *Arabidopsis*. *Proceedings of the National Academy of Sciences of the United States of America*, Vol.107, No.45, (November 2010), pp. 19591-19596, ISSN 0027-8424

Kodama, Y.; Suetsugu, N. & Wada, M. (2011). Novel protein-protein interaction family proteins involved in chloroplast movement response. *Plant Signaling & Behavior*, Vol.6, No.4, (April 2011), pp. 483-490, ISSN 1559-2316

Kondo, A.; Kaikawa, J.; Funaguma, T. & Ueno, O. (2004). Clumping and dispersal of chloroplasts in succulent plants. *Planta*, Vol.219, No.3, (July 2004), pp. 500-506, ISSN 0032-0935

Lehmann, P.; Bohnsack, M.T. & Schleiff, E. (2011). The functional domains of the chloroplast unusual positioning protein 1. *Plant Science,* Vol.180, No.4, (April 2011), pp. 650-654, ISSN 0168-9452

Loreto, F.; Tsonev, T. & Centritto, M. (2009). The impact of blue light on leaf mesophyll conductance. *Journal of Experimental Botany,* Vol.60, No.8, (May 2009), pp. 2283-2290, ISSN 0022-0957

Luesse, D.R.; DeBlasio, S.L. & Hangarter, R.P. (2006). Plastid movement impaired 2, a new gene involved in normal blue-light-induced chloroplast movements in Arabidopsis. *Plant Physiology,* Vol.141, No.4, (August 2006), pp. 1328-1337, ISSN 0032-0889

Oikawa, K.; Kasahara, M.; Kiyosue, T.; Kagawa, T.; Suetsugu, N.; Takahashi, F.; Kanegae, T.; Niwa, Y.; Kadota, A. & Wada, M. (2003). CHLOROPLAST UNUSUAL POSITIONING1 is essential for proper chloroplast positioning. *The Plant Cell,* Vol.15, No.12, (December 2003), pp. 2805-2815, ISSN 1040-4651

Oikawa, K.; Yamasato, A.; Kong, S.-G.; Kasahara, M.; Nakai, M.; Takahashi, F.; Ogura, Y.; Kagawa, T. & Wada, M. (2008). Chloroplast outer envelope protein CHUP1 is essential for chloroplast anchorage to the plasma membrane and chloroplast movement. *Plant Physiology,* Vol.148, No.2, (October 2008), pp. 829-842, ISSN 0032-0889

Park, Y.-I.; Chow, W.S. & Anderson, J.M. (1996). Chloroplast movement in the shade plant *Tradescantia albiflora* helps protect photosystem II against light stress. *Plant Physiology,* Vol.111, No.3, (July 1996), pp. 867-875, ISSN 0032-0889

Psaras, G.K.; Diamantopoulos, G.S. & Makrypoulias, C.P. (1996). Chloroplast arrangement along intercellular air spaces. *Israel Journal of Plant Sciences,* Vol.44, No.1, (1996), pp. 1-9, ISSN 0792-9978

Sakai, T.; Kagawa, T.; Kasahara, M.; Swartz, T.E.; Christie, J.M.; Briggs, W.R.; Wada, M. & Okada, K. (2001). *Arabidopsis* nph1 and npl1: Blue light receptors that mediate both phototropism and chloroplast relocation. *Proceedings of the National Academy of Sciences of the United States of America,* Vol.98, No.12, (June 2001), pp. 6969-6974, ISSN 0027-8424

Sato, Y.; Wada, M. & Kadota, A. (2001). Choice of tracks, microtubules and/or actin filaments for chloroplast photo-movement is differentially controlled by phytochrome and a blue light receptor. *Journal of Cell Science,* Vol.114, No.2, (January 2001), pp. 269-279, ISSN 0021-9533

Schmidt von Braun, S. & Schleiff, E. (2008). The chloroplast outer membrane protein CHUP1 interacts with actin and profilin. *Planta,* Vol.227, No.5, (April 2008), pp. 1151-1159, ISSN 0032-0935

Senn, G. (1908). *Die Gestalts- und Lageveränderung der Pflanzen-Chromatophoren,* Engelmann, Stuttgart, Germany

Suetsugu, N.; Dolja, V.V. & Wada, M. (2010b). Why have chloroplasts developed a unique motility system? *Plant Signaling & Behavior,* Vol.5, No.10, (October 2010), pp. 1190-1196, ISSN 1559-2316

Suetsugu, N.; Kagawa, T. & Wada, M. (2005a). An auxilin-like J-domain protein, JAC1, regulates phototropin-mediated chloroplast movement in Arabidopsis. *Plant Physiology,* Vol.139, No.1, (September 2005), pp. 151-162, ISSN 0032-0889

Suetsugu, N.; Mittmann, F.; Wagner, G.; Hughes, J. & Wada, M. (2005b). A chimeric photoreceptor gene *NEOCHROME,* has arisen twice during plant evolution.

Proceedings of the National Academy of Sciences of the United States of America, Vol.102, No.38, (September 2005), pp. 13705-13709, ISSN 0027-8424

Suetsugu, N.; Takano, A.; Kohda, D. & Wada, M. (2010c). Structure and activity of JAC1 J-domain implicate the involvement of the cochaperone activity with HSC70 in chloroplast photorelocation movement. *Plant Signaling & Behavior,* Vol.5, No.12, (December 2010), pp. 1602-1606, ISSN 1559-2316

Suetsugu, N.; Yamada, N.; Kagawa, T.; Yonekura, H.; Uyeda, T.Q.P.; Kadota, A. & Wada, M. (2010a). Two kinesin-like proteins mediate actin-based chloroplast movement in *Arabidopsis thaliana. Proceedings of the National Academy of Sciences of the United States of America,* Vol.107, No.19, (May 2010), pp. 8860-8865, ISSN 0027-8424

Suetsugu, N. & Wada, M. (2005). Photoreceptor gene families in lower plants, In: *Handbook of Photosensory Receptors,* W.R. Briggs & J.L. Spudich, (Eds.), 349-369, Wiley-VCH Verlag, ISBN 3-527-31019-3, Weinheim, Germany

Suetsugu, N. & Wada, M. (2007a). Phytochrome-dependent photomovement responses mediated by phototropin family proteins in cryptogam plants. *Photochemistry and Photobiology,* Vol.83, No.1, (January-February 2007), pp. 87-93, ISSN 0031-8655

Suetsugu, N. & Wada, M. (2007b). Chloroplast photorelocation movement mediated by phototropin family proteins in green plants. *Biological Chemistry,* Vol.388, No.9, (September 2007), pp. 927-935, ISSN 1431-6730

Suetsugu, N. & Wada, M. (2009). Chloroplast photorelocation movement, In: *The Chloroplasts. Plant Cell Monographs Series,* A.S. Sandelius & H. Aronsson, (Eds.), 349-369, Springer Berlin, ISBN 978-3-540-68692-7, Heidelberg, Germany

Sztatelman, O.; Waloszek, A.; Banas, A.K. & Gabrys, H. (2010). Photoprotective function of chloroplast avoidance movement: *In vivo* chlorophyll fluorescence study. *Journal of Plant Photobiology,* Vol.167, No.9, (June 2010), pp. 709-716, ISSN 0176-1617

Takano, A.; Suetsugu, N.; Wada, M. & Kohda, D. (2010). Crystallographic and functional analyses of J-domain of JAC1 essential for chloroplast photorelocation movement in *Arabidopsis thaliana. Plant & Cell Physiology,* Vol.51, No.8, (August 2010), pp. 1372-1376, ISSN 0032-0781

Terashima, I. & Hikosaka, K. (1995). Comparative ecophysiology of leaf and canopy photosynthesis. *Plant, Cell & Environment,* Vol.18, No.10, (October 1995), pp. 1111-1128, ISSN 0140-7791

Tholen, D.; Boom, C.; Noguchi, K.; Ueda, S.; Katase, T. & Terashima, I. (2008). The chloroplast avoidance response decreases internal conductance to CO_2 diffusion in *Arabidopsis thaliana* leaves. *Plant, Cell & Environment,* Vol.31, No.11, (November 2008), pp. 1688-1700, ISSN 0140-7791

Tsuboi, H.; Yamashita, H. & Wada, M. (2009). Chloroplasts do not have a polarity for light-induced accumulation movement. *Journal of Plant Research,* Vol.122, No.1, (January 2009), pp. 131-140, ISSN 0918-9440

Tsuboi, H. & Wada, M. (2010a). Speed of signal transfer in the chloroplast accumulation response. *Journal of Plant Research,* Vol.123, No.3, (May 2010), pp. 381-390, ISSN 0918-9440

Tsuboi, H. & Wada, M. (2010b). The speed of intracellular signal transfer for chloroplast movement. *Plant Signaling & Behavior,* Vol.5, No.4, (April 2010), pp. 433-435, ISSN 1559-2316

Tsuboi, H. & Wada, M. (2011a). Chloroplasts can move in any direction to avoid strong light. *Journal of Plant Research*, Vol.124, No.1, (January 2011), pp. 201-210, ISSN 0918-9440

Tsuboi, H. & Wada, M. (2011b). Distribution changes of actin filaments during chloroplast movement in *Adiantum capillus-veneris*. *Journal of Plant Research*, (July 2011), doi:10.1007/s10265-011-0444-8, ISSN 0918-9440

von Caemmerer, S & Furbank, R.T. (2003). The C_4 pathway: an efficient CO_2 pump. *Photosynthesis Research*, Vol.77, No.2-3, (August 2003), pp. 191-207, ISSN 0166-8595

Wada, M. (2007). The fern as a model system to study photomorphogenesis. *Journal of Plant Research*, Vol.120, No.1, (January 2007), pp. 3-16, ISSN 0918-9440

Wada, M.; Grolig, F. & Haupt, W. (1993). New trends in photobiology: Light-oriented chloroplast positioning. Contribution to progress in photobiology. *Journal of Photochemistry and Photobiology B: Biology*, Vol.17, No.1, (January 1993), pp. 3-25, ISSN 1011-1344

Wada, M. & Suetsugu, N. (2004). Chloroplasts can move in any direction to avoid strong light. *Current Opinion in Plant Biology*, Vol.7, No.6, (December 2004), pp. 626-631, ISSN 1369-5266

Whippo, C.W.; Khurana, P.; Davis, P.A.; DeBlasio, S.L.; DeSloover, D.; Staiger, C.J. & Hangarter, R.P. (2011). THRUMIN1 is a light-regulated actin-bundling protein involved in chloroplast motility. *Current Biology*, Vol.21, No.1, (January 2011), pp. 59-64, ISSN 0960-9822

Yamada, M.; Kawasaki, M.; Sugiyama, T.; Miyake, H. & Taniguchi, M. (2009). Differential positioning of C_4 mesophyll and bundle sheath chloroplasts: Aggregative movement of C_4 mesophyll chloroplasts in response to environmental stresses. *Plant & Cell Physiology*, Vol.50, No.10, (October 2009), pp. 1736-1749, ISSN 0032-0781

Yamashita, H.; Sato, Y.; Kanegae, T.; Kagawa, T.; Wada, M. & Kadota, A. (2011). Chloroplast actin filaments organize meshwork on the photorelocated chloroplasts in the moss *Physcomitrella patens*. *Planta*, Vol.233, No.2, (February 2011), pp. 357-368, ISSN 0032-0935

Yatsuhashi, H. (1996). Photoregulation systems for light-oriented chloroplast movement. *Journal of Plant Research*, Vol.109, No.2, (June 1996), pp. 139-146, ISSN 0918-9440

Yatsuhashi, H.; Kadota, A. & Wada, M. (1985). Blue- and red-light action in photoorientation of chloroplasts in *Adiantum* protonemata. *Planta*, Vol.165, No.1, (July 1985), pp. 43-50, ISSN 0032-0935

Yatsuhashi, H.; Wada, M. & Hashimoto, T. (1987). Dichroic orientation of phytochrome and blue-light photoreceptor in *Adiantum* protonemata as determined by chloroplast movement. *Planta*, Vol.9, No.3, (September 1987), pp. 27-63, ISSN 0137-5881

Zurzycki, J. (1955). Chloroplast arrangement as a factor in photosynthesis. *Acta Societasis Botanicorum Poloniae*, Vol.24, (1955), pp. 163-173, ISSN 0001-6977

Zurzycki, J. (1957). The destructive effect of intense light on the photosynthetic apparatus. *Acta Societasis Botanicorum Poloniae*, Vol.26, (1957), pp. 157-175, ISSN 0001-6977

Zurzycki, J. (1980). Blue light-induced intracellular movement, In: *Blue Light Syndrome*, H. Senger, (Ed.), 50-68, Springer-Verlag, ISBN 3-540-10075-X, Heidelberg, Germany

12

Light Harvesting and Photosynthesis by the Canopy

Mansour Matloobi

Department of Horticulture, Faculty of Agriculture, University of Tbariz, Tabriz,
Iran

1. Introduction

Photosynthesis is a life-sustaining process driven mostly by green plants to support not only life of plants, but also life on earth in general. The estimated dry matter produced by photosynthesis of land plants reaches as much as 125×10^9 per year (Field et al., 1998). About 40% of this material is composed of C, fixed in photosynthesis. Light has long been recognized as a source of energy to convert atmospheric CO_2 into energetic chemical bands which finally appear as sucrose, starch and many other energy containing substances. This conversion will not happen until there is specialized light-harvesting system to capture and transfer light energy to low-energy compounds. Leaves are this specialized system with broad, laminar surface well suited to gather and absorb light. When a large number of leaves are arranged beside each other the canopy will be formed. Organization and spatial arrangement of leaves within the canopy directly affect the amount of light absorbed by this integrated system. Therefore, photosynthetic capacity at canopy level depends not only on factors affecting leaf level photosynthesis but also on factors which influence properties of canopy microclimate, particularly its light distribution profile. Estimating photosynthesis at canopy scale, however can be of great importance as it provides a tool to predict crop yield and help producer to make decision and planning of production. While photosynthesis mechanism in C_4 plants differs virtually from that of C_3 plants, there have been no significant differences in the methods implemented to investigate light harvesting and upscaling photosynthesis from leaf to canopy level, so in this chapter the issues related to the photosynthesis of C_3 plants will be emphasized and addressed.

2. Canopy: An integrated foliage structure

There are many factors that determine plant canopy architecture. Some of these factors are genetic and relate to the plant species while some are ecological and relate to the plant-environment interactions. Under the influence of these factors plants develop their canopy so that they reach a compromise between affecting factors and internal physiological requirements. The resultant will be a volume composed of numerous leaves varying in size, thickness, inclination and many other physical and physiological properties distributed in space and time. Naturally, plants attempt to construct their canopy in a way that the highest ambient irradiance could be absorbed. This process is usually done by developing special branching system, efficient leaf arrangement; appropriate canopy dimension and even sometimes by natural pruning and removing weak and underdeveloped organs. Consequently canopies appear to be a complex, dynamic and ever-changing volumes; being difficult to interpret and understand. The complexity of canopy becomes more apparent when we move from leaf level to pure stand to heterogeneous plant communities, since each level contains elements of the lower levels (Norman & Campbell 1994). In vast plant communities, when diverse plant species mixed together and form a very heterogeneous vegetation stand, description of canopy structure become much more difficult. Therefore canopies composed of single species or integrated of only a few species usually assumed to be homogeneous with uniform monotypic plant stand (Beyshlag & Ryel, 2007).

In order to interpret canopy in detail we may have to consider its components. Canopy structure can be defined in detail by including the size, shape, orientation and positional distribution of various plant organs such as leaves, stems, branches, flowers and fruits. Getting such information for each element in a canopy is not currently feasible, so quantitative description of the canopy by means of mathematical and statistical methods seems to be appropriate. Norman and Campbell (1994) summarized all the methods applied in describing canopy structure to two main groups: direct and indirect methods. They explained that direct methods involve usually much labor in the field and require very simple data reduction when compared to the indirect methods which use simple and rapid field measurements but complex algorithms for the reduction of data. In spite of recent considerable progresses achieved in 3D modeling by computers, this technique still requires a considerable effort to sample all the growing organs of a canopy. Because of this, only a few variables, such as the leaf area density, and the leaf inclination distribution function could be used to describe canopy structure (Weiss et al., 2004). Sound estimation of a crop whole canopy leaf area may be sufficient to predict crop productivity in large scale, but does not give an accurate estimation of vertical gradient of light or spatial distribution of materials applied to the plant canopy. Plant architectural models attempt to fill the gap caused by not considering the influence of plant functioning or environmental variables on the process of morphogenesis through including physiological processes of plant growth and development as well as the physical structure of plants. To do this, more precise and extensive data will be required than usually collected on the dynamics of production of individual organs of plants (Birch et al., 2003).

3. Light harvesting

Light harvesting by plants is influenced by many factors such as, diurnal variation in solar elevation and variation in leaf angle, leaf position in the canopy, sky cloudiness, degree of leaf clumping and amount of sunflecks penetrated through the canopy, and all the factors affecting gas flux properties of individual leaves. Photosynthesis occurs in leaves, the small-sized food factories constituting the majority of the canopy volume. Any disturbances in canopy microclimate such as variations occurred in ambient gas composition, light quantity and quality, temperature and humidity will clearly lead to corresponding changes in C uptake by the leaves. Therefore studying leaves as the primary light harvesting organ within the canopy could merit first priority.

3.1 Light harvesting at the leaf scale
Before being intercepted by leaves, light travels a long distance between the sun and the earth, passing through the atmosphere according to its composition and physical features, it

experiences some quantitative and qualitative alterations which favor life sustaining processes occurring on the planet. Upon reaching leaf surface light transferred and distributed through the leaf by a phenomenon called **lens effect** created by the planoconvex nature of epidermal cells covering leaf surface. The consequent of this effect is efficient redirecting of incoming radiation to the chloroplasts confined in mesophyll cells. The mesophyll tissue consisted of two distinct cells: palisade and spongy cells. Palisade cells are elongated and cylindrical with the long axis perpendicular to the surface of the leaf, while spongy cells situated below this layer and surrounded by the prominent air spaces (Hopkins & Huner, 2004). Although a large number of chloroplasts occupy the cell volume of palisades, there is still a significant proportion of cell volume that does not contain chloroplast. This chloroplast-free portion of the cell helps to distribute incoming light and maximize absorption by chlorophyll. Consequently, some of the incident light may pass through the first palisade layer without being absorbed, but more likely will be intercepted through successive layers by **the sieve effect.** Additionally, palisade cells help efficient distribution of incoming light by **light-guide effect**, a feature that assists light reaching the cell-air interfaces to be reflected and channeled through these layers to the spongy mesophyll below (Hopkins & Huner, 2004).

A large portion of the light reaching the leaf surface then finally targets the chloroplasts, where photochemical reactions occur. Although the mesophyll layer is the main place hosting chloroplasts, these organelles may be also found in other organs such as; buds, the bark of stems and branches, flowers and fruits. Light interception in chloroplasts is carried out specifically by antenna complex or light harvesting complex (LHC), mainly consisted of chlorophylls (i.e. chlorophyll a and b) and several hundred accessory pigments clustered together in the thylakoid membrane. Carotenoides are one of the most important accessory pigments in green plants which absorb light at wavelength different from that of chlorophyll and so act together to maximize the light harvested. When a pigment molecule absorbed incoming photon energy and excited, it transfers the energy to two special chlorophyll molecules in the photosynthetic reaction center. The reaction center then passes on the energy as a high-energy electron to a chain of electron carriers in the thylakoid membrane. The high energy electrons are then exploited to produce high energy molecules which are eventually used to reduce RuBP by CO_2.

In response to changes of environmental conditions chloroplast may undergo some modifications in structure and biochemical composition in order to cope with new environment. Some of these environmental factors negatively affect chloroplast activity and therefore directly limit the photosynthetic rate. The consequence of most of these factors, such as high light intensity, UV radiation, air pollutants, herbicides, water and heavy metal stress will usually appear as oxidative stress and often leads to the symptoms of structural damage which emerges as swelling of thylakoids, plastoglobule and starch accumulation, photodestruction of pigments, and inhibition of photosynthesis (Mostowska, 1997). It was shown similarly that chloroplast property changes in accordance with the light gradient within a bifacial leaf (Terashima & Inoue, 1985). That is, near leaf surface facing to ambient light, the chloroplasts have higher rates of electron transport and Rubisco activities per unit of chlorophyll than chloroplasts farther away from the surface. Moreover, in plants acclimated to shade conditions, it was shown that chloroplasts migrate in response to inducing ambient light (lambers et al., 1998).

Plants try to increase their light absorption at leaf level by adjusting leaf weight to plant weight or leaf weight to leaf area. One of the parameters which can be very helpful in giving good understanding of the plant manner of investment on light harvesting complexes is specific leaf area (SLA). It is defined as projected leaf area per unit leaf dry mass. This parameter relates with the other plant growth parameters as follows:

$$LAR=LWR×SLA \tag{1}$$

LWR is the ratio of leaf weight to plant weight (gg^{-1}), LAR is the ratio of leaf area to plant weight (m^2g^{-1}). The equation that links LAR to RGR is:

$$RGR=NAR×LAR \tag{2}$$

Where RGR is relative growth rate ($gg^{-1}d^{-1}$) and NAR denotes net assimilation rate ($gm^{-2}d^{-1}$). This relationship implies that transferring a sun-acclimated plant to a shade environment will result in a reduction in RGR caused by a lowering in NAR, reflecting the effect of PAR on photosynthesis. In order to keep RGR unchanged, plant has to increase LAR with the assumption that there is no change in the light dependence of photosynthesis. LAR directly changes with any changes in LWR and/or SLA. It has been revealed that LWR may proportionally change in accordance with plant light regime alterations, having tendency to increase in shade-adapted plants, while showing decline in non-adapted plants in shade (Fitter & Hay, 2002). Studying with many plants indicated that SLA seems to change faster than LWR, playing an important role in acclimation process to varying environmental light regimes. Plants developed under high light usually have thick leaves with a low SLA (Bjorkman, 1981, as cited in Fitter & Hay, 2002). Light-saturated photosynthesis remained unchanged in plants acclimated to shade environment because of doubling SLA (Evans & Poorter, 2001). It can be deduced that SLA is more variable than LWR, or, leaf area is more plastic than leaf weight. Studying with *Cucumis sativa* , a light-demanding species, showed that leaf area changes proportionally with the total ambient light, with a maximum at about 4.2 $Mjm^{-2}d^{-1}$ (Newton, 1963, as cited in Fitter & Hay, 2002). Instantaneous light variations do not exert any immediate changes in SLA, while these changes generally occur in response to total radiation load; this is probably the case for *Impatients parviflora* which shows an almost threefold increase in SLA when grown in 7% of full daylight (Evans & Hughes, 1961, as cited in Fitter & Hay, 2002). Findings of Evans and Poorter (2001) indicated that increasing SLA is a very important means applied by plants to maximize carbon gain per unit leaf mass under different environmental light conditions.

3.2 Light harvesting at the canopy scale

Foliage density distribution and leaves orientation highly impact sunlight attenuation through the canopy. As described before, canopies normally are not solid sheets, but are loosely stacked formation of leaves which help plants to effectively absorb most of the incident light, with leaves near the top of the canopy absorbing near maximum solar radiation and the lower leaves perceiving sunlight of a reduced intensity and also an altered spectral composition. Therefore the amount of photosynthetically active radiation intercepted by a leaf usually depends on its position in the canopy and the angle it faces incoming solar radiation. Leaves within the canopy are generally subject to three types of radiation: light beam, reflected and transmitted radiation. Light beam penetrates through the gaps created in the canopy probably by instantaneous fluttering of leaves caused by wind, or sparse leaf arrangement which naturally forms gaps within the canopy. While passing through the canopy light beam is usually trapped by the lower leaf layers, however, depending on the canopy architecture some may reach the most lower layers and form "sunflecks". These packages of high light intensity are not generally stable, but dynamically change their location due to movements of branches, and the changing angle of the sun. Their duration may range from less than a second to minutes. Small-sized sunflecks typically carry lower light energy than direct sunlight because of penumbral effects, but large ones can approach irradiances of direct sunlight (Lambers et al., 1998). Direct beam light predominantly absorbed by the leaves at the top of the canopy, some portion transmitted down with altered spectral quality, due to action of the various leaf pigments. Leaves typically transmit only a few percent of incident PAR in the green band at around 550 nm, and are otherwise efficiently opaque in the visible range. Transmittance of PAR is normally less than 10%, whereas transmittance of far-red light is substantial (Terashima & Hikosala, 1995). This spectral alteration affects the phytochrome photoequilibrum and allows plants to perceive shading by other plants to adjust their photomorphogenesis activities (Lambers et al., 1998). Leaves like to other biological surfaces not only transmit light but reflect a proportion. The amount of reflection depends on morphological and physical properties of leaves such as, leaf shape, thickness and shininess of the cuticle. However, it should be noted that reflected light then may be absorbed or transmitted by the lower leaves similar to the radiation reaching the canopy surface.

Rundel and Gibson (1996) found that leaf angle and orientation are the main factors which control daily integrated radiation, maximum irradiance and diurnal distribution of irradiance. Orientation of leaves at the top of the canopy is usually at oblique (acute or obtuse) to incident light. When leaves in the uppermost layer of the canopy arrange obliquely, they allow a given amount of light to distribute over a greater total leaf area of the plant than when they arrange at right angles to the direction of incoming light. While leaves on the canopy surface are most efficient at utilizing full sunlight when at an oblique angle to the sun's ray, the leaves located in lower parts do best in lower irradiance if the leaf area is at right angle to the light, intercepting the greatest sunlight per leaf surface. Ontogenetically change from sessile juvenile leaves to petiolate adult leaves is accompanied by a change in leaf orientation from horizontal to vertical (king, 1997). Research by Shelley and Bell (2000) on the heteroblastic species *Eucalyptus globules* Labill. ssp. Globules showed that there was no active diurnal orientation between juvenile and adult leaves twoard or away from incident radiation. They concluded that greater interception of light by juvenile leaves compared with vertical adult leaves, may be due to their high adaptation capacity to high incident light.

3.2.1 Light profile within the canopy

Beer's law has long been used by many authors to describe light penetration in plant canopy. With the assumption that the gaps are randomly distributed horizontally, the area of direct-beam irradiance penetrating to any depth in the canopy is an exponential function of the cumulative LAI from the top of the canopy (Boote & Loomis, 1991):

$$I = I_O \, exp(-KLAI) \tag{3}$$

I and I_o are respectively the irradiance beneath and above the canopy (umolm^{-2}s^{-1}), K is the extinction coefficient and LAI denotes leaf area index. The extinction coefficient is actually the ratio of horizontally projected shadow area per unit ground area per unit leaf area. Both leaf angle and solar elevation angle (β) affect the shadow projection of leaves. At any point within the canopy, radiation is composed of contributions from all directions. The angle between leaf surface and incident radiation depend on leaf orientation and the radiation direction. However, for horizontally positioned leaves, the fraction of radiation intercepted by any leaf will be proportional to the leaf area itself, independent of the radiation direction (Marcelis et al., 1998). Consequently, the extinction coefficient is high for horizontally inclined leaves, but low for vertical leaf arrangements. When all leaves distributed randomly in the horizontal plane and are perpendicular to the direct beam with solar elevation of 90°, the value of K is 1. Solar position changes during the day influence the value of K by the factor $1/\sin(\beta)$. Variations occurred in leaf angle also change K value dramatically, as vertically oriented leaves intercept less light than horizontal leaves.(Boote & Loomis, 1992). For greenhouse roses trained by arching system K ranged from 0.58 to 0.66 at different hours of day, with a daily average value of about 0.63 (Gonzalez –Real et al., 2007). Typical values for K are in the range of 0.5 to 0.8 (Marcelis et al., 1998). A canopy with low extinction coefficient allows more effective light reaches lower leaves. Some crops tend to arrange upper leaves at oblique angles to incident radiation to minimize the probability of photoinhibition and increase light penetration to lower leaves in high light environments, thereby maximizing whole-canopy photosynthesis. (Terashima & Hikosaka, 1995). It should be noted that direct and diffuse light have different extinction profiles in the canopy and due to light saturation of photosynthesis, direct beam should be singled out from the rest of the incoming radiation (Spitters, 1986). For this reason, experimentally determined values for total light extinction would not necessarily be the same as K. Leaf area index varies with the number and density of leaves within the canopy. In sparsely vegetated communities like deserts or tundra LAI value is less than 1, while for crops it is about 5 to 7 and for dense forests it estimated to range between 5 to 10 (Schulze et al. 1994). About 90% of PAR is absorbed by the canopy when LAI exceeds a value of 3. At leaf level absorption of PAR is approximately 80%-85% (Marcelis et al., 1998).

4. Whole canopy photosynthesis

Photosynthesis is a fundamental process occurring in green plants, algae and photosynthetic bacteria. During the process solar energy is trapped and utilized to drive the synthesis of carbohydrate from carbon dioxide and water. There are two distinct phases in the reactions of photosynthesis: the light reactions and the dark reactions. Light reactions use light energy to synthesize NADPH and ATP, which then transfer the energy to produce carbohydrate from CO_2 and H_2O during dark reactions. Chloroplast is an organelle in which photosynthesis takes place and has highly permeable outer membrane and an inner membrane that is impermeable to most molecules and ions. Light reactions occur in thylakoids, stacks of flattened chloroplast membranes extended into stroma, the place where the dark reactions are taken place. Two photosystems are involved in light reactions: photosystem I (PSI) and photosystm II (PS II). The difference is that PSI contains chlorophylls which have an absorption peak at 700 nm and so is called P700 but chlorophylls in the reaction center of PSII absorb light mostly at 680 nm and so is referred to as P680. The two photosystems are linked by a chain of electron carriers and when arranged in order of their redox potentials they form so-called Z scheme. Electrons released in PSII flow through Z scheme to reduce $NADP^+$ in PSI. Pigments involved in the light harvesting complex (LHC) have already been discussed and here a brief explanation of overall photosynthesis reactions is presented:

Photosynthesis process begins with absorbing light-energy photos and transferring them to the reaction centers of the photosystems where the second process starts. In the thylakoid membrane water splits to release electrons which are then transported along an electron-transport chain to produce NADPH and ATP. All the reactions occurred up to this point are called light reactions of photosynthesis as they depend on light energy to proceed. The produced NADPH and ATP during the light reactions then enter the carbon-reduction cycle (Calvin cycle), in which CO_2 assimilated, leading to the synthesis of C_3 compounds (triose-phosphates). This reaction does not need light to proceed and therefore is referred to as the dark reaction of photosynthesis. In Calvin cycle, CO_2 molecules are condensed with ribulose 1,5-bisphoshate (a five-carbon molecule) to produce a transient six-carbon intermediate that immediately hydrolyzes to two molecules of 3-phosphoglycerate. This important process will not be complete without the mediation of the key enzyme ribulose bisphosphate carboxylase/oxygenase (often called Rubisco). Rubisco is known as a slow enzyme as it only fixes three molecules of its substrate every second and hence plants need a large amount of this enzyme to assimilate enough CO_2 to sustain plant life. Approximately 50% of chloroplast protein content is Rubisco, probably the most abundant protein on the earth.

4.1 Factors affecting leaf level photosynthesis

The rate of photosynthesis at leaf level varies widely and is influenced not only by leaf internal biochemical and physiological conditions, but also by many environmental variables such as, CO_2 concentration, light intensity, temperature and humidity fluctuations. Temperature is an essential factor that control enzymatically catalyzed reactions and membrane processes and in this way it controls photosynthesis (Lambers et al., 1998). Photosynthetic response to temperature varies among species because of the different activation energy required by different reactions processed in various plants. Consequently, temperature-dependent photosynthesis of plants range widely from temperatures near freezing to over 40°C, implying that the specific range depends on species and genotype, plant age, plant origin, and season (Pallardy, 2008). Optimal temperature for any plant is usually defined by the temperature that plant has experienced and adapted to during the entire growing period. High temperatures increase affinity of Rubisco to oxygen than carbon dioxide, consequently leading to enhancing photorespiration. In addition, the solubility of CO_2 declined with increasing temperature more strongly than does that of O_2. At temperatures below about 15°C the rate of photosynthesis is often reduced in many (sub)tropical plants. For example, after exposing coffee trees to 4°C at night, the rate of photosynthesis was reduced by more than half (Pallardy, 2008). This kind of damage is called chilling injury and differs from frost damage, a type of damage that only happens below 0°C. Chilling injury generally results from a precipitate decrease in the activity of metabolic processes, notably respiration, which can be fatal within a few hours or days (Fitter & Hay, 2002). Part of the chilling injury is due to the depression of photosynthetic metabolism caused by: (i) decrease in membrane fluidity, (ii) changes in processes and activities of the enzymes related to the membrane, such as the photosynthetic electron transport, (iii) decline in activity of cold-sensitive enzymes (Lambers et al., 1998). Freezing injury commonly occurs to the woody plants of north temperate, subarctic, and alpine regions. It happens not because of low temperatures *per se*, but due to ice formation within tissues. If ambient temperature falls with intermediate rates (10°C to 100°C min^{-1}), it will cause intracellular ice formation which disrupts the fine structure of the cells and invariably results death.

Leaves absorb approximately about 85% to 90% of incident PAR (Nobel, 1999). The rate of absorption depends on leaf morphology and structure, especially on the number of palisade and spongy mesophyll layers (Vogelmann & Martin, 1993). In darkness there is no photosynthesis and leaves continue respiration, releasing CO_2 to the atmosphere. In accordance with increasing light intensity, the rate of photosynthesis starts to increase until it reaches compensation point where the uptake of CO_2 in photosynthesis equals releasing of CO_2 in respiration. At this point there is no CO_2 exchange between leaves and the atmosphere. When light intensity goes beyond the compensation point the photosynthetic rate starts to increase linearly. The initial slope of this line, located between compensation point and light saturation point, is referred to as quantum efficiency. Quantum efficiency describes the efficiency with which light is converted into fixed carbon. With further increase in light intensity, photosynthesis became saturated and is limited by the carboxylation rate. Increasing irradiance beyond the upper limits may even cause a decline in photosynthesis due to occurrence of photoinhibition, particularly in shade-adapted leaves. Photoinhibition may take place in both shade intolerant and shade tolerant plants. However, shade tolerant species and plants grown under shade are especially prone to photoinhibition (Pallardy, 2008). It was shown when willow leaves that previously developed in the shade were exposed to full sunlight, they showed more photoinhibition than leaves developed in the light (Ogren, 1988). Photoinhibition reduces plant quantum efficiency, therefore negatively influence photosynthetic productivity. Nevertheless, plants develop mechanisms to recover from photoinhibition, and it was indicated that the level of recovery is partly related to the duration of exposure to higher light environment.

Although under normal conditions the probability of photosynthesis reduction caused by decreased levels of enzyme Rubisco is very low, there are nevertheless circumstances under which Rubisco concentration exerts strong control over photosynthetic capacity, for example, in low plants transferred to high light (Lauerer et al., 1993). In addition, it was proven that sufficient amount of Rubsico may effectively regulate other components of photosynthetic apparatus. Anthisense plants with greatly reduced levels of Rubisco often suffer imbalances in electron transport and decreased water-use efficiency (Quick et al., 1991).

Availability of carbon dioxide at the carboxylation site within the chloroplast highly affects photosynthesis capacity. This availability is strongly limited by resistances in its diffusion path twoards the mesophyll cells. Resistance may be generated by boundary layer of air, cuticle, stomata, and mesophyll air space and liquid diffusion resistance. Regarding predicted atmospheric CO_2 elevation up to 700-1000 µmol mol^{-1} by the end of the 21th century (Houghton et al., 2001), many researches have been conducted over the past decades on the effects of rising atmospheric CO_2 on the physiological aspects of higher plants. These researches showed that leaf-level photosynthesis was often increased in plants developed under long-term exposure to increased levels of CO_2 (Curtis, 1996; Gonzalez-Real & Baille, 2000; Tissues et al., 1997). Since Rubisco uses CO_2 and RuBP as the principal substrate to catalyze the carboxylation reactions, it could be expected that any increase in the environmental CO_2 concentration may cause increases in the rate of photosynthesis, assuming that there is no other limiting factor . The rate of carboxylation per unit leaf area can be governed by elevated CO_2 through at least two fundamentally different ways: biochemical mechanisms and leaf morphological modifications (Peterson et al., 1999). The biochemical mechanism consists of three levels, as stated by Peterson et al. (1999):" (i) a reduction in substrate limitation of Rubsico catalysis (Farquahr, von Caemerer & Berry, 1980) , (ii) competitive reduction of RuBP oxygenation (Farquhar et al., 1980), and (iii) any adjustments in the photosynthetic apparatus (from light capture through starch and sucrose synthesis) that change the RuBP limitation of Rubsico (Sage, Sharkey & Seemann,1989; Sage, 1990)". Modifications in leaf morphology and anatomy are the second way that influences the rate of leaf carboxylation. These alterations appear as changes in mesophyll cell number, carbohydrate concentration and leaf mass per unit area (Lambers et al., 1998).

Photosynthetic capacity at leaf level also depends highly on stomata density per unit leaf area and their gas exchange behavior controlled by environmental factors. Stomatal opening is bordered by a pair of unique guard cells which actively regulate the rate of aperture opening by means of swelling and shrinking mechanism, controlled by proton pump and potassium ion uptake processes. Outer surface of epidermis is coated with CO_2-impermeable cuticle, therefore nearly all of the CO_2 taken into the leaf for photosynthesis must enter only by diffusion through stomatal pores. The degree of stomata opening determines the rate of gas exchange between the leaf and environment which in turn results in direct influence on the rate of transpiration and CO_2 assimilation. Of the environmental variables affecting stomatal movements, CO_2 and light appear to make a substantial contribution to the rate of opening. Stomatal pores tend to be open when the leaves experience low CO_2 concentration or light, and gradually begin to close when face high CO_2 concentrations (Fig. 1). Although high environmental CO_2 concentration may gradually stimulate closure of stomata, there are a number of studies which show elevated ambient CO_2 enhances plant photosynthesis [Curtis, 1996; Gonzalez- Real & Baille, 2000; Tissues et al., 1997]. Gas concentration gradient between leaf intercellular air spaces and leaf boundary layer, together with the size of aperture determine the rate of gas movement across the stomata, referred to as stomatal conductivity. Water vapor and CO_2 are the two main gases crossing stomata which directly influence the rate of transpiration and photosynthesis respectively. As the concentration gradient across the stomata differs considerably for H_2O and CO_2, and since they are not equal in the coefficient of diffusion, rate of gas exchange through the stomata will be different for them. As a result, gas exchange would affect photosynthesis and transpiration almost independently (Fig. 2). It has been recognized that elevating ambient CO_2 increases plant water use efficiency. This term is defined as the ratio of CO_2 molecules assimilated by photosynthesis to the number of water molecules lost *via* transpiration. Efficient water use by crops will result in increased agricultural products per liter of water consumed and therefore it can be highly beneficial to agriculture in arid and semi-arid regions with elevated CO_2.

Investigations have shown that all the stomata distributed over the entire leaf do not respond homogeneously to environmental factors at least in stressful condition (Pospisilova & Santrucek, 1997)). This is called stomata patchiness which occurs when some stomata over the leaf close completely, whereas others are almost open. Meyer and Genty (1999) documented that inhibition of photosynthesis was mainly mediated through stomatal closure, when leaves undergo stress caused by dehydration or ABA treatment.

Water stress is another important factor that controls the rate of leaf photosynthesis. It is carried out partially through regulating the size of stomatal aperture and thus limiting CO2 diffusion to the leaf air spaces, and partly by means of increasing diffusional resistance to CO_2 movement from intercellular spaces to the chloroplast. Since water stress in drought conditions usually coincide with high solar radiation and higher temperatures, the mechanism of this down-regulation of photosynthesis in response to water stress is not fully understood (Lambers et al., 1998).

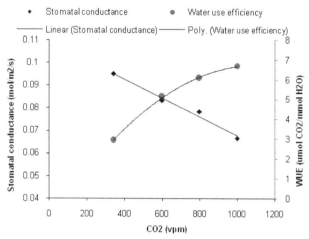

Fig. 1. Effect of CO_2 on stomatal conductance and water use efficiency of *Rosa hybrida* 'Habari'

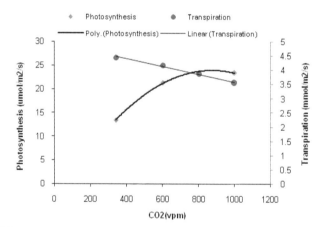

Fig. 2. Effect of CO_2 on the rate of photosynthesis and transpiration of *Rosa hybrida* 'Habari'

4.1.1 Modeling photosynthesis

Models in plant science can be divided into two main groups: mechanistic and empirical. Mechanistic models are descriptive and developed usually based on exhaustive and comprehensive studies which led to globally accepted findings. However, empirical models sometimes referred to as statistical, regression or black-box models are limited to time, location and species on which that model developed and cannot accurately be extrapolated to other conditions and species. Empirical models could be valuable in that they implicitly take into account all unknown effects (Marcelis et al., 1998). In modeling photosynthesis processes both mechanistic and empirical approaches have been considered. Photosynthetic light response curve was one of the cases that has been noted by many authors to be described by several mathematical functions. Three functions have been used to describe photosynthesis light response cure:

This model was initially proposed by Rabinowitch in 1951, and later reviewed by Johnson and Thornley (1984) in order to describe photosynthesis light response. Non-rectangular hyperbola seems to be one of the best equations in prediction of leaf photosynthetic light response. The function is as follows (Fig. 3):

$$P = \frac{\alpha I + P_{max} - \sqrt{(\alpha I + P_{max})^2 - 4\alpha\theta I P_{max}}}{2\theta} \tag{4}$$

where P and P_{max} are respectively the rate of leaf gross photosynthesis ($\mu molm^{-2}s^{-1}$) and light-saturated photosynthesis ($\mu molm^{-2}s^{-1}$), a is quantum efficiency (mol CO_2 mol quanta^{-1}), I is irradiance quantity (μmol qunta $m^{-2}s^{-1}$) and θ denotes curvature (convexity) of the light-photosynthesis relationship (dimensionless).

This model have three parameters to be estimated: (i) the quantum efficiency (a),the initial slope of the curve which relates the rate of CO_2 uptake to absorbed or incident light at very low light intensity. Values of this parameter change with the species, leaf history of stress such as; low temperatures, drought and high irradiance and usually range between 0.040 to 0.075 mol CO_2 mol quanta^{-1} at ambient CO_2 concentrations (Cannell & Thornley, 1998), (ii) the light –saturated photosynthetic rate (P_{max}), which varies extremely among species and is affected by the temperature and life history of leaves which influence leaf morphological and physiological properties like N content and leaf thickness, (iii) and the curvature factor (θ), which indicates how quickly the transition of the curve is made from Rubisco-limited rgion to RuBP-regeneration-limited region. There are many studies, indicating that non-rectangular hyperbola usually best fitted with θ ranging from 0.7 to 0.9 (Matloobi, 2007; Marshall & Biscoe, 1980; Cannel & Thornley, 1998; Kim et al., 2004).

In non-rectangular hyperbola when θ closes zero, the equation appears to be a rectangular hyperbola (Fig. 4):

$$P = \frac{\alpha I P_{max}}{\alpha I + P_{max}} \tag{5}$$

This equation recalls the famous Michaelis-Menten equation. With $\theta=1$ the equation will be Blackman response with two intersecting straight lines.

Another equation that is used to model photosynthesis light response is asymptotic exponential equation (Fig. 5):

$$P = P_{max}(1 - e^{(-\alpha I/P_{max})}) \tag{6}$$

In an experiment we measured leaf gas exchange parameters in a rose crop (*Rosa hybrid* "Habari") by a portable photosynthesis measurement system (Matloobi,2008). Obtained data then were used to estimate the models parameters by non-linear least squares regression method. A linear regression was fitted to the data obtained by direct measurement and those estimated by the models. Results showed that all the models had the potential to present good estimations of the leaf photosynthetic light response with roughly high R^2 (coefficient of determination), but the non-rectangular hyperbola with the highest R^2 ($R^2=0.968$) was the best as it predicted values more closer to the observed ones

(Fig. 3). Non-rectangular hyperbola has been frequently used to describe observed leaf photosynthetic responses to environmental variables (e.g. Pasian, 1989; kim et al., 2004; Cannel & Thornley, 1998)

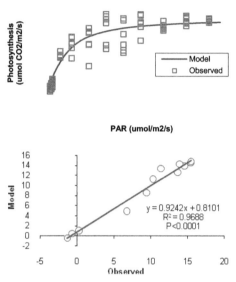

Fig. 3. Photosynthetic light response of *Rosa hybrida* 'Habari', non-rectangular hyperbola model was fitted to the observed data (top), regression between the model and observed data estimates model efficiency (bottom)

The model proposed by Farquahr et al. in 1980 for leaf photosynthesis of C_3 plants is the only mechanistic model which is accepted and widely used for determination of the leaf CO_2 assimilation capacity. This model developed based on the amount and kinetic properties of Rubisco and the ratio of RuBP to enzyme (Rubisco) active site (Harley & Tenhunen, 1991). In this model two limiting factors were assumed to control the leaf photosynthetic capacity:

$$A_n = min\{A_v, A_j\} - R_d \tag{7}$$

where A_v and A_j are the rate of gross photosynthesis limited by Rubisco activity and the rate of RuBP regeneration through electron transport, respectively, and R_d is the rate of mitochondrial respiration. Rubisco limited photosynthesis is given by:

$$A_v = V_{cmax} \frac{C_i - \Gamma^*}{C_i + K_c\left((1 + O / K_o)\right)} \tag{8}$$

where V_{cmax} is the maximum carboxylation rate, with K_c and K_o the Michaelis constants for carboxylation and oxygenation, respectively, and C_i and O are the partial pressure of CO_2 and O_2 in the intercellular air spaces, and Γ^* is the CO_2 compensation point in the absence of mitochondrial respiration. The rate of photosynthesis limited by RuBP regeneration is given by:

$$A_j = (\frac{J}{4}) \times \frac{(C_i - \Gamma^*)}{(C_i + 2\Gamma^*)}$$ (9)

where J is the rate of electron transport, J is related to irradiance usefully absorbed by photosystem II, I_2 by:

$$? J^2 - (I_2 + J_{max})J + I_2 J_{max} = 0$$ (10)

where J_{max} is the potential rate of electron transport, I_2 is related to the incident PAR, I_o by the following equation:

$$I_2 = I_o(1-f)(1-r)/2$$ (11)

where f is spectral correction factor ($\sim$ 0.15) and r is the reflectance plus any small transmittance of the leaf to PAR ($\sim$ 0.12).

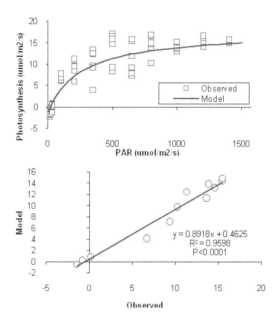

Fig. 4. Photosynthetic light response of *Rosa hybrida* 'Habari', rectangular hyperbola model was fitted to the observed data (top), regression between the model and observed data estimates model efficiency (bottom)

The two key parameters in this model which vary among species are V_{cmax} and J_{max}. The potential rate of electron transport, J_{max}, is a property of the thylakoids that varies depending on growth conditions (Farquhar & Evans, 1991). Factors affecting the chlorophyll content per unit leaf area determine the rate of J_{max}. There are many studies showing that the chlorophyll content of leaves dynamically change according to the environmental light availability [Kitajamia & Hogan, 2003; Matloobi et al., 2009; Walters, 2005). Plants acclimated to low irradiance are enriched in the light-harvesting chlorophyll a/b protein complex and deplete in the photosystem II reaction-center complexes, therefore the electron-transport capacity per unit of chlorophyll is less in leaves acclimated to low irradiance (Farquhar & Evans, 1991).

It was found that there is a good correlation between the leaf N content and the photosynthetic maximum carboxylation rate, V_{cmax}. Gonzalez-Real and Baille (2000) documented that in rose crop there is a gradient in the leaf photosynthetic N concentration from the top of canopy down to the bottom layers according to the amount of light absorbed. The value of V_{cmax} decreased from 66 $\mu molm^{-2}s^{-1}$ for leaves situated at the top of the canopy to 44 $\mu molm^{-2}s^{-1}$ for leaves located at the bottom layers. The ratio of J_{max}/V_{cmax} for all leaf layers within the canopy was almost constant and resulted 2.3. It should be noted that photosynthetic key parameters (J_{max} and V_{cmax}) change proportionally with seasonal variations in soil water content, air temperature and VPD (Xu & Baldocchi, 2003). Parameterization of the photosynthetic models for several plants have been previously done (Gonzalez-Real & Baille, 2000; Kim & Lieth, 2002; Kim & Lieth, 2003; Matloobi, 2007) and still it is noted and under research by many authors around the world.

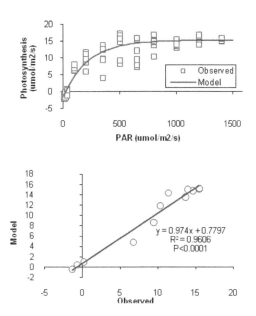

Fig. 5. Photosynthetic light response of *Rosa hybrida* 'Habari', asymptotic exponential model was fitted to the observed data (top), regression between the model and observed data estimates model efficiency (bottom)

4.2 Photosynthesis at canopy level
Canopy photosynthesis at the first step highly depends on the rate of photosynthesis at leaf level which is controlled by interaction of many internal and some external factors, outlined earlier. Therefore in order to obtain an estimation of the plant whole-canopy photosynthesis one must first consider the factors affecting leaf level CO_2 assimilation and then accurately

incorporate them to upscale photosynthesis from individual leaves to the canopy. To achieve this, the first challenge will be calculating the amount of radiation absorbed by individual leaves and finally by the plant whole canopy.

Beer's law (equation 3) has been used as a basis to develop more precise models in order to get a clear profile of light distribution within the canopy. Three approaches have been considered in modeling light absorption by the canopy: (i) big leaf model (ii) multi-layer model (iii) sun and shade model.

Big leaf model tries to simplify rather than increase canopy structural complexity (Beyschlag & Ryel, 2007). The concept comes from the findings of Farquhar (Farquhar, 1989, as cited in Evans & Farquhar, 1991) who demonstrated that the equation for whole-leaf photosynthesis would be the same form as for individual chloroplast provided that chloroplast photosynthetic capacity distributes in proportion to the profile of absorbed irradiance and that in all layers the shape of the response to irradiance become identical. This approach applies to predict canopy light absorption by reduction of properties of all leaves within the canopy to a single leaf. However, this prediction will not be accurate enough to ignore developing alternative models. While Beer's law describes time-averaged profile of absorbed irradiance and the spatially averaged instantaneous profiles in a canopy, it doesn't describe the actual instantaneous distribution of absorbed irradiance. In fact, some leaves located deep in the canopy receive much higher radiation than the amount that Beer's law would predict when they are subject to sunflecks. Generally Beer's law does not represent the instantaneous profiles of absorbed irradiance in canopies because of errors created by both sunflecks and leaf angles (De Pury & Farquhar, 1997).

Multi layer model of light penetration through the canopies was proposed by Goudriaan (Goudriaan,1977, as cited in De Pury & Farquhar, 1997). In this model the plant canopy is divided into multiple leaf layers (increments in L of 0.1) distributed horizontally and assumed to be homogeneous with respect to leaf angels. Two groups of leaves are identified in each layer: shade and sunlit leaves, and sunlit section is divided into nine leaf angle classes. Irradiance absorption by each leaf group (sunlit and shade leaves) and also by each angle class of sunlit leaves is then calculated separately and integrated to give the whole canopy light absorption profile.

Sun-shade model initially introduced by Sinclair et al. (Sinclair et al., 1976, as cited in De Pury & Farquhar, 1997) and then applied by Norman (Norman,1980, as cited in De Pury & Farquhar, 1997), recently improved by De Pury and Farquhar (1997). This model gives predictions of canopy photosynthesis that closely match estimations of multi layer model with far fewer calculations. Sun-shade model divides canopy into large foliage groups: sunlit foliage which receives direct beam, and shade foliage which is subject to diffuse and/or transmitted irradiation. Amount of irradiance absorbed by each of these parts is calculated as an integral of absorbed light and the corresponding leaf area fraction.

Regardless of the way one calculates the rate of irradiance absorbed by the canopy, the next step in prediction of canopy photosynthesis will be estimation of the rate of photosynthesis undertaken by each group of leaves. In big leaf model an averaged value of light intercepted by the whole canopy enters photosynthesis model to calculate entire plant CO_2 assimilation rate. The method may be quite complex with multi layer model and somehow with sun-shade model as in these cases the calculations should be done in detail and more accurately for each leaf class. The performance of big leaf model in estimation of canopy photosynthesis depends in part on the accuracy by which the nitrogen distribution was predicted in proportion to the daily irradiance. While sun-shade model gives predictions of canopy photosynthesis with best approximation to those predicted by multi layer model, the big leaf model usually shows deviations ranged from 10% to 45% (De Pury & Farquhar, 1997). Each model accompanies advantages and disadvantages, differing in the rate of accuracy and degree of complexity in calculations. Presently, computer software makes it so feasible to integrate many mathematical formulas into one distinctive program, facilitating calculations of even more complex equations.

5. Training systems and canopy photosynthesis

Pruning and training techniques are professional horticultural practices developed not only to control plant growth in some circumstances but also to modify plant canopy in such a way that increases the amount of light absorption. Fruit trees commonly are subject to training systems during their juvenile period when the plant canopy is being formed. Depend on the type of buds (vegetative or reproductive), abundance and method of distribution within tree crown, pruning and training practices are carefully adopted so that it ensures maximum light penetration through the canopy, and provides plants the highest growth and productivity. In an experiment with two cultivars of apple trees Mierowska et al. (2002) indicated that summer pruning enhances photosynthetic acclimation of spur leaves, previously developed under shade, by rapid increasing of the chlorophyll a/b ratio. Similarly, in *phalaenopsis* , it was shown that providing the lower shade-developed leaves with higher rates of light intensities caused increased rate of photosynthesis (Lin & Hsu, 2004). Pruning resulted in changes in light harvesting complexes of rose plants, showing that rose leaves are very plastic and acclimate rapidly to any changes in light intensity (Calatayud, et al., 2007).

Training systems alter canopies light harvesting hehaviour through changing the foliage density, spatial form; the ratio of sun/shade leaves, leaf angles and finally the canopy leaf distribution pattern. There are several training systems developed for fruit trees based on the tree reproductive biology such as, central leader and modified leader particularly appropriate for pome fruits (apple, pear and quince trees) and open center specifically developed for prunes (peach, plum, and cherry trees). Recently, most greenhouse cut rose producers apply a type of training system, called arching technique recognized as an effective method to improve marketable qualities of cut flowers (Lieth & Kim, 2001; Sarkka & Rita, 1999). In this system most weak and blind shoots (shoots without flower bud) are bent toward the aisle instead of being pruned, a common practice traditionally performed before introducing bending method. This training system divides the rose canopy into two different parts: upright shoots which comprise the crop harvesting stems and bent stems which consisted of unmarketable shoots devoted to extend plant leaf area facing high solar radiation and to act as a pool to store and reserve assimilates in order to be used in future production of high quality flower shoots. It was found that K value (equation 3) for bent layer of the rose canopy is higher than the value determined for upright canopy (Gonalez-Real, et al. 2007). Additionally, bent layer showed lower rate of photosynthesis than upright shoots (Kim et al., 2004; Gonzalez-Real et al., 2007).

In an experiment we examined effects of 5 training systems on the rate of canopy light absorption and photosynthesis of *Rosa hybrida* 'Habari'(Matloobi et al., 2009). Treatments were combinations of bending height on the mother stem and height of harvesting on the successive flower shoots: (i) T.S. 1-1: bending at the base of the primary shoot and harvesting all flowering shoots above the first bud, (ii) T.S. 3-3: bending above the third bud, and harvesting above third bud for the first-order flowering shoot and above first bud for the following flowering shoots, (iii) T.S. 3-3-2: after bending primary shoot above third bud, the first-order flower shoot was harvested above the third bud and the second-order one above the second bud, (iv) T.S. 5-1: primary shoot was bent above fifth bud and the bearing flower shoot was harvested above first bud, (v): T.S. 5-3: primary shoot was bent above fifth bud and the bearing shoot was harvested above third bud. Leaf photosynthetic measurements have been performed for three layers of upright shoots (top, middle and bottom layer), and bent layer. Results exhibited that training system did not affect whole canopy light absorption significantly, but affected photosynthetic rate at canopy level (Table 1). This implies that photosynthetic rate at canopy level was influenced particularly by the interaction between light distribution profile through the canopy and canopy leaf area distribution rather than the

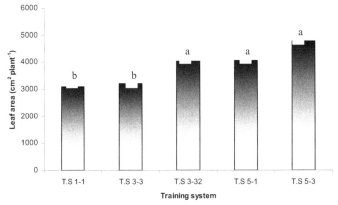

Fig. 6. Effect of different types of training system on the leaf area of *Rosa hybrida* 'Habari'. Different letters above columns indicate significantly difference according to the Duncan's multiple range test (p<0.05).

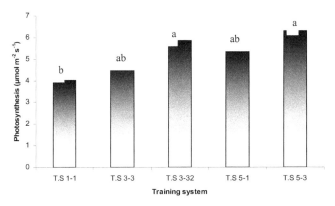

Fig. 7. Effect of different types of training system on the whole photosynthetic rate of *Rosa hybrida* 'Habari'. Different letters above columns indicate significantly difference according to the Duncan's multiple range test (p<0.05).

amount of light incident on the canopy. In other words, interaction between light intensity, light quality, leaf age, leaf area and many other intrinsic factors related to the leaf photosynthetic capacity have determined the canopy entire photosynthetic rate (Fig. 6 & Fig. 7). However, it was clearly deduced that canopy training system affected plant leaf area distribution between different leaf layers and accordingly impacted the rate of photosynthesis in each leaf layer. Plants developed by T.S. 1-1 method showed the lowest rate of canopy photosynthesis because they produced much less leaf area among the other training systems (Fig. 6). Although photosynthetic rate of the bent-shoots layer per unit leaf area was lower owing to the lower incident PAR, this layer accounted for about 40% of the whole plant photosynthetic capacity as a result of increased leaf area. Increasing leaf area does not enhance canopy assimilation rate unlimitedly due to leaves mutual shading caused by clumping effect. Lower layer of the bent shoots contributes negatively to the total canopy carbohydrate balance if its leaf area exceeds an optimal range (Pien et al., 2001). As a consequence, before adopting any type of training system or pruning strategy one should consider the results of *in situ* researches and try to optimize the canopy architecture and morphology based on environmental conditions and plant physiological and phonological characteristics.

Leaf layers	Incident PAR μmol m^{-2} s^{-1}	Photosynthesis μmol m^{-2}leaf s^{-1}	Photosynthesis μmol m^{-2}layer s^{-1}	Leaf area cm^2
Top layer	630.1 a	18.25 a	0.979 c	534.8 c
Middle layer	502.6 b	16.91 a	1.495 b	874.4 b
Bottom layer	343.6 d	10.78 b	0.747 c	664.6 c
Bent layer	411.1 c	11.90 b	2.102 a	1769.5 a

Table 1. Means comparison of the measured properties within different leaf layers of *Rosa hybrida* 'Habari'

Training systems or pruning methods may influence canopy photosynthetic rate by altering source-sink relationship. This alteration may lead to negative feedback control of leaf photosynthesis capacity. In *citrus unshiu*, girldling and defruiting induced leaf starch accumulation and reduced photosynthesis, whereas partial defoliation induced the opposite effect (Iglesias et al., 2002). Partial defoliated apple trees have shown similar results (Zhou & Quebedeaux, 2003). Matloobi et al. (2008) indicated that in cut roses the leaf attached to the bud immediately below the harvesting place, actively contributes in assimilate supply to the new growing shoot. Photosynthetic rate of the leaf attached to the bud, above which the shoot was pruned, was more or less constant from time of harvest until the growth of axillary bud. After the bud started to grow, the photosynthetic rate began to decline sharply, showing that the leaf had been degrading photosynthesis-related enzymes and other chloroplast proteins in order to support the growing young shoot (strong sink). This reduction in carbon fixation may arise from N depletion due to remobilization of N towards the growing point. Surprisingly, removing flower bud (another strong sink) did not significantly affect carbon assimilation rate of the leaf nearest to the flower bud over one week of gas exchange measurements. This implies that sink removal might have contrasting responses regarding plant species, type of the sink organ to be removed and its spatial position in relation to the sources.

6. References

Beyschlag, W. and Ryel, R.J. (2007). Canopy Photosynthesis Modeling, In: *Functional Plant Ecology*, Pugnaire, F.I. & Valladares, F., pp. 627-647, CRC Press, ISBN 9780849374883, NY, USA

Birch, C.J., Andrieu, B., Fournier, C., Vos, J. & Room, P. (2003). Modelling Kinetics of Plant Canopy Architecture – Concepts and Applications, *European Journal of Agronomy*, Vol. 19, pp. 519-533

Boote, K.J., & Loomis, R.S. (1991). The Prediciton of Canopy Assimilation, In: *Modeling Crop Photosynthesis – from Biochemsitry to Canopy*, Boote, K.J. & Loomis, R. S., pp. 109-137, CSSA, No. 19, Wisconsin, USA

Calatayud, A., Roca, D., Gorbe, E. & Martinez F.P. (2007). Light Acclimation in Rose (*Rosa hybrida* cv. Grand Gala) Leaves after Pruning: Effects on Chlorophyll a Fluorescence, Nitrate Reductase, Ammonium and Carbohydrates. *Sci. Hort.* Vol. 111, pp. 152-159.

Cannell, M. G. R. & Thornley, J. H. M. (1998). Temperature and CO_2 Response of Leaf and Canopy Photosynthesis: a Clarification Using the Non-rectangular Hyperbola Model of Photosynthesis, *Annals of Botany*, Vol. 82, pp. 883-892

Curtis, P. S., Vogel, C. S. Pregitzer, K. S., Zak, D. R. & Teeri, J. A. (1995). Interacting Effects of Soil Fertility and Atmospheric CO_2 on Leaf Area Growth and Carbon Gain Physiology in Populus x euramericana (Dode) Guinier. New Phytologist, Vol. 129, pp. 253-263

De Pury D.G.G. & Farquhar G.D. (1997). Simple Scaling of Photosynthesis from Leaves to Canopies without the Errors of Big-leaf Models, *Plant, Cell and Environment*, Vol. 20, pp. 537–557.

Evans, J.R. & Poorter, H. (2001). Photosynthetic Acclimation of Plants to Growth Irradiance: the Relative Importance of Specific Leaf Area and Nitrogen Partitioning in Maximizing Carbon Gain, *Plant, Cell & Environment*, Vol. 24, pp. 755-767

Farquhar, G. D., Caemmerer, S. Von. & Berry, J. A. (1980). A Biochemical Model of Photosynthetic CO_2 Assimilation in Leaves of C_3 Species, Planta, Vol. 149, pp. 78-90

Farquhar, G. D. & Evans, J. R. (1991). Modeling Canopy Photosynthesis from the Biochemistry of the C_3 Chloroplast, In: *Modeling Crop Photosynthesis – from Biochemsitry to Canopy*, Boote, K.J. & Loomis, R. S., pp. 109-137, CSSA, No. 19, Wisconsin, USA

Field, C. B., Behrenfeld, M. J., Randerson, J. T., & Falkowski, P. (1998). Primary Production of the Biosphere—Integrating Terrestrial and Oceanic Components. *Science*, Vol. 281, pp. 237–240.

Fitter, A.H. & Hay, R. KM. (2002). Environmental Physiology of Plants. Academis Press, ISBN 0122577663, London, UK

Gonzalez-Real, M.M., Baille, A. (2000). Changes in Leaf Photosynthetic Parameters with Leaf Position and Nitrogen Content within a Rose plant Canopy (Rosa hybrida), *Plant Cell Environ*, Vol. 23, pp. 351–363.

Gonzalez-Real, M.M., Baille, A. & Gutierrez Colomer, R.P. (2007). Leaf Photosynthesis Properties and Radiation Profiles in a Rose Canopy (*Rosa hybrida* L.) with Bent Shoots, Scientia Horticulturae, Vol. 114, pp. 177-187

Harley, P. C. & Tenhunen, J. D. (1991). Modeling the Photosynthetic Response of C₃ Leaves to Environmental Factors, In: *Modeling Crop Photosynthesis – from Biochemsitry to Canopy*, Boote, K.J. & Loomis, R. S., pp. 109-137, CSSA, No. 19, Wisconsin, USA

Hopkins, W.G. & Huner, N. P. A. (2004). Introduction to Plant Physiology,John Wiley & Sons, ISBN 0471389153, USA

Houghton, J.T., Ding, Y., Griggs, D.J., Noguer, M., van der Linden, P.J. & Xiaosu, D. (2001). Climate Change 2001. The Scientific Basis. Contribution of Working Group I to the Third Assessment Report of the Intergovernmental Panel on Climate Change (IPCC). Cambridge University Press, Cambridge.

Iglesias, D.J., Liso, I., Tadeo, F.R. & Talon, M. (2002). Regulation of Photosynthesis through Source–sink Imbalance in Citrus is Mediated by Carbohydrate Content in Leaves. *Physiol. Plant,* Vo. 116, pp. 563–572.

Johnson, I. R., & Thornley, J. H. M. (1984). A Model of Instantaneous and Daily Canopy Photosynthesis, *J. Theor. Biol.,* Vol. 107, pp. 531-545

Kim, S. H. & Lieth, H. (2002). Parameterization and Testing of a Coupled Model of Photosynthesis Stomatal Conductance for Greenhouse Rose Crop. *Acta Hort.,* Vol. 593, pp. 113-120.

Kim, S-H & Lieth, H. (2003). A Coupled Model of Photosynthesis, Stomatal Conductance and Transpiration for a Rose Leaf (*Rosa hybrida* L.). *Anl. of Bot.* 91: 771-781.

Kim, S. H., Shackel & Lieth, K. A. (2004). Bending Alters Water Balance and Reduces Photosynthesis of Rose Shoots, *J. Amer. Soc. Hort. Sci.,* Vol. 129, pp. 896-901

King, D.A. (1997). The Functional Significance of Leaf Angle in Eucalyptus. *Australian Journal of Botany*, Vol. 45, pp. 619-639

Kitajima, K. & Hogan, K. P. (2003). Increases of Chlorophyll a/b Ratios during Acclimation of Tropical Woody Seedlings to Nitrogen Limitation and High Light, *Plant, Cell and Environment*, Vol. 26, pp. 857-865

Laurerer, M., Saftic, D., Quick, WP., Labate, C., Fichtner, K., Schulze, ED., Rodermel, SR., Bogorad, L. & Stitt, M. (1993). Decreased Ribuloxe-1, 5-bisphosphate Carboxylase-oxygenase in Transgenic Tobacco Transformed with Antisense rbcS. 6. Effect on Photosynthesis in Plants Grown at Different Irradiance. *Planta*, Vol. 190, pp. 332-345

Lambers, H., Chapin, F.S. III. & Pons, T.L. (1998). Plant Physiological Ecology, Springr-Verlag, ISBN 0387983260, NY, USA

Lieth, J. H. & Kim, S. H. (2001). Effects of Shoot-bending in Relation to Root Media on Cut-flower Production, *Acta Horticulturae,* Vol. 547 pp. 303-310.

Lin, M. J. & Hsu, B. D. (2004). Photosynthetic Plasticity of Phalaenopsis in Response to Different Light Environment, *Journal of Plant Physiology*, Vol. 161, pp. 1259-1268

Marshall, B. & Biscoe, P. V. (1980). A Model for C₃ Leaves describing the Dependence of Net Photosynthesis on Irradiance, *J. Exp. Bot.,* Vol. 31, pp. 29-39

Matloobi, M. (2007). Possibilty of Optimizing *Rosa hybrida* L. 'Habari' Canopy in order to Increase Yield and Quality of Cut Flowers, Ph.D. Thesis, Tarbiat Modares University, Tehran, Iran

Matloobi, M., Baille, A., Gonzalez-Real, M. M. & Guiterrez Colomer, R.P. (2008). Effects of Sink Removal on Leaf Photosynthesis Attributes of Rose Flower Shoots (*Rosa hybrida* L., cv. Dallas), *Scientia Horticulturae*, Vol. 118, pp. 321-327

Matloobi, M., Ebrahimzadeh, A., Khalighi, A. & Hassndokhot, M. (2009). Training System Affect Whole Canopy Photosynthesis of the Greenhouse Roses (Rosa hybrida L. 'Habari'), *Journal of Food, Agriculture & Environment*, Vol. 7 (1), pp. 114-117

Marcelis, L.F.M., Heuvelink, E. & Goudriaan, J. (1998). Modelling Biomass Production and Yield of Horticultural Crops: a Review, *Scientia Horticulturae*, Vol. 74, pp. 83-111

Meyer, S., Genty, B. (1999). Heterogeneous Inhibition of Photosynthesis over the Leaf Surface of *Rosa ubiginosa* L. During Water Stress and Abscisic Acid Treatment: Induction of a Metabolic Component by Limitation of CO_2 Diffusion, *Planta*, Vol. 210, pp. 126-131

Mierowska, A., Keutgen, N., Huysamer, M. & Smith, V. (2002). Photosynthetic Acclimation of Apple Spur Leaves to Summer-pruning, *Scientia Horticulturae*, Vol. 92, pp. 9-27

Mostowska, A. (1997). Environmental Factors Affecting Chloroplasts, In: *Handbook of Photosynthesis*, Pessarakli, M., pp. 407-426, Marcel Dekker, ISBN 0824797086, NY, USA

Nobel, P.S. (1999). Physicochemical and Environmental Plant Physiology. Academic Press, San Diego

Norman, J.M. & Campbell, G.S. (1994). Canopy Structure, In: *Plant Physiological Ecology*, Pearcy, R.W., Ehleringer, J.R., Mooney, H.A., & Rundel, P.W., pp. 301-323, Chapman & Hall, ISBN 0412407302, London, UK

Ogren, E. (1988). Photoinhibition of Photosynthesis in Willow Leaves under Field Conditions. *Planta*, Vol.175, pp. 229–236

Pallardy, S. G. (2008). Physiology of Woody Plants, Academis Press, ISBN 9780120887651, California , USA

Pasian, C. C. & Lieth, J. H. (1989). Analysis of the Response of Net Photosynthetically Active Radiation and Temperature, *J. Amer. Soc. Hort. Sci.*, Vol. 114, pp. 581-586

Peterson, A. G., Ball, J. T., Luo, Y., Field, C. B., Curtis, P. S., Griffin, K. L., Gunderson, C. A., Norby, R.J., Tissue, D. T., Forstreuter, M., Rey, A., Vogel, C.S., & Participan, C. (1999). Quantifying the Response of Phyotosynthesis to Changes in Leaf Nitrogen Content and Leaf Mass per Area in Plants Grown under Atmospheric CO_2 , Enrichment, *Plant, cell Environment*, Vol. 22, pp. 1109-1119

Pien, H., Bobelyn, E., Lemeur, R. & Van Labeke, M. C. (2001). Optimising LAI in bent Rose Shoots. *Acta Horticulturae*, Vol. 547, pp. 319-327.

Pospisilova, J. & Santrucek, J. (1997). Stomatal Patchiness: Effects on Photosynthesis, In: Handbook of Photosynthesis, Pessarakli, M., Marcel Dekker, ISBN 0824797086, NY, USA

Quick, WP., Schurr, U., Scheibe, R., Schulze, ED., Rodermel, SR., Bogorad, L. & Stitt, M. (1991). Decreased Ribuloxe-1, 5-bisphosphate Carboxylase-oxygenase in Transgenic Tobacco Transformed with Antisense rbcS. 1. Impact on Photosynthesis in Ambient Growth-conditions. *Planta*, Vol. 183, pp. 542-554

Rundel, P.W. & Gibson, A. C. (1996). Adaptations of Mojave Desert Plants. In: *Ecological Commuities and Processes in a Mojave Desert Ecosystem*, Rock Vlley, Nevada, Cambridge Unviversity Press, Cambridge, pp. 55-83

Sarkka, L.E. & Rita, H.J. (1999) . Yield and Quality of Cut Roses Produced by Pruning or Bending down Shoots. *Gartenbauwissenschaft*, Vol. 64, pp. 173–176.

Schulze, E.D., Kelliher, F.M., Lorner, C., Lioyd, J., & Leuning, R. (1994). Relationship among Maximum Stomatal Conductacnce, Ecosystem Surfce Conductance, Carbon

Assimilation Rate, and Plant Nitrogen Nutrition: A global Ecology Scaling Exercise. *Annu. Rev. Ecol. Syst.* Vol. 25, pp. 629-660

Shelley, A.J. & Bell, D.T. (2000). Leaf Orientation, Light Interception and Stomatal Conductance of *Eucalyptus globulus* ssp. *Globulus* Leaves, *Tree Physiology*, Vol. 20, pp. 815-823

Spitters, C.J.T.(1986). Separating the Diffuse and Direct Component of Global Radiation and its Implications for

Modelling Canopy Photosynthesis: II. Calculation of Canopy Photosynthesis, *Agricultural and Forestry Meteorology.* Vo.38, pp. 231-242

Terashima, I. & Hilosaka, K. (1995). Comparative Ecophysiology of Leaf and Canopy Photosynthesis, *Plant Cell Environment*, Vol. 18, pp. 1111-1128

Terashima, I., & Saeki, T. (1985). Vertical Gradients in Photosynthetic Properties of Spinach Chloroplasts Dependent on Intraleaf Light Environment. *Plant Cell Physiology*, Vol. 24, pp. 1493-1501

Tissue D.T., Thomas, R.B. & Strain B.R. (1997) Atmospheric CO_2 enrichment Increases Growth and Photosynthesis of *Pinus taeda*: a 4 Year Experiment in the Field. *Plant, Cell and Environment*, Vol. 20, pp. 1123-1134

Vogelmann, T.C. & Martin, G. (1993). The Functional Significance of Palisade Tissue: Penetration of Directional *versus* Diffuse Light, *Plant, Cell and Environment*, Vol. 16, pp. 65-72

Walters, R. G. (2005). Towards an Understandign of Photosynthetic Acclimation, *J. Exp. Bot*, Vol. 56, pp. 435-447

Weiss, M., Baret, F., Smith, G.J., Joncheere, I. & Coppin, P. (2004). Review of Methods for *in situ* Leaf Area Index (LAI) Determination Part II. Estimation of LAI, Errors and Sampling. *Agricultural and Forest Meteorology*, Vol. 121, pp. 37-53

Wullschleger, S. D. (1993). Biochemical Limitations to Carbon Assimilation in C_3 Plants – A Retrospective Analysis of the A/C_i Curves from 109 Species, *Journal of Experimental Botany*, Vol. 44, pp. 907-920

Xu, L. & Baldocchi, D. (2003). Seasonal Trends in Photosynthetic Parameters and Stomatal Conductance of Blue Oak (*Quercus douglasii*) under Prolonged Summer Drought and High Temperature, *Tree Physiology*, Vol. 23, pp. 865-877

Zhou, R. & Quebedeaux, B. (2003). Changes in Photosynthesis and Carbohydrate Metabolism in Mature Apple Leaves in Response to Whole Plant Source-sink Manipulation, J. Amer. Soc. Hort. Sci. Vol. 128, 113-119

Permissions

The contributors of this book come from diverse backgrounds, making this book a truly international effort. This book will bring forth new frontiers with its revolutionizing research information and detailed analysis of the nascent developments around the world.

We would like to thank Mohammad Mahdi Najafpour, for lending his expertise to make the book truly unique. He has played a crucial role in the development of this book. Without his invaluable contribution this book wouldn't have been possible. He has made vital efforts to compile up to date information on the varied aspects of this subject to make this book a valuable addition to the collection of many professionals and students.

This book was conceptualized with the vision of imparting up-to-date information and advanced data in this field. To ensure the same, a matchless editorial board was set up. Every individual on the board went through rigorous rounds of assessment to prove their worth. After which they invested a large part of their time researching and compiling the most relevant data for our readers. Conferences and sessions were held from time to time between the editorial board and the contributing authors to present the data in the most comprehensible form. The editorial team has worked tirelessly to provide valuable and valid information to help people across the globe.

Every chapter published in this book has been scrutinized by our experts. Their significance has been extensively debated. The topics covered herein carry significant findings which will fuel the growth of the discipline. They may even be implemented as practical applications or may be referred to as a beginning point for another development. Chapters in this book were first published by InTech; hereby published with permission under the Creative Commons Attribution License or equivalent.

The editorial board has been involved in producing this book since its inception. They have spent rigorous hours researching and exploring the diverse topics which have resulted in the successful publishing of this book. They have passed on their knowledge of decades through this book. To expedite this challenging task, the publisher supported the team at every step. A small team of assistant editors was also appointed to further simplify the editing procedure and attain best results for the readers.

Our editorial team has been hand-picked from every corner of the world. Their multi-ethnicity adds dynamic inputs to the discussions which result in innovative outcomes. These outcomes are then further discussed with the researchers and contributors who give their valuable feedback and opinion regarding the same. The feedback is then collaborated with the researches and they are edited in a comprehensive manner to aid the understanding of the subject.

Apart from the editorial board, the designing team has also invested a significant amount of their time in understanding the subject and creating the most relevant covers. They scrutinized every image to scout for the most suitable representation of the subject and create an appropriate cover for the book.

The publishing team has been involved in this book since its early stages. They were actively engaged in every process, be it collecting the data, connecting with the contributors or procuring relevant information. The team has been an ardent support to the editorial, designing and production team. Their endless efforts to recruit the best for this project, has resulted in the accomplishment of this book. They are a veteran in the field of academics and their pool of knowledge is as vast as their experience in printing. Their expertise and guidance has proved useful at every step. Their uncompromising quality standards have made this book an exceptional effort. Their encouragement from time to time has been an inspiration for everyone.

The publisher and the editorial board hope that this book will prove to be a valuable piece of knowledge for researchers, students, practitioners and scholars across the globe.

List of Contributors

Mohammad Mahdi Najafpour and Babak Pashaei
Chemistry Department, Institute for Advanced Studies in Basic Sciences (IASBS), Zanjan, Iran

Claudia Stange and Carlos Flores
Universidad de Chile, Chile

Kevin M. Folta
Horticultural Sciences Department, Graduate Program in Plant Molecular and Cellular Biology, University of Florida, Gainesville, FL, USA

Felipe Caycedo-Soler
Ulm University, Institute of Theoretical Physics, Ulm, Germany
Departamento de Física, Universidad de los Andes, Bogotá, Colombia

Ferney J. Rodríguez and Luis Quiroga
Departamento de Física, Universidad de los Andes, Bogotá, Colombia

Guannan Zhao and Neil F. Johnson
Department of Physics, University of Miami, Coral Gables, Miami, Florida, USA

Sabina Jodłowska and Adam Latała
Department of Marine Ecosystems Functioning, Institute of Oceanography, University of Gdańsk, Gdynia, Poland

Ana Serrano and Milagros Medina
Department of Biochemistry and Molecular and Cellular Biology and Institute of Biocomputation and Physics of Complex Systems, University of Zaragoza, Spain

Maja Berden-Zrimec and Alexis Zrimec
Institute of Physical Biology, Slovenia

Marina Monti
Istituto Nazionale di Oceanografia e Geofisica Sperimentale, Italy

Joaquín Herrero, Alberto Esteban-Carrasco, José Miguel Zapata and Alfredo Guéra
University of Alcalá, Spain

Francisco Gasulla and Eva Barreno
University of Valencia, Spain

Alfonso Ros-Barceló
University of Murci, Spain

Sabrina Iñigo, Mariana R. Barber, Maximiliano Sánchez-Lamas, Francisco M. Iglesias and Pablo D. Cerdán
Fundación Instituto Leloir, IIBBA-CONICET and FCEN-UBA, Argentina

Lea Vojta and Hrvoje Fulgosi
Rudjer Boskovic Institute, Croatia

Noriyuki Suetsugu and Masamitsu Wada
Kyushu University, Japan

Mansour Matloobi
Department of Horticulture, Faculty of Agriculture, University of Tbariz, Tabriz, Iran

Printed in the USA
CPSIA information can be obtained
at www.ICGtesting.com
JSHW011443221024
72173JS00004B/917